Advanced Spectroscopy Techniques in Food Analysis: Qualitative and Quantitative Chemometric Approaches

Advanced Spectroscopy Techniques in Food Analysis: Qualitative and Quantitative Chemometric Approaches

Editors

Mourad Kharbach
Samuli Urpelainen

MDPI

Basel • Beijing • Wuhan • Barcelona • Belgrade • Novi Sad • Cluj • Manchester

Editors
Mourad Kharbach
University of Helsinki
Helsinki
Finland

Samuli Urpelainen
University of Oulu
Oulu
Finland

Editorial Office
MDPI
St. Alban-Anlage 66
4052 Basel, Switzerland

This is a reprint of articles from the Special Issue published online in the open access journal *Foods* (ISSN 2304-8158) (available at: https://www.mdpi.com/journal/foods/special_issues/Advanced_Spectroscopy_Techniques_Food_Analysis).

For citation purposes, cite each article independently as indicated on the article page online and as indicated below:

Lastname, A.A.; Lastname, B.B. Article Title. *Journal Name* **Year**, *Volume Number*, Page Range.

ISBN 978-3-0365-8580-2 (Hbk)
ISBN 978-3-0365-8581-9 (PDF)
doi.org/10.3390/books978-3-0365-8581-9

Contents

foods

Editorial

Special Issue "Advanced Spectroscopy Techniques in Food Analysis: Qualitative and Quantitative Chemometric Approaches"

Mourad Kharbach [1,2,*] and Samuli Urpelainen [3,*]

1 Department of Food and Nutrition, University of Helsinki, 00014 Helsinki, Finland
2 Department of Computer Sciences, University of Helsinki, 00560 Helsinki, Finland
3 Nano and Molecular Systems Research Unit, University of Oulu, 90014 Oulu, Finland
* Correspondence: mourad.kharbach@helsinki.fi (M.K.); samuli.urpelainen@oulu.fi (S.U.)

The globalization of the food market has created a pressing need for food producers to meet the ever-increasing demands of consumers while ensuring adherence to stringent food safety and quality standards [1]. The comprehensive analysis of food quality encompasses numerous aspects, such as chemical characterization, physical properties, sensory evaluation, authentication, traceability, processing, storage, and microbiological safety [2]. Traditional analytical techniques have long been employed in food analysis, but they often involve destructive procedures that are labor-intensive, time-consuming, costly, and environmentally burdensome [3].

In response to these challenges, the field of food analysis has witnessed remarkable advancements through the utilization of advanced spectroscopic techniques. These cutting-edge methods, including X-ray-based approaches, hyperspectral and multispectral imaging, NMR, Raman, IR, mass, UV, visible, and fluorescence spectroscopy, offer non-destructive, rapid, solvent-efficient, eco-friendly, and cost-effective alternatives to conventional methods [4]. Leveraging these techniques in tandem with statistical analysis, particularly through chemometric approaches, allows for the extraction and exploration of vital information hidden within spectral fingerprints or image data. Furthermore, this extracted information can be utilized to construct calibration models for qualitative and quantitative analysis of various food samples. The integration of advanced spectroscopy and chemometrics holds immense potential in the field of food science and technology, bolstering consumer confidence and contributing to overall food quality assurance [3].

It is with great pleasure that we present this Special Issue, which focuses on recent developments and applications of advanced spectroscopic techniques in food analysis, quality evaluation, safety assessment, and practical industrial implementations, with a specific emphasis on chemometric approaches. The collection of papers included in this Issue offers a valuable insight into the diverse range of research and applications in this field, shedding light on the potential of these techniques to revolutionize food analysis.

The accepted papers cover a broad spectrum of topics within the scope of this Special Issue. The first paper presents a comprehensive review of the current applications of advancing spectroscopy techniques in food analysis, focusing on the data handling aspect with chemometric approaches [3]. This review offers an overview of the progress made in the field and identifies avenues for further research and development.

Furthermore, one paper details an innovative application of laser-induced breakdown spectroscopy coupled with variable selection algorithms and chemometrics for the detection of heavy metals in *Fritillaria thunbergia* [5]. Another paper delves into the phenotypic analysis of Fourier-transform infrared milk spectra in dairy goats, providing valuable insights into the characterization and quality assessment of dairy products [6]. Additionally, the utilization of spatial frequency domain imaging and machine learning for the rapid and accurate detection of bruised tissue in pears is explored, highlighting the potential of these techniques for quality control purposes [7].

Citation: Kharbach, M.; Urpelainen, S. Special Issue "Advanced Spectroscopy Techniques in Food Analysis: Qualitative and Quantitative Chemometric Approaches". *Foods* **2023**, *12*, 2831. https://doi.org/10.3390/foods12152831

Received: 19 July 2023
Revised: 24 July 2023
Accepted: 24 July 2023
Published: 26 July 2023

In another study, the discrimination of Brazilian stingless bee honey based on its iron-based biogeographical origin is investigated, showcasing the applicability of discriminant analysis in ensuring the authenticity and traceability of food products [8]. The quality evaluation of fair-trade cocoa beans from different origins using portable near-infrared spectroscopy (NIRS) is also examined, illustrating the potential of NIRS as a non-destructive tool for rapid quality assessment in the cocoa industry [9].

Additionally, the effect of moisture content on the analysis of quality attributes of red pepper powder is explored using a hyperspectral system, providing valuable insights into the impact of moisture on food analysis outcomes [10]. Moreover, time-resolved laser-induced breakdown spectroscopy is employed for the accurate qualitative and quantitative analysis of brown rice flour adulteration, offering a promising approach to combat food fraud and adulteration [11].

Furthermore, the classification of *Prunus* genus by botanical origin and harvest year based on carbohydrates profiles is investigated, shedding light on the application of spectroscopic techniques for the authentication of botanical products [12]. The chemical authentication and speciation of *Salvia* botanicals are explored using GC/Q-ToF and chemometrics, providing crucial insights into the identification and characterization of herbal products [13].

Lastly, the detection of pesticide residue levels in grapes is studied using hyperspectral imaging and machine learning, illustrating the potential of these techniques for ensuring food safety [14].

In conclusion, this Special Issue brings together a collection of research papers that highlight the immense potential of advanced spectroscopic techniques in the field of food analysis and quality evaluation. By presenting a range of innovative applications, these studies demonstrate the power of these techniques to enhance food safety, authenticity, and overall quality. We hope that the papers in this Special Issue provide valuable insights, inspire further research, and encourage the adoption of advanced spectroscopic techniques in the food industry.

Author Contributions: M.K. and S.U. contributed equally to this Editorial. All authors have read and agreed to the published version of the manuscript.

Conflicts of Interest: The authors declare no conflict of interest.

References

1. Rivera, D.; Toledo, V.; Reyes-Jara, A.; Navarrete, P.; Tamplin, M.; Kimura, B.; Wiedmann, M.; Silva, P.; Switt, A.I.M. Approaches to empower the implementation of new tools to detect and prevent foodborne pathogens in food processing. *Food Microbiol.* **2018**, *75*, 126–132. [CrossRef] [PubMed]
2. González-Domínguez, R.; Sayago, A.; Fernández-Recamales, A. An Overview on the Application of Chemometrics Tools in Food Authenticity and Traceability. *Foods* **2022**, *11*, 3940. [CrossRef] [PubMed]
3. Kharbach, M.; Mansouri, M.A.; Taabouz, M.; Yu, H. Current Application of Advancing Spectroscopy Techniques in Food Analysis: Data Handling with Chemometric Approaches. *Foods* **2023**, *12*, 2753. [CrossRef]
4. Franca, A.S.; Nollet, L.M. (Eds.) *Spectroscopic Methods in Food Analysis*; CRC Press: Boca Raton, FL, USA, 2017.
5. Kabir, M.H.; Guindo, M.L.; Chen, R.; Luo, X.; Kong, W.; Liu, F. Heavy Metal Detection in *Fritillaria thunbergii* Using La-ser-Induced Breakdown Spectroscopy Coupled with Variable Selection Algorithm and Chemometrics. *Foods* **2023**, *12*, 1125. [CrossRef]
6. Villar-Hernández, B.d.J.; Amalfitano, N.; Cecchinato, A.; Pazzola, M.; Vacca, G.M.; Bittante, G. Phenotypic Analysis of Fourier-Transform Infrared Milk Spectra in Dairy Goats. *Foods* **2023**, *12*, 807. [CrossRef] [PubMed]
7. Xing, S.; Zhang, J.; Luo, Y.; Yang, Y.; Fu, X. Extracting Tissue Optical Properties and Detecting Bruised Tissue in Pears Quickly and Accurately Based on Spatial Frequency Domain Imaging and Machine Learning. *Foods* **2023**, *12*, 238. [CrossRef] [PubMed]
8. Lavinas, F.C.; Gomes, B.A.; Silva, M.V.T.; Nunes, R.M.; Leitão, S.G.; Moura, M.R.L.; Simas, R.C.; Carneiro, C.S.; Rodrigues, I.A. Discriminant Analysis of Brazilian Stingless Bee Honey Reveals an Iron-Based Biogeographical Origin. *Foods* **2023**, *12*, 180. [CrossRef] [PubMed]
9. Forte, M.; Currò, S.; Van de Walle, D.; Dewettinck, K.; Mirisola, M.; Fasolato, L.; Carletti, P. Quality Evaluation of Fair-Trade Cocoa Beans from Different Origins Using Portable Near-Infrared Spectroscopy (NIRS). *Foods* **2022**, *12*, 4. [CrossRef] [PubMed]
10. Choi, J.Y.; Cho, J.S.; Park, K.J.; Choi, J.H.; Lim, J.H. Effect of Moisture Content Difference on the Analysis of Quality At-tributes of Red Pepper (*Capsicum annuum* L.) Powder Using a Hyperspectral System. *Foods* **2022**, *11*, 4086. [CrossRef] [PubMed]

11. Ma, H.; Shi, S.; Zhang, D.; Deng, N.; Hu, Z.; Liu, J.; Guo, L. Time-resolved laser-induced breakdown spectroscopy for accurate qualitative and quantitative analysis of brown rice flour adulteration. *Foods* **2022**, *11*, 3398. [CrossRef] [PubMed]
12. Miricioiu, M.G.; Ionete, R.E.; Costinel, D.; Botoran, O.R. Classification of *Prunus* Genus by Botanical Origin and Harvest Year Based on Carbohydrates Profile. *Foods* **2022**, *11*, 2838. [CrossRef] [PubMed]
13. Lee, J.; Wang, M.; Zhao, J.; Avula, B.; Chittiboyina, A.G.; Li, J.; Wu, C.; Khan, I.A. Chemical authentication and speciation of *Salvia* botanicals: An investigation utilizing GC/Q-ToF and chemometrics. *Foods* **2022**, *11*, 2132. [CrossRef] [PubMed]
14. Ye, W.; Yan, T.; Zhang, C.; Duan, L.; Chen, W.; Song, H.; Zhang, Y.; Xu, W.; Gao, P. Detection of Pesticide Residue Level in Grape Using Hyperspectral Imaging with Machine Learning. *Foods* **2022**, *11*, 1609. [CrossRef] [PubMed]

foods

MDPI

Article

Heavy Metal Detection in *Fritillaria thunbergii* Using Laser-Induced Breakdown Spectroscopy Coupled with Variable Selection Algorithm and Chemometrics

Muhammad Hilal Kabir [1,2], Mahamed Lamine Guindo [1], Rongqin Chen [1], Xinmeng Luo [3], Wenwen Kong [3] and Fei Liu [1,4,*]

1 College of Biosystems Engineering and Food Science, Zhejiang University, 866 Yuhangtang Road, Hangzhou 310058, China
2 Department of Agricultural and Bio-Resource Engineering, Abubakar Tafawa Balewa University, Bauchi PMB 0248, Nigeria
3 College of Mathematics and Computer Science, Zhejiang A&F University, Hangzhou 311300, China
4 Key Laboratory of Spectroscopy Sensing, Ministry of Agriculture and Rural Affairs, Hangzhou 310058, China
* Correspondence: fliu@zju.edu.cn; Tel.: +86-571-88982825

Abstract: Environmental and health risks associated with heavy metal pollution are serious. Human health can be adversely affected by the smallest amount of heavy metals. Modeling spectrum requires the careful selection of variables. Hence, simple variables that have a low level of interference and a high degree of precision are required for fast analysis and online detection. This study used laser-induced breakdown spectroscopy coupled with variable selection and chemometrics to simultaneously analyze heavy metals (Cd, Cu and Pb) in *Fritillaria thunbergii*. A total of three machine learning algorithms were utilized, including a gradient boosting machine (GBM), partial least squares regression (PLSR) and support vector regression (SVR). Three promising wavelength selection methods were evaluated for comparison, namely, a competitive adaptive reweighted sampling method (CARS), a random frog method (RF), and an uninformative variable elimination method (UVE). Compared to full wavelengths, the selected wavelengths produced excellent results. Overall, RC^2, RV^2, RP^2, RSMEC, RSMEV and RSMEP for the selected variables are as follows: 0.9967, 0.8899, 0.9403, 1.9853 mg kg^{-1}, 11.3934 mg kg^{-1}, 8.5354 mg kg^{-1}; 0.9933, 0.9316, 0.9665, 5.9332 mg kg^{-1}, 18.3779 mg kg^{-1}, 11.9356 mg kg^{-1}; 0.9992, 0.9736, 0.9686, 1.6707 mg kg^{-1}, 10.2323 mg kg^{-1}, 10.1224 mg kg^{-1} were obtained for Cd Cu and Pb, respectively. Experimental results showed that all three methods could perform variable selection effectively, with GBM-UVE for Cd, SVR-RF for Pb, and GBM-CARS for Cu providing the best results. The results of the study suggest that LIBS coupled with wavelength selection can be used to detect heavy metals rapidly and accurately in Fritillaria by extracting only a few variables that contain useful information and eliminating non-informative variables.

Keywords: laser-induced breakdown spectroscopy; *Fritillaria thunbergii*; heavy metals; chemometrics; variable selection; machine learning

Citation: Kabir, M.H.; Guindo, M.L.; Chen, R.; Luo, X.; Kong, W.; Liu, F. Heavy Metal Detection in *Fritillaria thunbergii* Using Laser-Induced Breakdown Spectroscopy Coupled with Variable Selection Algorithm and Chemometrics. *Foods* **2023**, *12*, 1125. https://doi.org/10.3390/foods12061125

Academic Editors: Mourad Kharbach and Samuli Urpelainen

Received: 29 January 2023
Revised: 3 March 2023
Accepted: 5 March 2023
Published: 7 March 2023

1. Introduction

The World Health Organization (WHO) reports that herbal medicines remain the primary treatment for a number of diseases in developing countries [1]. The use of nutraceuticals and medicinal products derived from medicinal herbs is increasing even in developed countries [1]. The consumption of healthy herbs is currently receiving considerable attention, and there is a focus on consuming products that are as natural as possible and have as little contamination as possible. The demand for nutraceuticals and herbal dietary supplements has increased significantly in recent years. Plants and other natural materials are used to make these products. There is, therefore, a high probability of heavy metal contamination [2]. The food chain may introduce heavy metals to humans through the accumulation of heavy metals in the environment [3–6]. There are numerous anthropogenic sources and activities that lead to heavy metal contamination of the environment, such as mining, traffic, agriculture, and industrial processes. However, heavy metals are naturally present in the Earth's crust [6,7]. Risks associated with heavy metals can be represented in a variety of ways [8]. They can be found in the air we breathe, the food we eat, and the water we drink. They can also be found in soil and dust, which can be inhaled or ingested. As a result, they can enter the body and cause health problems.

In traditional Chinese medicine (TCM), heavy metals such as arsenic (As), mercury (Hg), lead (Pb), copper (Cu) and cadmium (Cd) are of particular concern [2]. Human health is at risk from these pollutants. The intake of excessive amounts of heavy metals is detrimental to the human body because it results in neurotoxicity, organic damage, and diseases of the skin and blood [9,10].

Fritillaria thunbergii Miq is a perennial herbaceous plant that is native to the provinces of Zhejiang, Jiangsu, and Anhui in China [11]. A dry bulb of Thunbergii fritillaria (Zhebeimu), a plant from the family of Thunbergii, is frequently used in Chinese medical clinical practice for treating coughs caused by wind and phlegm heats, as well as bronchitis, inflammation, hypertension, gastric ulcer, diarrhea, and bacterial infections [12]. Furthermore, Zhebeimu is extensively used to treat leukemia that is resistant to drugs [13]. Heavy metals have become one of the most serious safety concerns due to increasingly stringent TCM regulations [14–16]. Thus, accurate detection of heavy metal concentrations in TCM is crucial. Among the methods commonly used for heavy metal detection are atomic absorption spectroscopy (AAS), atomic fluorescence spectroscopy (AFS), X-ray fluorescence spectroscopy (XRFS), inductively coupled plasma optical emission spectroscopy (ICP-OES) [17,18], and electrochemical methods, especially stripping and cyclic voltammetry, which are commonly used methods for detection. Voltammetry is a sensitive electrochemical method that is widely used for heavy metal detection [19–21]. A traditional heavy metal detection method involves sampling, pretreatment, and laboratory chemical analysis; all of which are time-consuming, costly, and require extensive preparation [22,23].

Multi-elemental detection can be achieved through laser-induced breakdown spectroscopy (LIBS) [24]. LIBS uses pulsed laser ablation to create plasma on a sample and then detects and analyzes the emission light emanating from the plasma. With LIBS, a sample does not need to be prepared prior to analysis, thus allowing for rapid results [25], minimum requirements for small samples [26], and cost-effective instrumentation [27]. LIBS is widely used in a wide variety of industries [28], including mining [29], plastics [30], biomedicine [31,32], food [33], and the environment [34]. LIBS can provide information on the composition of samples within a short period of time, as well as the element content of samples. In comparison with other detection technologies, LIBS has many advantages, such as the requirement for fewer samples, the lack of complex pretreatment, the ability to measure multiple elements simultaneously, and the possibility of rapid implementation [35].

A number of studies have been conducted in recent years that focus on using LIBS to detect heavy metal pollution. Wang et al. [36] applied LIBS to detect cadmium content in rice. The LIBS analysis of rice stems demonstrated that it is an effective method for detecting cadmium. Su et al. [37] simultaneously and quantitatively analyzed the heavy metals in *Sargassum fusiforme* by using laser-induced breakdown spectroscopy. Liu et al. (2020) [38] analyzed cosmetics for trace lead and cadmium through laser-induced breakdown spectroscopy and ultrasound-assisted extraction. Rehan et al. [39] analyzed henna paste, and fresh leaves and soils were tested with LIBS to detect lead and nutrients. Wang et al. [40] conducted an analysis of lead and copper in Ligusticum wallichii using LIBS. Lead (Pb) levels in soil were quantitatively analyzed by Zhao et al. [41]. It was demonstrated that the dual-pulse laser-induced breakdown spectroscopy (DP-LIBS) was an efficient spectroscopic tool for improving the quantitative analysis of Pb heavy metal in soil. Zhu et al. [42] performed an analysis of the content of arsenic in traditional Chinese medicine using laser-induced breakdown spectroscopy (LIBS). Rehan et al. [43] used LIBS to assess the amount of toxic heavy metals (Pb, Cr, Ni) present in different brands of face foundation powder. Zhu et al. [44] detected lead in rhododendron leaves using laser-induced breakdown spectroscopy assisted by laser-induced fluorescence. Peng et al. [45] analyzed rice leaves using collinear DP-LIBS to determine their chromium content. Yang et al. [46] determined the content of lead and cadmium in rice using LIBS. These studies are indicative of the increasing consolidation of LIBS associated with chemometrics methods for the analysis of heavy metals.

However, despite the increase of fast and clean methods for TCM analysis [47], the application of gradient-boosting machine learning algorithms has rarely been used for heavy metal prediction in TCM. Thus, detecting heavy metals in different varieties of Fritillaria based on LIBS technology combined with a gradient boosting machine is unique and important for monitoring human exposure and establishing effective environmental control strategies.

Considering the potential risks associated with heavy metals in traditional Chinese medicine (TCM), the present work investigated the feasibility of LIBS combined with chemometrics in measuring cadmium (Cd), copper (Cu) and lead (Pb) simultaneously in twelve (12) different varieties of *Fritillaria thunbergii*. However, the specific objectives were as follows: (a) to test the feasibility of the gradient boosting machine (GBM) as a method of measuring heavy metals in different varieties of Fritillaria using both full and extracted variables, (b) to verify the effectiveness of using three variable selection methods, namely, competitive adaptive reweighed sampling (CARS), random frog (RF) and uninformative variable elimination (UVE) by comparing model (GBM, SVR and PLSR) performances, and (c) to establish a quantitative analysis model for heavy metals based on full and extracted variables.

2. Materials and Methods

2.1. Sample Collection and Preparation

A total of twelve (12) different varieties of Fritillaria thunbergia were used in the experiment. The varieties were provided by the Faculty of Biosystems Engineering and Food Science (Zhejiang University, Hangzhou, China). Different copper (Cu), cadmium (Cd) and lead (Pb) samples were prepared in the laboratory using $Cu(NO_3)_2$, $Cd(NO_3)_2 \cdot 4H_2O$ and Pb $(NO_3)_2$ [40,44,46,48], respectively. Fritillaria varieties were randomly divided into eight groups to obtain samples with varying levels of Cu, Cd, and Pb. The first group was designated as a control group, whereas the remaining seven groups were designated as treatment groups. In order to accurately quantify Cu, Cd, and Pb concentrations in Fritillaria samples, the solution was artificially contaminated for 48 h at 4 °C and rinsed three times with super-pure water. This was done to simulate the effects of heavy metal pollution in the environment and to determine the best course of action for mitigating the environmental damage caused by these pollutants. Following drying at room temperature, all samples were milled at high speeds using a high-speed pulverizer (FW100, TAISITE, Tianjin, China). In order to produce pellets from these ground samples, they were compressed using a tablet compressor (FY-24, SCJS, Tianjin, China), of 1.5 cm in diameter, at a pressure of 25 kN for the duration of one minute. In total, 288 pellets were prepared.

2.2. Experimental Setup

A LIBS self-assembled schematic diagram is available in [49]; Figure 1 illustrates the methods used in this experiment. Laser pulses were generated at 532 nm with a maximum energy of 200 mJ and a pulse width of 8 ns using a Q-switched Nd: YAG pulse laser (Vlite 200, Beamtech, Beijing, China). A planoconvex lens (f = 100 mm) finally focused the laser onto the sample surface after passing through the optical system. The laser ablation generated plasma which emitted electromagnetic waves that diffused outward. In order to measure the waves, a light collector was used, and the waves were captured by a spectrometer (SR-500i-A-R, Andor Technology, Belfast, UK) along with an intensified charge-coupled device (ICCD) camera (DH334T-18F-03, Andor Technology, Belfast, UK).

The laser Q-switch and ICCD camera were delayed using a delay generator (DG645, Stanford Research Systems, Sunnyvale, CA, USA). Several parameters were optimized before the experiment, including the laser energy of 60 mJ, the delay time of 1.5 μs, and the gate width of 10μs. The Fritillaria pellets were automatically placed, and the laser ablation path was controlled using a 4 × 4 array of craters designed using automatic x-y-z translation. Laser pulses accumulated five times faster in each crater. For each sample, an average of 80 spectra (4 × 4 × 5) were taken in order to reduce fluctuation between the laser points. Approximately one minute was required to collect LIBS information for one sample.

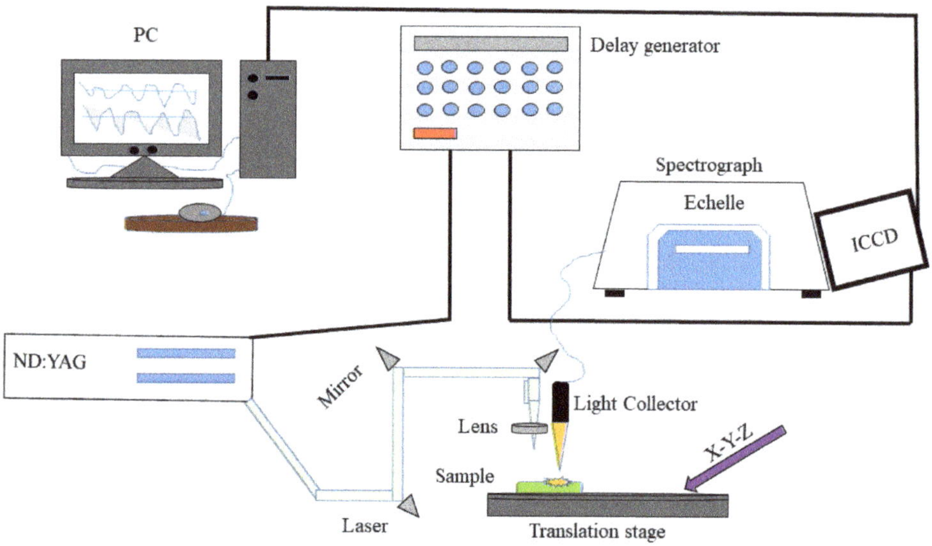

Figure 1. An illustration of the LIBS experimental setup.

2.3. Determination of Heavy-Metals Reference Value

Inductively coupled plasma mass spectrometry (ICP-MS) was used to determine the contents of three (3) heavy metals (Cu, Cd, and Pb) in *Fritillaria thunbergii*. Table 1 lists the statistics of heavy metal contents in different samples. A number of procedures are involved in the ICP-MS analysis of the samples, including digestion of the sample, filtration and purification of the digestion solution, and detection of the digestion solution using ICP-MS (ELAN DRC-e, Perkin Elmer, USA). The pH value as well as all steps in the ICP-MS analysis were carried out by experimental technicians at the Zhejiang College of Biosystems Engineering and Food Science, Zhejiang University. This was a preliminary attempt to detect multiple heavy metals (Cd, Cu and Pb) in Fritillaria thubergii using the LIBS technique. The pH value was determined by measuring the electrical conductivity of the sample solution with a pH meter. The ICP-MS analysis was performed using a quadrupole-based ICP-MS system which allowed the technicians to accurately measure the concentrations of elements in the sample solution. Similar steps are described by Su et al. [37]. Unscrambler X, version 10.1 (CAMO Software AS, Oslo, Norway, 2011) was used for the descriptive statistics (file imported in MATLAB format).

Table 1. Heavy metals (Cu, Cd and Pb) contents of Fritillaria obtained by ICP-MS.

Heavy Metal	Groups	CK	1	2	3	4	5	6	7
Cu	Number	36	36	36	36	36	36	36	36
	Min.	1.50	13.80	33.01	42.51	53.76	66.49	155.41	169.66
	Max.	4.88	27.99	55.97	61.28	69.33	82.97	242.04	245.36
	Mean	2.50	20.06	37.31	52.48	65.95	79.66	172.57	215.06
	Range	3.37	14.18	22.95	18.76	15.57	16.47	86.63	75.69
	Var.	1.20	13.14	36.71	37.67	16.87	20.33	517.21	773.97
	Std	1.09	3.62	6.05	6.13	4.10	4.50	22.74	27.82
Cd	Min.	0.17	5.21	9.64	20.81	24.73	44.85	63.06	83.66
	Max.	1.19	7.08	11.70	25.34	29.38	97.26	117.89	100.65
	Mean	0.43	5.86	10.64	22.87	26.39	60.32	82.48	93.91
	Range	1.02	1.87	2.05	4.52	4.65	52.41	54.82	16.98
	Var.	0.07	0.21	0.31	2.34	1.72	394.02	175.37	35.38
	Std.	0.28	0.46	0.56	1.53	1.31	19.85	13.24	5.94
Pb	Min.	0.13	4.12	21.28	43.13	58.61	63.14	102.16	143.12
	Max.	0.76	7.33	23.90	102.03	71.15	90.51	112.26	219.07
	Mean	0.29	6.06	22.75	51.58	65.62	85.00	112.26	199.24
	Range	0.63	3.21	2.61	58.90	12.53	27.36	30.33	75.95
	Var.	0.03	0.56	0.64	265.51	14.55	59.54	63.84	604.43
	Std.	0.19	0.74	0.80	16.29	3.81	7.71	7.99	24.58

CK: Control group.

2.4. Data Analysis

2.4.1. Spectral Modeling

The gradient boosting technique (GB) is a machine-learning method used in regression and classification problems. As a result, a prediction model is produced as an ensemble of weak prediction models. Every step evaluates the model values at each training data point, using the residuals of previous steps to minimize the loss function [50]. A GBM utilizes the best practices to avoid overfitting the classification machine. A subsample of the training data is randomly selected (without replacement) from the full dataset for each iteration in order to fit the base learner for that iteration. Figure 2 illustrates the main processes involved in gradient boosting through a simplified flow chart.

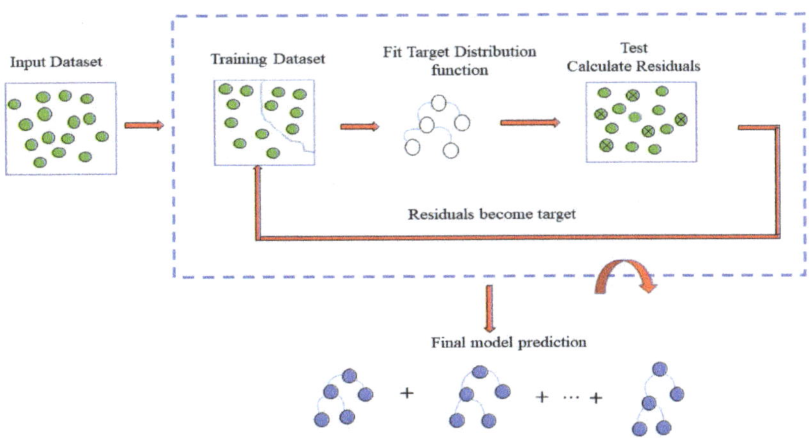

Figure 2. Gradient Boosting Machine Flow Chart.

In comparison to other machine learning methods, gradient boosting machines (GBMs) have several advantages. GBM, in addition to its complex classification capability, allows soft classification, which entails calculating the probability of each sample being a member of each class, rather than labeling every sample as part of a single category (hard classification). It facilitates the assessment of the reliability of the statistical model (the potential for overfitting) and the study of the chemical-physical properties of the model, thus fostering the development of further qualitative and quantitative research studies.

In addition, it provides a natural measure of how significant each spectral feature is for classifying data, something that is usually lost in the black-box nature of many machine learning algorithms, which include artificial neural networks. The relative importance of variables is computed in GBM. A refinement of the split criterion is computed at each split in each tree (MSE for regression). Averaging the improvements made by each variable over every tree that uses that variable follows. In the split criterion, the variables with the greatest average decrease are listed as the most significant [50].When it comes to GBM modeling, there are a variety of tuning parameters available. The following variables were used in this study: (boosting_type = 'gbdt', num_leaves = 31, max_depth = −1, learning_rate = 0.1, n_estimators = 100). In order to forecast discrete values, SVR is an approach that uses supervised learning. By comparison, it aims to determine the hyperplane with the most points, or the line of best fit [51,52]. PLS uses projections to build linear regression models using variables and observables [53–55]. A significant amount of collinearity can be analyzed by this algorithm, unlike previous algorithms.

2.4.2. Variable Selection Methods

Generally, LIBS data display high covariance due to the capability of the technique to measure multiple emission lines associated with the same element or species. Moreover, each peak is a result of a combination of many factors. It is therefore possible to reduce the covariance and complexity of the model by selecting variables. Selection of more explanatory variables improves the understanding of the multivariate system [56]. The removal of noisy areas, such as the extremes of each spectrometer and variables without analytical information, is also likely to increase the explained variance and enhance the accuracy of the model [56]. To select the optimal variable, variables should be selected and eliminated. In this study, three methods of variable selection were used to simplify the calculation process and improve model performance (CARS, RF, and UVE).

In CARS, wavelengths are selected using a survival-of-the-fittest principle [57]. Firstly, wavelengths with small regression coefficients are removed by using an exponentially decreasing function (EDF). An EDF equation is then used to calculate the wavelength ratio. The following steps are involved in each sampling run: (a) the Monte Carlo (MC) principle is used to sample models; (b) EDF-based wavelength selection is performed; (c) adaptive reweighted sampling is used for competitive wavelength selection and (d) cross-validation evaluation of the subset is performed [57]. A subset of wavelengths with the lowest root mean squared error of cross-validation (RMSECV) is retained as the effective wavelength, and wavelengths with little or no effective information are eliminated [58].

The RF method is iterative in nature. There are primarily three steps in the random frog algorithm: (1) A random selection of features is used to create an initialized feature subset. (2) Iteration is performed until a candidate feature subset is selected. This is accepted with a certain probability, then replaced, and this step is looped until the desired number of iterations is achieved. (3) As a measure of feature importance, the selection probability of each feature is calculated [59]. The total number of feature subsets can be determined after multiple iterations. Its selection probability can be calculated as

$$P_j = \frac{N_j}{N}, \ j = 1, 2, \ldots\ldots n \quad (1)$$

N_j denotes the frequency of the j_{th} feature, j = 1, 2, . . . , n, selected from the features. For each feature, N_j is the feature subsets after iteration, and N is the feature frequency.

Accordingly, the more optimal a feature, the more likely it is to be selected for inclusion in these subsets of features. As a result, features can be selected according to feature importance. The UVE method uses regression coefficients from a PLS model to select variables. This method is useful in eliminating non-informative variables, and the remaining variables can be used to analyze and classify chemicals [60,61].

2.4.3. Model Evaluation and Calculation

In order to assess the performance of the calibration model, several evaluation indices were calculated. The point-to-point fluctuations in the spectra were reduced by area normalization [45]. The data were all normalized prior to modeling. The accuracy of the model was assessed by determining the root mean square error of calibration (RMSEC), the root mean square error of validation (RMSEV), the root mean square error of prediction (RMSEP) and the coefficient of calibration (RC^2), coefficient of validation (RV^2) and coefficient of prediction (RP^2) based on the predicted results. In summary, a good calibration model should have a small RMSEC, and RMSEP, as well as large values for (RC^2) and (RP^2) [37,48,62,63]. The calculations were performed using Python with Scikit-Learn and the figures were generated using Origin 2022.

3. Results and Discussion

3.1. Spectra Analysis

In LIBS, atoms and ions are expelled from a generated plasma as a result of their excited state [64]. Figure 3 illustrates the normalized spectra of 12 Fritillaria varieties. In accordance with the National Institute of Standards and Technology, USA, Electronic Database, characteristic lines for Cd, Cu, and Pb were identified. In this study, however, the purpose is to quantitatively analyze the content of three heavy metals (Cd, Cu and Pb), and because the LIBS spectra of the different varieties have similar curves, it is difficult to observe the LIBS spectra simultaneously and quantitatively to analyze the heavy metal content of Fritillaria simultaneously and quantitatively. Hence, the LIBS data must be further analyzed using chemometric methods.

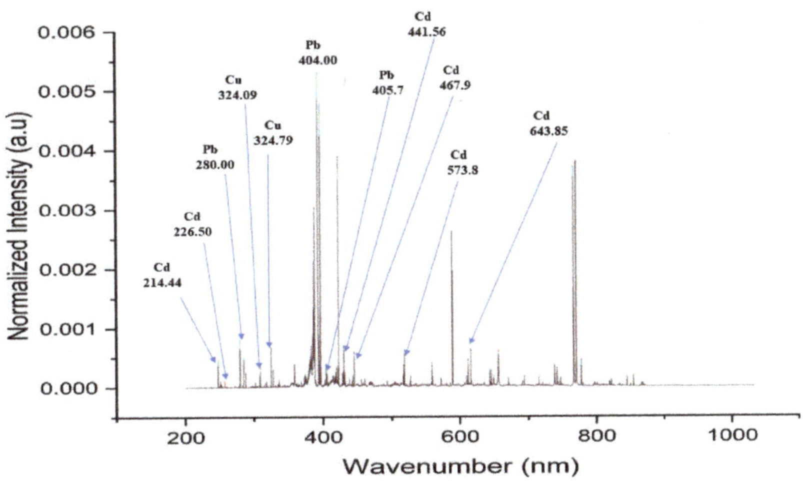

Figure 3. Normalized spectra of 12 Fritillaria varieties.

3.2. Heavy Metals Prediction Using Full and Selected Variables

We performed LIBS analyses in accordance with best practices [65]. The full spectrum was divided into three parts: calibration (60%), validation (28%), and prediction (12%). A summary of the GBM results for the full wavelength and the selected wavelength is presented in Tables 1 and 2. The peaks in full LIBS data usually exhibit high covariance due to the capability of measuring multiple emission lines of the same element or species and the fact that a number of x variables are required to generate a single peak. A useful range of wavelengths should be selected from the entire range of wavelengths in order to simplify the LIBS calibration models and improve prediction accuracy. As a result, three methods of variable selection (CARS, RF, and UVE) were used to select the informative variables from the whole wavelength range to simplify the calculation process and to improve the performance of the model.

Table 2. Prediction results of the different models using full variables.

	Models	Variables	RC^2	RMSEC	RV^2	RMSEV	RP^2	RMSEP
Cd	SVR	Full variables	0.9999	0.0995	0.8957	11.0874	0.8276	14.5054
	PLSR	Full variables	0.9923	3.0637	0.6791	19.4539	0.5499	23.4405
	GBM	Full variables	0.9984	1.3693	0.8924	11.2625	0.9139	10.2494
Cu	SVR	Full variables	0.9999	0.0995	0.9606	13.9359	0.9280	17.5096
	PLSR	Full variables	0.9936	5.7691	0.6902	39.1138	0.5622	43.1827
	GBM	Full variables	0.9979	3.2885	0.9308	18.4815	0.9596	13.1156
Pb	SVR	Full variables	0.9999	0.0996	0.9304	16.6308	0.8876	19.1794
	PLSR	Full variables	0.9938	4.9443	0.6925	34.9791	0.6272	34.9311
	GBM	Full variables	0.9998	0.8488	0.9673	11.3933	0.9635	10.9220

RC^2: Coefficient of determination for calibration; RV^2: Coefficient of determination for validation; RP^2: Coefficient of determination for prediction; RMSEC: Root mean square error for calibration; RMSEV: Root mean square error for validation; RMSEP: Root mean square error for prediction.

3.2.1. Cd Content Prediction Using Full and Selected Variables

All three variable selection methods (CARS, RF and UVE) were computed with SVR, PLSR and GBM to predict Cd (Table 2), and the best result was achieved with UVE-GBM. The calibration RC^2 achieved by the model was 0.9967, the RMSEC was 1.9853 mg kg^{-1}, the validation set RV^2 achieved 0.8899, the RMSEV was 11.3934 mg kg^{-1}, the prediction RP^2 achieved 0.9403 and the RMSEP was 8.5344 mg kg^{-1}. Figure 4b displays the scatter plots of the reference value and the prediction value for the Cd using the UVE-GBM.

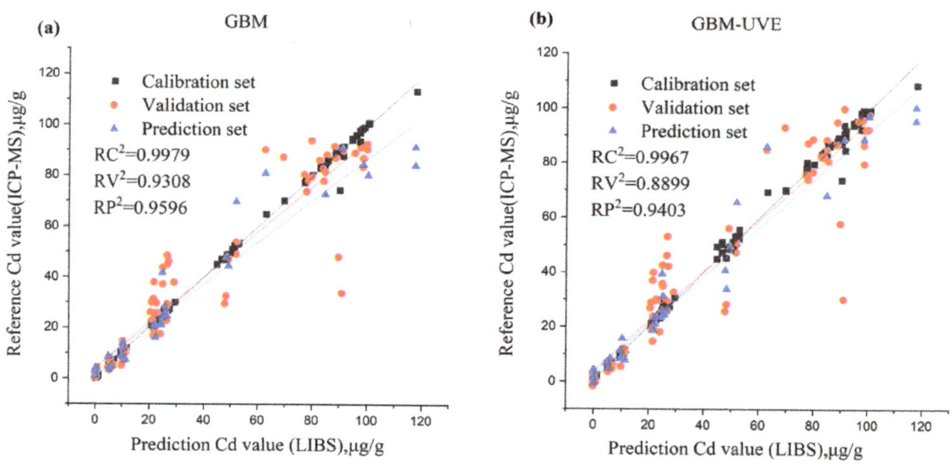

Figure 4. (**a**) Scatter plots of the full variables and reference value (**b**) Scatter plots of GBM combined with UVE feature selection method and reference value for the content of Cd.

RF-SVR and RF-PLSR also produced comparatively good results. For the RF-SVR calibration set, RC^2 = 0.9999, RMSEC = 0.1000 mg kg^{-1}, the validation set RV^2 achieved 0.9287, the RMSEV was 9.1671 mg kg^{-1} and the prediction set was RP^2 = 0.9322, RMSEP = 9.0933 mg kg^{-1}. For the RF-PLSR calibration set RC^2 = 0.9825, RMSEC = 4.6313 mg kg^{-1}, the validation set RV^2 achieved 0.9313, the RMSEV was 9.0014 mg kg^{-1} and the prediction set was RP^2 = 0.9182, RMSEP = 9.9903 mg kg^{-1}. On the other hand, CARS-SVR, UVE-SVR and CARS-GBM, RF-GBM produced lower prediction outcomes (Table 3). Meanwhile, for the full spectra, GBM was the best (Table 2); for the calibration set, RC^2 was 0.9984, RMSEC = 1.3693 mg kg^{-1}, the validation set RV^2 achieved 0.8924, the RMSEV was 11.2625 mgkg^{-1}, and the prediction set was RP^2 = 0.9139, RMSEP = 10.2494 mg kg^{-1}, as shown in Figure 4a,b below. GBM combined with UVE offers superior predictions than full wavelength GBM alone. GBM-UVE performs better with Cd because it is able to handle the large number of variables and select the most important ones.

Table 3. Prediction results of the different models using three variable selection (CARS, RF and UVE) methods.

	Elements	Methods	N	RC^2	RMSEC	RV^2	RMSEV	RP^2	RMSEP
SVR	Cd	CARS	37	0.9707	5.9935	0.9235	9.4984	0.8139	15.0703
		RF	151	0.9999	0.1000	0.9287	9.1671	0.9322	9.0933
		UVE	192	0.9525	7.6277	0.9066	10.4929	0.9116	10.3842
	Cu	CARS	66	0.9803	10.1823	0.9622	13.6476	0.9430	15.5721
		RF	120	0.9860	8.5699	0.9640	13.3319	0.9658	12.0658
		UVE	231	0.9715	12.2395	0.9412	17.0324	0.9648	12.2429
	Pb	CARS	17	0.9869	7.1947	0.9710	10.7251	0.9341	14.6841
		RF	124	0.9992	1.6707	0.9736	10.2323	0.9686	10.1224
		UVE	93	0.9799	8.9112	0.9585	12.8355	0.9395	14.0633
PLSR	Cd	CARS	37	0.9315	9.1659	0.9098	10.3132	0.9060	10.7075
		RF	151	0.9825	4.6313	0.9313	9.0014	0.9182	9.9903
		UVE	192	0.9659	6.4658	0.8910	11.3386	0.9072	10.6405
	Cu	CARS	66	0.9758	11.2746	0.9520	15.3841	0.9568	13.5625
		RF	120	0.9769	11.0291	0.9385	17.4200	0.9457	15.1970
		UVE	231	0.9856	8.7040	0.9411	17.0462	0.9575	13.4396
	Pb	CARS	17	0.9683	11.2039	0.9718	10.5798	0.9599	11.4563
		RF	124	0.9844	7.8436	0.9303	16.6519	0.9129	16.8794
		UVE	93	0.9720	10.5175	0.9556	13.2832	0.9381	14.2331
GBM	Cd	CARS	37	0.9907	3.3649	0.9009	10.8113	0.9146	10.2061
		RF	151	0.9982	1.4753	0.8474	13.4134	0.8909	11.5370
		UVE	192	0.9967	1.9853	0.8899	11.3934	0.9403	8.5344
	Cu	CARS	66	0.9933	5.9332	0.9316	18.3779	0.9665	11.9356
		RF	120	0.9929	6.0952	0.9371	17.6235	0.9545	13.9099
		UVE	231	0.9964	4.3168	0.9304	18.5337	0.9648	12.2323
	Pb	CARS	17	0.9970	3.4434	0.9469	14.5335	0.9429	13.6631
		RF	124	0.9982	2.6103	0.9329	16.3340	0.9136	16.8175
		UVE	93	0.9992	1.7248	0.9562	13.1967	0.9609	11.3113

N: Number of features selected; RC^2: Coefficient of determination for calibration; RV^2: Coefficient of determination for validation; RP^2: Coefficient of determination for prediction; RMSEC: Root mean square error for calibration; RMSEV: Root mean square error for validation; RMSEP: Root mean square error for prediction.

3.2.2. Cu Content Prediction Using Full and Selected Variables

For the prediction of Cu content, GBM combined with CARS achieved the best outcome with the following calibration set: RC^2 = 0.9933, RMSEC = 5.9332 mg kg^{-1}, the validation set RV^2 achieved 0.9316, the RMSEV was 18.3779 mg kg^{-1}, and the prediction set was RP^2 = 0.9665, RMSEP = 11.9356 mg kg^{-1}. This was followed by RF-SVR: for the calibration set RC^2 = 0.9860, RMSEC = 8.5699 mg kg^{-1}, the validation set RV^2 achieved 0.9640, the RMSEV was 13.3319 mg kg^{-1}, and the prediction set was RP^2 = 0.9658, RMSEP = 12.0658 mg kg^{-1}; UVE-SVR: for the calibration set RC^2 = 0.9715, RMSEC = 12.2395 mg kg^{-1}, the validation set RV^2 achieved 0.9412, the RMSEV was 17.0324 mg kg^{-1}, and the prediction set was RP^2 = 0.9648, RMSEP = 12.2429 mg kg^{-1} and UVE-GBM: for the calibration set (RC^2 = 0.9964, RMSEC = 4.3168 mg kg^{-1}), the validation set RV^2 achieved 0.9304, the RMSEV was 18.5337 mg kg^{-1}, and the prediction set was RP^2 = 0.9648, RMSEP = 12.2323 mg kg^{-1}. Figure 5 shows a scatter plot of the reference value and prediction value for Cu content using the CARS-GBM model. Although for the full spectra, GBM was the best (Table 2), for the calibration set RC^2 = 0.9979, RMSEC = 3.2885 mg kg^{-1}, the validation set RV^2 achieved 0.9308, the RMSEV was 18.4815 mg kg^{-1}, and the prediction set was RP^2 = 0.9596, RMSEP = 13.1156 mg kg^{-1}. It can be seen in Figure 5a,b that the prediction result from GBM combined with CARS is better than the result from the full wavelength GBM-CARS is better for Cu because it is able to capture interactions between variables.

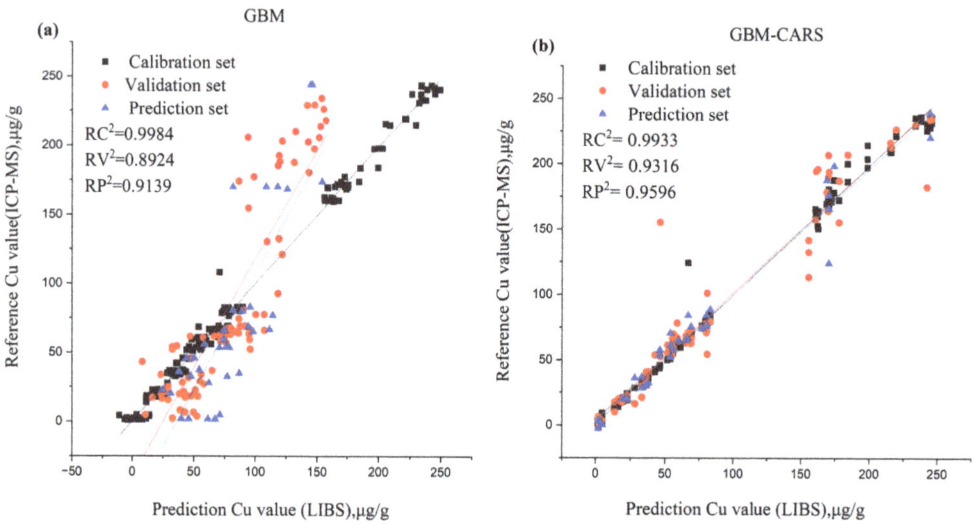

Figure 5. (**a**) Scatter plots of the full variables and reference value (**b**) Scatter plots of GBM combined with CARS feature selection method and reference value for the content of Cu.

3.2.3. Pb Content Prediction Using Full and Selected Variables

As shown in Table 3, RF combined with SVR produced the best result for the variable selection methods for Pb content prediction with the following calibration set: RC^2 = 0.9992, RMSEC = 1.6707 mg kg^{-1}; the validation set RV^2 achieved 0.9736, the RMSEV was 10.2323 mg kg^{-1}, and the prediction set was RP^2 = 0.9686, RMSEP = 10.1224 mg kg^{-1}. This was followed by UVE-GBM: calibration set RC^2 = 0.9992, RMSEC = 1.7248 mg kg^{-1}, the validation set RV^2 achieved 0.9562, the RMSEV was 13.1967 mg kg^{-1}, and the prediction set was RP^2 = 0.9609, RMSEP = 11.3113 mg kg^{-1} and CARS-PLSR: calibration set RC^2 = 0.9683, RMSEC = 11.2039 mg kg^{-1}, the validation set RV^2 achieved 0.9718, the RMSEV was 10.5798 $mgkg^{-1}$, and the prediction set was RP^2 = 0.9599, RMSEP = 11.4563 mg kg^{-1}. Although for the full spectra, GBM was also the best (Table 2), for the calibration set RC^2 = 0.9998, RMSEC = 0.8488 mg kg^{-1}, the validation set RV^2 achieved 0.9673, the RMSEV was 11.3933 mg kg^{-1}, and the prediction set was RP^2 = 0.9635, RMSEP = 10.9220 mg kg^{-1}. It can be noticed that the prediction result for SVR combined with RF, not GBM, as it is with Cd and Cu, as shown in the Figure 6a,b, it is slightly better than that for the full wavelength. SVR-RF is better for Pb because it is able to capture non-linear relationships between variables.

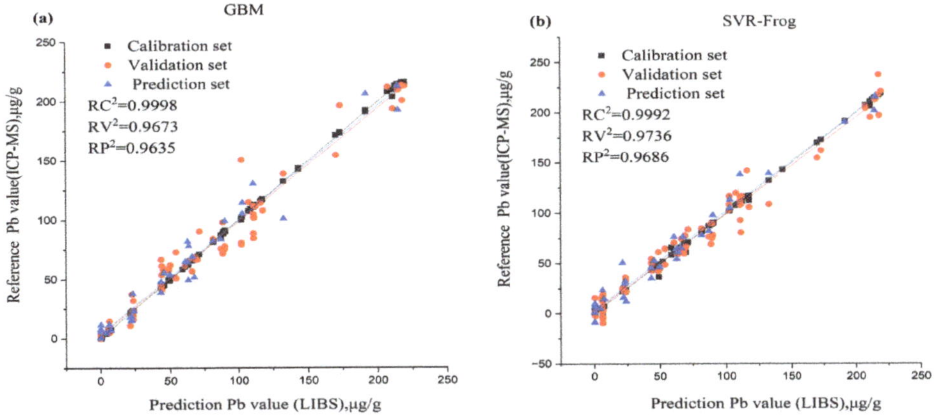

Figure 6. (**a**) Scatter plots of the full variables and reference value (**b**) Scatter plots of SVR combined with RF feature selection method and reference value for the content of Pb.

4. Discussion

Among the perennial herbaceous plants found mainly in the Zhejiang, Jiangsu, and Anhui provinces of China, *Fritillaria thunbergii* Miq. [11] was selected as the research object for collecting LIBS data and assessing its heavy metal contents (Cd, Cu, and Pb). The presence of heavy metals in excess of the standard poses a significant health risk. Due to increasing regulation of (TCM), heavy metals have become a priority pollutant in TCM and a serious safety concern [14–16]. Thunbergii fritillaria bulbs (Zhebeimu) are most commonly used in Chinese medical clinical practice to treat coughs caused by wind-heat and phlegm-heat, bronchitis, inflammation, hypertension, gastric ulcers, diarrhea, and bacterial infections [12].

In addition, Zhebeimu is widely used in the treatment of leukemia that is resistant to drugs [13]. Despite the development of fast and clean methods for TCM analysis [47]. Using LIBS to detect heavy metals in Fritillaria Thunbergii has still not been reported. Therefore, the ability to detect heavy metals in different varieties of Fritillaria using LIBS technology is a unique and vital process for establishing effective environmental control strategies and monitoring human exposure to heavy metals. In light of the above considerations and the advantages of the LIBS technique, in this study, a quantitative and simultaneous analysis of the contents of three heavy metals in *Fritillaria thunbergii* was performed. LIBS data, however, typically show a high degree of covariance due to the ability of LIBS to measure multiple emission lines from the same species or element; besides, several variables are responsible for creating each peak. Therefore, selecting variables may result in a reduction in the covariance and complexity of the model. It is also helpful to select more explanatory variables because this allows a better understanding of the multivariate system in terms of its chemical characteristics [56]. As well as improving the model fit and increasing the explained variance, removing noisy areas, such as extreme regions of the spectrometer and x variables with no analytical information, can also help to remove noisy regions [56].

In multivariate analysis, the matrix effect and fluctuations in the LIBS spectrum can be taken into account, in addition to the fluctuation in the LIBS spectrum from shot to shot. Recent years have seen an extensive use of chemometric methods, such as partial least squares (PLSR) and support vector machines (SVM) in the analysis of LIBS spectra for multivariable analysis [66,67]. An analysis of three variable selection methods (CARS, RF, and UVE) was conducted in order to simplify the calculation process and improve model performance. PLSR, SVM, and GBM models were run on the full spectra and spectral variables from CARS, RF, and UVE, respectively. A comparison of the performance of PLSR, SVM and GBM models was made by determining the root mean square error of calibration and prediction (RMSEC and RMSEP), as well as the greatest correlation coefficient square (R^2) between calibration and prediction sets (RC^2 and RP^2). It has been found that the combination of SVR and CARS yielded the smallest number of variables selected.

The combination of GBM with UVE yielded the greatest number of variables for Cd and Cu, and for Pb, GBM with RF yielded a limited number of variables (Table 3). The results of the different feature selection methods differ when compared when calibration, validation, and prediction sets are analyzed. Different feature selection methods were best for the different heavy metals. For Cd: GBM combined with UVE obtained the best performance; for Cu: GBM combined with CARS obtained the best performance, whereas for Pb: SVR combined with RF obtained the best performance. Several studies have demonstrated that there is no single technique for selecting features that is universally optimal [68], and multiple subsets of features are usually equally effective in predicting the data [69–71]. There is no doubt that full spectra contain essential information for elemental analysis but inevitably contain irrelevant information and noise, which weaken the model's capability [63,72,73]. The data analysis step in LIBS, as in other fields of spectroscopic analysis, is heavily constrained both by the high-dimensional input spaces and their inherent sparsity. In addition to reducing the measurement and storage requirements for LIBS data, properly selecting spectral features can facilitate the visualization and understanding of the data and enable more timely and cost-effective classification methods to be developed. The selection of feature variables needs to be applied to reduce computational complexity. As described above, it is worth noting that the number of variables selected by different feature selection methods varies widely (Table 3). For Cd, CARS showed the least variables (37), followed by RF (151) and UVE (192) and for Cu, CARS also showed the least variables (66), followed by RF (120), UVE (231) and finally for Pb, CARS showed the least variables (17), followed by UVE (93), RF (124), as can be seen from Figure 7. All methods select the informative region around 200–1000 nm, which is consistent with the Cd, Cu and Pb results in previous literature [1,37,39,40,48,63,74,75], indicating that these variable intervals are the informative variables.

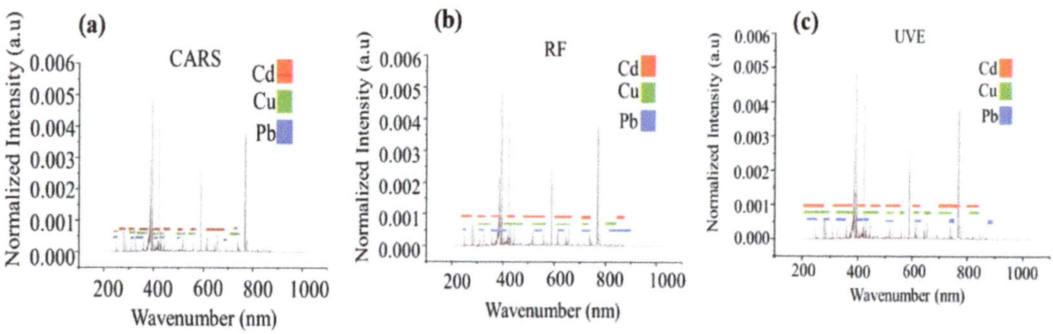

Figure 7. (**a**) Distributions of selected wave number by CARS (**b**) Distributions of selected wave number by RF (**c**) Distributions of selected wave number by UVE.

The different feature selection algorithms indicate some essential variables, so the variables in those intervals must be supported to improve the model's prediction ability, indicating those regions are informative variable intervals. Table 3 summarizes the results of CARS, RF, and UVE variable selection methods. Based on Figure 5, all selection methods perform significantly better in the test set than when compared with the full spectrum. The spectral matrix was then transformed by reducing the spectral information of the LIBS measurements to the most relevant variables that contained the most relevant spectral information of the respective heavy metals in Fritillaria. Multivariate analysis could address laser-to-sample interaction, experimental parameter variance, and matrices, among other factors [76].

5. Conclusions

The study indicated that when using the wavelength selection method, only a limited number of useful variables were extracted, and non-informative variables were eliminated. Therefore, the study explored effective variables by variable selection methods. The experimental results showed that all three methods applied could accomplish variable selection effectively, among which GBM-UVE for Cd, SVR-RF for Pb, and GBM-CARS for Cu produced the best results. Results of some recent heavy metal detection using LIBS are presented in Table 4 for comparison with the current work. Table 4 provides a comprehensive comparison of the results from the current work with those obtained from recent LIBS studies. It allows for a direct comparison of the accuracy of the detection methods. This study demonstrated the potential of LIBS coupled with variable selection and chemometrics as a tool for the rapid detection of heavy metals in varieties of *Fritillaria thunbergii*. It is essential to select variables that are associated with spectral information, so that subsequent modeling analysis can be based on more concise and effective spectral data. Additionally, the wavelengths selected could provide a theoretical foundation for the development of new instruments.

Table 4. Results of some recent heavy metals (Pb, Cd and Cu) detection using LIBS of various samples and current work.

Heavy-Metal	Sample	Spectral Line (nm)	Reference
Cd	Lettuce	214.44, 226.50, 228.80	[63]
	Sargassum fusiforme	441.56, 643.85	[37]
	Lipstick	467.9, 573.80	[23]
	This work	214.44, 226.50, 441.56, 467.90, 573.80, 643.85	
Cu	Traditional Chinese medicinal materials	324.79, 327.35	[74]
	Glycyrrhiza	324.70	[75]
	Ligusticum wallichii	324.46, 327.09	[40]
	Sargassum fusiforme	324.75, 327.40	[37]
	Rice	324.754, 327.396	[48]
	This work	324.09, 324.79	
Pb	Paint samples	405.70	[77]
	Henna paste	405.78	[39]
	Ligusticum wallichii	405.80	[40]
	Medicinal herbs	405.78, 404.00	[1]
	This work	280.00, 404.00, 405.70	

Author Contributions: Conceptualization, F.L. and M.H.K.; data curation, M.H.K., X.L. and R.C.; formal analysis, M.H.K. and R.C.; funding acquisition, F.L.; investigation, M.H.K.; methodology, M.H.K.; project administration, F.L.; resources, F.L. and W.K.; software, M.H.K. and M.L.G.; validation, M.H.K. and F.L.; visualization, M.H.K. and F.L.; writing—original draft, M.H.K.; writing—review and editing, F.L., M.H.K., M.L.G., R.C., X.L. and W.K. All authors have read and agreed to the published version of the manuscript.

Funding: This study was supported by National Natural Science Foundation of China (61975174) and Science and Technology Department of Zhejiang Province (2021C02023).

Data Availability Statement: The data presented in this study are available on request from the corresponding author.

Acknowledgments: The authors wish to acknowledge the funding agency stated above.

Conflicts of Interest: The authors declare no conflict of interest.

References

1. Jiang, Y.; Kang, J.; Wang, Y.; Chen, Y.; Li, R. Rapid and sensitive analysis of trace leads in medicinal herbs using laser-lnduced breakdown spectroscopy-laser-induced fluorescence (libs-lif). *Appl. Spectrosc.* **2019**, *73*, 1284–1291. [PubMed]
2. Gyamfi, E.T. Metals and metalloids in traditional medicines (ayurvedic medicines, nutraceuticals and traditional chinese medicines). *Environ. Sci. Pollut. Res. Int.* **2019**, *26*, 15767–15778. [CrossRef] [PubMed]
3. Dudka, S.; Miller, W.P. Accumulation of potentially toxic elements in plants and their transfer to human food chain. *J. Environ. Sci. Health Part B* **1999**, *34*, 681–708. [CrossRef] [PubMed]
4. Dube, A.; Zbytniewski, R.; Kowalkowski, T.; Cukrowska, E.; Buszewski, B. Adsorption and migration of heavy metals in soil. *Pol. J. Environ. Stud.* **2001**, *10*, 1–10.
5. Ercal, N.; Gurer-Orhan, H.; Aykin-Burns, N. Toxic metals and oxidative stress part i: Mechanisms involved in metal-induced oxidative damage. *Curr. Top Med. Chem.* **2001**, *1*, 529–539. [CrossRef]
6. Duruibe, J.; Ogwuegbu, M.O.C.; Egwurugwu, J. Heavy metal pollution and human biotoxic effects. *Int. J. Phys. Sci.* **2007**, *2*, 112–118.
7. Wei, B.; Yang, L. A review of heavy metal contaminations in urban soils, urban road dusts and agricultural soils from china. *Microchem. J.* **2010**, *94*, 99–107. [CrossRef]
8. Orecchio, S.; Amorello, D.; Barreca, S.; Valenti, A. Wood pellets for home heating can be considered environmentally friendly fuels? polycyclic aromatic hydrocarbons (pahs) in their ashes. *Microchem. J.* **2016**, *124*, 267–271. [CrossRef]
9. Tang, Y.; Wang, X.; Jia, J. Mercury poisoning presenting as sporadic creutzfeldt-jakob disease: A case report. *Ann. Intern. Med.* **2015**, *162*, 462–463. [CrossRef]
10. Chan, N.C.N.; Chan, K.P. Coarse basophilic stippling in lead poisoning. *Blood* **2017**, *129*, 3270. [CrossRef]
11. Hao, D.-C.; Gu, X.-J.; Xiao, P.-G.; Peng, Y. Phytochemical and biological research of fritillaria medicine resources. *Chin. J. Nat. Med.* **2013**, *11*, 330–344. [CrossRef]
12. Hill, C.B.; Roessner, U. Metabolic Profiling of Plants by GC–MS. In *The Handbook of Plant Metabolomics*; John Wiley & Sons: Hoboken, NJ, USA, 2013.
13. Ju, W. Clinical observation on effects of adjuvant chemotherapy with fritillary thunbergii bulb complex granules on symptoms of refractory acute leukemia patients. *Beijing J. Tradit. Chin. Med.* **2013**, *32*, 833–835.
14. Tian, M.; Fang, L.; Yan, X.; Xiao, W.; Row, K.H. Determination of heavy metal ions and organic pollutants in water samples using ionic liquids and ionic liquid-modified sorbents. *J. Anal. Methods Chem.* **2019**, *2019*, 1948965. [CrossRef]
15. Yin, H.; Truskewycz, A.; Cole, I. Quantum dot (qd)-based probes for multiplexed determination of heavy metal ions. *Microchim. Acta* **2020**, *187*, 336. [CrossRef]
16. Fu, F.; Wang, Q. Removal of heavy metal ions from wastewaters: A review. *J. Environ. Manag.* **2011**, *92*, 407–418. [CrossRef]
17. Fortes, F.J.; Moros, J.; Lucena, P.; Cabalín, L.M.; Laserna, J.J. Laser-induced breakdown spectroscopy. *Anal. Chem.* **2013**, *85*, 640–669. [CrossRef]
18. Kim, G.; Kwak, J.; Kim, K.-R.; Lee, H.; Kim, K.-W.; Yang, H.; Park, K. Rapid detection of soils contaminated with heavy metals and oils by laser induced breakdown spectroscopy (libs). *J. Hazard. Mater.* **2013**, *263*, 754–760. [CrossRef]
19. Sawan, S.; Maalouf, R.; Errachid, A.; Jaffrezic-Renault, N. Metal and metal oxide nanoparticles in the voltammetric detection of heavy metals: A review. *TrAC Trends Anal. Chem.* **2020**, *131*, 116014. [CrossRef]
20. Malik, L.A.; Bashir, A.; Qureashi, A.; Pandith, A.H. Detection and removal of heavy metal ions: A review. *Environ. Chem. Lett.* **2019**, *17*, 1495–1521. [CrossRef]
21. Amorello, D.; Barreca, S.; Gulli, E.; Orecchio, S. Platinum and rhodium in wine samples by using voltammetric techniques. *Microchem. J.* **2017**, *130*, 229–235. [CrossRef]
22. Rai, N.K.; Rai, A.K. Libs—An efficient approach for the determination of cr in industrial wastewater. *J. Hazard. Mater.* **2008**, *150*, 835–838. [CrossRef] [PubMed]
23. Gondal, M.; Seddigi, Z.; Nasr, M.; Gondal, B. Spectroscopic detection of health hazardous contaminants in lipstick using laser induced breakdown spectroscopy. *J. Hazard. Mater.* **2010**, *175*, 726–732. [CrossRef]
24. Hahn, D.W.; Omenetto, N. Laser-induced breakdown spectroscopy (libs), part i: Review of basic diagnostics and plasma–particle interactions: Still-challenging issues within the analytical plasma community. *Appl. Spectrosc.* **2010**, *64*, 335A–366A. [CrossRef] [PubMed]
25. Ma, S.; Tang, Y.; Ma, Y.; Chu, Y.; Chen, F.; Hu, Z.; Zhu, Z.; Guo, L.; Zeng, X.; Lu, Y. Determination of trace heavy metal elements in aqueous solution using surface-enhanced laser-induced breakdown spectroscopy. *Opt. Express* **2019**, *27*, 15091–15099. [CrossRef] [PubMed]
26. Ma, S.; Tang, Y.; Ma, Y.; Dong, D.; Guo, L.; Zhu, H.; Liu, J.; Lu, Y. The pH effect on the detection of heavy metals in wastewater by laser-induced breakdown spectroscopy coupled with a phase transformation method. *J. Anal. At. Spectrom.* **2020**, *35*, 198–203. [CrossRef]
27. Hahn, D.W.; Omenetto, N. Laser-induced breakdown spectroscopy (libs), part ii: Review of instrumental and methodological approaches to material analysis and applications to different fields. *Appl. Spectrosc.* **2012**, *66*, 347–419. [CrossRef]
28. Noll, R.; Fricke-Begemann, C.; Connemann, S.; Meinhardt, C.; Sturm, V. Libs analyses for industrial applications—An overview of developments from 2014 to 2018. *J. Anal. At. Spectrom.* **2018**, *33*, 945–956. [CrossRef]

29. Zou, X.H.; Guo, L.B.; Shen, M.; Li, X.Y.; Hao, Z.Q.; Zeng, Q.D.; Lu, Y.F.; Wang, Z.M.; Zeng, X.Y. Accuracy improvement of quantitative analysis in laser-induced breakdown spectroscopy using modified wavelet transform. *Opt. Express* **2014**, *22*, 10233–10238. [CrossRef]
30. Gondal, M.A.; Siddiqui, M.N. Identification of different kinds of plastics using laser-induced breakdown spectroscopy for waste management. *J. Environ. Sci. Health Part A* **2007**, *42*, 1989–1997. [CrossRef]
31. Unnikrishnan, V.K.; Choudhari, K.S.; Kulkarni, S.D.; Nayak, R.; Kartha, V.B.; Santhosh, C.; Suri, B.M. Biomedical and environmental applicationsof laser-induced breakdown spectroscopy. *Pramana* **2014**, *82*, 397–401. [CrossRef]
32. Ahmed, I.; Yang, J.; Law, A.W.L.; Manno, F.A.M.; Ahmed, R.; Zhang, Y.; Lau, C. Rapid and in situ optical detection of trace lithium in tissues. *Biomed. Opt. Express* **2018**, *9*, 4459–4471. [CrossRef]
33. Tang, S.S.; Hao, Z.Q.; Ma, Y.Y.; Ma, S.X.; Guo, L.B.; Guo, Y.M.; Zeng, X.Y.; Chu, Y.W.; Lu, Y.F. Accuracy and stability improvement for meat species identification using multiplicative scatter correction and laser-induced breakdown spectroscopy. *Opt. Express* **2018**, *26*, 10119.
34. Pandhija, S.; Rai, N.K.; Rai, A.K.; Thakur, S.N. Contaminant concentration in environmental samples using libs and cf-libs. *Appl. Phys. B* **2010**, *98*, 231–241. [CrossRef]
35. Gondal, M.A.; Hussain, T.; Yamani, Z.H.; Baig, M.A. On-line monitoring of remediation process of chromium polluted soil using libs. *J. Hazard. Mater.* **2009**, *163*, 1265–1271. [CrossRef]
36. Wang, W.; Kong, W.; Shen, T.; Man, Z.; Zhu, W.; He, Y.; Liu, F.; Liu, Y. Application of laser-induced breakdown spectroscopy in detection of cadmium content in rice stems. *Front. Plant Sci.* **2020**, *11*, 599616. [CrossRef]
37. Su, L.; Shi, W.; Chen, X.; Meng, L.; Yuan, L.; Chen, X.; Huang, G. Simultaneously and quantitatively analyze the heavy metals in Sargassum fusiforme by laser-induced breakdown spectroscopy. *Food Chem.* **2021**, *338*, 127797. [CrossRef]
38. Liu, Y.; Chu, Y.; Hu, Z.; Zhang, S.; Ma, S.; Khan, M.S.; Chen, F.; Zhang, D.; Guo, L.; Lau, C. High-sensitivity determination of trace lead and cadmium in cosmetics using laser-induced breakdown spectroscopy with ultrasound-assisted extraction. *Microchem. J.* **2020**, *158*, 105322. [CrossRef]
39. Rehan, I.; Gondal, M.A.; Rehan, K.; Sultana, S.; Dastageer, M.A.; Al-Adel, F.F. Libs for the detection of lead in ready to use henna paste and nutrients in fresh henna leaves and cultivated soils. *Talanta* **2019**, *199*, 203–211. [CrossRef]
40. Wang, J.; Xue, S.; Zheng, P.; Chen, Y.; Peng, R. Determination of lead and copper in ligusticum wallichii by laser-induced breakdown spectroscopy. *Anal. Lett.* **2017**, *50*, 2000–2011. [CrossRef]
41. Zhao, S.; Song, C.; Gao, X.; Lin, J. Quantitative analysis of pb in soil by femtosecond-nanosecond double-pulse laser-induced breakdown spectroscopy. *Results Phys.* **2019**, *15*, 102736. [CrossRef]
42. Zhu, C.; Lv, J.; Liu, K.; Chen, J.; Liu, K.; Li, G.; Lu, B.; Li, X. Rapid determination of arsenic in traditional chinese medicine by laser-induced breakdown spectroscopy (libs). *Anal. Lett.* **2022**, *55*, 2531–2541. [CrossRef]
43. Rehan, I.; Gondal, M.A.; Rehan, K.; Sultana, S. Spectral diagnosis of health hazardous toxins in face foundation powders using laser induced breakdown spectroscopy and inductively coupled plasma-optical emission spectroscopy (icp-oes). *Talanta* **2020**, *217*, 121007. [CrossRef] [PubMed]
44. Zhu, C.; Tang, Z.; Li, Q.; Zhou, R.; Lv, J.; Zhang, W.; Zhan, K.; Li, X.; Zeng, X. Lead of detection in rhododendron leaves using laser-induced breakdown spectroscopy assisted by laser-induced fluorescence. *Sci. Total Environ.* **2020**, *738*, 139402. [CrossRef] [PubMed]
45. Peng, J.; He, Y.; Jiang, J.; Zhao, Z.; Zhou, F.; Liu, F. High-accuracy and fast determination of chromium content in rice leaves based on collinear dual-pulse laser-induced breakdown spectroscopy and chemometric methods. *Food Chem.* **2019**, *295*, 327–333. [CrossRef] [PubMed]
46. Yang, P.; Zhou, R.; Zhang, W.; Yi, R.; Tang, S.; Guo, L.; Hao, Z.; Li, X.; Lu, Y.; Zeng, X. High-sensitivity determination of cadmium and lead in rice using laser-induced breakdown spectroscopy. *Food Chem.* **2019**, *272*, 323–328. [CrossRef] [PubMed]
47. Li, K.; Yang, H.; Yuan, X.; Zhang, M. Recent developments of heavy metals detection in traditional chinese medicine by atomic spectrometry. *Microchem. J.* **2021**, *160*, 105726. [CrossRef]
48. Liu, F.; Ye, L.; Peng, J.; Song, K.; Shen, T.; Zhang, C.; He, Y. Fast detection of copper content in rice by laser-induced breakdown spectroscopy with uni-and multivariate analysis. *Sensors* **2018**, *18*, 705. [CrossRef]
49. Kabir, M.H.; Guindo, M.L.; Chen, R.; Sanaeifar, A.; Liu, F. Application of laser-induced breakdown spectroscopy and chemometrics for the quality evaluation of foods with medicinal properties: A review. *Foods* **2022**, *11*, 2051. [CrossRef]
50. Huffman, C.; Sobral, H.; Terán-Hinojosa, E. Laser-induced breakdown spectroscopy spectral feature selection to enhance classification capabilities: A t-test filter approach. *Spectrochim. Acta Part B At. Spectrosc.* **2019**, *162*, 105721. [CrossRef]
51. Gholizadeh, A.; Žižala, D.; Saberioon, M.; Borůvka, L. Soil organic carbon and texture retrieving and mapping using proximal, airborne and Sentinel-2 spectral imaging. *Remote Sens. Environ.* **2018**, *218*, 89–103. [CrossRef]
52. Guindo, M.L.; Kabir, M.H.; Chen, R.; Liu, F. Particle swarm optimization and multiple stacked generalizations to detect nitrogen and organic-matter in organic-fertilizer using vis-nir. *Sensors* **2021**, *21*, 4882. [CrossRef]
53. Douglas, R.K.; Nawar, S.; Cipullo, S.; Alamar, M.C.; Coulon, F.; Mouazen, A.M. Evaluation of vis-nir reflectance spectroscopy sensitivity to weathering for enhanced assessment of oil contaminated soils. *Sci. Total Environ.* **2018**, *626*, 1108–1120. [CrossRef]
54. Shi, T.; Chen, Y.; Liu, H.; Wang, J.; Wu, G. Soil organic carbon content estimation with laboratory-based visible-near-infrared reflectance spectroscopy: Feature selection. *Appl. Spectrosc.* **2014**, *68*, 831–837. [CrossRef]

55. Rossel, R.A.V.; Behrens, T. Using data mining to model and interpret soil diffuse reflectance spectra. *Geoderma* **2010**, *158*, 46–54. [CrossRef]
56. Xiaobo, Z.; Jiewen, Z.; Povey, M.J.W.; Holmes, M.; Hanpin, M. Variables selection methods in near-infrared spectroscopy. *Anal. Chim. Acta* **2010**, *667*, 14–32. [CrossRef]
57. Li, H.; Liang, Y.; Xu, Q.; Cao, D. Key wavelengths screening using competitive adaptive reweighted sampling method for multivariate calibration. *Anal. Chim. Acta* **2009**, *648*, 77–84. [CrossRef]
58. Vohland, M.; Ludwig, M.; Thiele-Bruhn, S.; Ludwig, B. Determination of soil properties with visible to near- and mid-infrared spectroscopy: Effects of spectral variable selection. *Geoderma* **2014**, *223–225*, 88–96. [CrossRef]
59. Wang, S.-L.; Li, J.; Fang, J. Predicting progression of als disease with random frog and support vector regression method. In *Intelligent Computing Methodologies*; Springer International Publishing: Cham, Switzerland, 2016; pp. 160–170.
60. Pan, L.; Lu, R.; Zhu, Q.; Tu, K.; Cen, H. Predict compositions and mechanical properties of sugar beet using hyperspectral scattering. *Food Bioprocess Technol.* **2016**, *9*, 1177–1186. [CrossRef]
61. Dong, J.; Guo, W.; Wang, Z.; Liu, D.; Zhao, F. Nondestructive determination of soluble solids content of 'fuji' apples produced in different areas and bagged with different materials during ripening. *Food Anal. Methods* **2016**, *9*, 1087–1095. [CrossRef]
62. Guindo, M.L.; Kabir, M.H.; Chen, R.; Liu, F. Potential of vis-nir to measure heavy metals in different varieties of organic-fertilizers using boruta and deep belief network. *Ecotoxicol. Env. Saf.* **2021**, *228*, 112996. [CrossRef]
63. Shen, T.; Kong, W.; Liu, F.; Chen, Z.; Yao, J.; Wang, W.; Peng, J.; Chen, H.; He, Y. Rapid determination of cadmium contamination in lettuce using laser-induced breakdown spectroscopy. *Molecules* **2018**, *23*, 2930. [CrossRef] [PubMed]
64. Tawfik, W.; Askar, A. Study of the matrix effect on the plasma characterization of heavy elements in soil sediments. *Prog. Phys.* **2007**, *3*, 46–52.
65. El Haddad, J.; Canioni, L.; Bousquet, B. Good practices in libs analysis: Review and advices. *Spectrochim. Acta Part B At. Spectrosc.* **2014**, *101*, 171–182. [CrossRef]
66. Zhang, T.; Wu, S.; Dong, J.; Wei, J.; Wang, K.; Tang, H.; Yang, X.; Li, H. Quantitative and classification analysis of slag samples by laser induced breakdown spectroscopy (libs) coupled with support vector machine (svm) and partial least square (pls) methods. *J. Anal. At. Spectrom.* **2015**, *30*, 368–374. [CrossRef]
67. Yan, C.; Qi, J.; Ma, J.; Tang, H.; Zhang, T.; Li, H. Determination of carbon and sulfur content in coal by laser induced breakdown spectroscopy combined with kernel-based extreme learning machine. *Chemom. Intell. Lab. Syst.* **2017**, *167*, 226–231. [CrossRef]
68. Pořízka, P.; Klus, J.; Hrdlička, A.; Vrábel, J.; Škarková, P.; Prochazka, D.; Novotný, J.; Novotný, K.; Kaiser, J. Impact of laser-induced breakdown spectroscopy data normalization on multivariate classification accuracy. *J. Anal. At. Spectrom.* **2017**, *32*, 277–288. [CrossRef]
69. Myakalwar, A.K.; Spegazzini, N.; Zhang, C.; Anubham, S.K.; Dasari, R.R.; Barman, I.; Gundawar, M.K. Less is more: Avoiding the libs dimensionality curse through judicious feature selection for explosive detection. *Sci. Rep.* **2015**, *5*, 13169. [CrossRef]
70. Wang, Q.; Teng, G.; Qiao, X.; Zhao, Y.; Kong, J.; Dong, L.; Cui, X. Importance evaluation of spectral lines in laser-induced breakdown spectroscopy for classification of pathogenic bacteria. *Biomed. Opt. Express* **2018**, *9*, 5837–5850. [CrossRef]
71. Menze, B.H.; Kelm, B.M.; Masuch, R.; Himmelreich, U.; Bachert, P.; Petrich, W.; Hamprecht, F.A. A comparison of random forest and its gini importance with standard chemometric methods for the feature selection and classification of spectral data. *BMC Bioinform.* **2009**, *10*, 213. [CrossRef]
72. Duan, F.; Fu, X.; Jiang, J.; Huang, T.; Ma, L.; Zhang, C. Automatic variable selection method and a comparison for quantitative analysis in laser-induced breakdown spectroscopy. *Spectrochim. Acta Part B At. Spectrosc.* **2018**, *143*, 12–17. [CrossRef]
73. Zhang, C.; Wang, Q.; Liu, F.; He, Y.; Xiao, Y. Rapid and non-destructive measurement of spinach pigments content during storage using hyperspectral imaging with chemometrics. *Measurement* **2017**, *97*, 149–155. [CrossRef]
74. Sun, D.; Yang, F.; Su, M.; Han, W.; Dong, C. Quantitative analysis of cu in traditional chinese medicinal materials using laser-induced breakdown spectroscopy. *Microw. Opt. Technol. Lett.* **2022**. [CrossRef]
75. Zhu, C.; Lv, J.; Liu, K.; Li, Q.; Tang, Z.; Zhou, R.; Zhang, W.; Chen, J.; Liu, K.; Li, X. Fast detection of harmful trace elements in glycyrrhiza using standard addition and internal standard method–Laser-induced breakdown spectroscopy (sais-libs). *Microchem. J.* **2021**, *168*, 106408. [CrossRef]
76. Liu, F.; Shen, T.; Kong, W.; Peng, J.; Zhang, C.; Song, K.; Wang, W.; Zhang, C.; He, Y. Quantitative analysis of cadmium in tobacco roots using laser-induced breakdown spectroscopy with variable index and chemometrics. *Front. Plant Sci.* **2018**, *9*, 1316. [CrossRef]
77. Gondal, M.A.; Nasr, M.M.; Ahmed, M.M.; Yamani, Z.H.; Alsalhi, M. Detection of lead in paint samples synthesized locally using-laser-induced breakdown spectroscopy. *J. Environ. Sci. Health Part A* **2011**, *46*, 42–49. [CrossRef]

foods MDPI

Article

Phenotypic Analysis of Fourier-Transform Infrared Milk Spectra in Dairy Goats

Bartolo de Jesús Villar-Hernández [1], Nicolò Amalfitano [2], Alessio Cecchinato [2], Michele Pazzola [3], Giuseppe Massimo Vacca [3] and Giovanni Bittante [2,*]

1 Departamento de Zootecnia, Universidad Autónoma Chapingo, Texcoco 56230, Mexico
2 Department of Agronomy, Food and Natural Resources, Animals and Environment (DAFNAE), University of Padova, Viale dell'Università 16, 35020 Legnaro, Italy
3 Department of Veterinary Medicine, University of Sassari, 07100 Sassari, Italy
* Correspondence: bittante@unipd.it

Abstract: The infrared spectrum of bovine milk is used to predict many interesting traits, whereas there have been few studies on goat milk in this regard. The objective of this study was to characterize the major sources of variation in the absorbance of the infrared spectrum in caprine milk samples. A total of 657 goats belonging to 6 breeds and reared on 20 farms under traditional and modern dairy systems were milk-sampled once. Fourier-transform infrared (FTIR) spectra were taken (2 replicates per sample, 1314 spectra), and each spectrum contained absorbance values at 1060 different wavenumbers (5000 to 930 $\times$ cm^{-1}), which were treated as a response variable and analyzed one at a time (i.e., 1060 runs). A mixed model, including the random effects of sample/goat, breed, flock, parity, stage of lactation, and the residual, was used. The pattern and variability of the FTIR spectrum of caprine milk was similar to those of bovine milk. The major sources of variation in the entire spectrum were as follows: sample/goat (33% of the total variance); flock (21%); breed (15%); lactation stage (11%); parity (9%); and the residual unexplained variation (10%). The entire spectrum was segmented into five relatively homogeneous regions. Two of them exhibited very large variations, especially the residual variation. These regions are known to be affected by the absorbance of water, although they also exhibited wide variations in the other sources of variation. The average repeatability of these two regions were 45% and 75%, whereas for the other three regions it was about 99%. The FTIR spectrum of caprine milk could probably be used to predict several traits and to authenticate the origin of goat milk.

Keywords: FTIR; mid-infrared; caprine milk; milk absorbance spectra; variance components; sources of variation

Citation: Villar-Hernández, B.d.J.; Amalfitano, N.; Cecchinato, A.; Pazzola, M.; Vacca, G.M.; Bittante, G. Phenotypic Analysis of Fourier-Transform Infrared Milk Spectra in Dairy Goats. *Foods* **2023**, *12*, 807. https://doi.org/10.3390/foods12040807

Academic Editors: Mourad Kharbach and Samuli Urpelainen

Received: 18 January 2023
Revised: 7 February 2023
Accepted: 9 February 2023
Published: 14 February 2023

1. Introduction

Fourier-transform infrared spectroscopy (FTIR) is a high-throughput method with multiple applications that has revolutionized the livestock sector [1]. FTIR technology measures the vibrations of the atoms in a molecule related to their bond strengths. When the frequency of the IR radiation directed at the bond is equal to the frequency of the bond's vibration, the bond absorbs the radiation. The frequencies absorbed constitute the molecule's IR spectrum. Analyzing infrared spectra can tell us what molecules (hence, what compounds) are present in a sample (of milk, cheese, meat, etc.) and at what concentrations.

According to Smith, 2011 [2], infrared spectroscopy is almost universal, in the sense that the infrared spectra of solids, liquids, and gases can all be measured. A second advantage concerns the richness of information obtained: the position of a spectral peak reveals the structure of the molecules, the peak intensity reveals the concentration of molecules, and the peak width is sensitive to the chemical matrix. Further key features of FTIR are that it is relatively easy, fast, and sensitive, i.e., it is a non-destructive method that requires only grams/milliliters of material to produce a good spectrum. The main disadvantage of FTIR with regard to milk samples is the presence of water, which has intense peaks that can mask the spectra of the milk components.

In milk analyses, transmittance is defined as $T = I_m / I_w$, where I_m and I_w represent the transmitted radiation of milk and water (reference or identity testing), respectively. Usually, $I_m < I_w$ (and $T < 1$) due to the presence of other milk components that affect the transmittance radiance, but it may also be that $I_m > I_w$ (and $T > 1$). Another related metric is absorbance, defined as $A = -\log_e T$. Note that, by construction, the values of T are centered at one, so when taking the base e logarithm, A is centered at zero. When conducting quantitative analyses, absorbance is preferred because it is linearly proportional to concentration, according to Beer's law [2].

FTIR spectrometry can be used—with different levels of accuracy—for:

i. the prediction of milk components than can be easily distinguished from each other due to the specific vibrational properties of their chemical bonds;
ii. the prediction of groups of milk components with similar chemical and vibrational properties;
iii. the prediction of the chemical components or physical-technological characteristics of milk that do not have specific vibrational properties;
iv. the prediction of the metabolic characteristics of animals affecting certain properties of milk;
v. the authentication of the origin of milk.

Fourier-transform infrared spectroscopy is, of course, most commonly used to predict milk components with specific vibrational properties with highly accurate results guaranteed. This is the case for the major chemical components of milk (fat, protein, and lactose), the predictions of which are regulated by the International Organization for Standardization (ISO) and certified by the International Committee of Animal Recording [3–5].

Although prediction of the total milk content of fat and protein is accurate, the quantification of individual milk fatty acids [6–8] or individual protein fractions [9–11] is much less accurate. In fact, the chemical and vibrational characteristics of the members of the same chemical family are very similar, so discriminating one milk fatty acid from another one is not simple, and the same is true when comparing a milk protein fraction with another one. In this case, prediction is based not only on the chemical bonds of each compound, but probably also on the relationships between the concentration of the compound in question and other characteristics of the milk [12].

Predicting chemical compounds without specific vibrational characteristics, such as minerals [13–15], and the physical-technological properties of milk [16–18] is based substantially on covariance between the compound or trait in question and others related in some way to it. Accuracy in such cases is never very high and depends on the closeness of association and the specificity of the vibrational properties of the associated compounds.

The structure of the covariance matrix is also fundamental to predicting traits defining the metabolism of animals using milk FTIR spectroscopy. Examples include the prediction of blood metabolites [19,20], animal dismetabolism [21], nutritional efficiency [22,23], animal energy balance [24,25], enteric methane emissions [26–28], fertility [29–31], etc.

Lastly, infrared spectra can be used to determine the fingerprint of milk for authentication purposes where adulteration is suspected [32–34], or to certify the area of origin [35], or the farming system in which the milk was produced [36,37]. Some other secondary method has been proposed for the rapid prediction of some substance or property of milk, but none has the versatility of FTIR spectrometry in that, with just one sample and one instrument, in a single passage, many components and characteristics of the milk, the dairy animal, and the dairy system can be predicted. Moreover, provided that the spectra are stored, new traits can be predicted a posteriori, simply using new calibration equations on old spectra.

Many studies have been carried out on milk from bovine species, but there has been very little research on the milk of other species, particularly goats [38–40].

The characteristics of bovine FTIR milk spectra have been extensively studied with the aim of understanding the properties of different fractions of the spectrum and of individual wavelengths, and in order to identify the areas related to specific chemical bonds. In a previous study on the FTIR spectrum of bovine milk, we clearly identified five major spectral fractions in the range of near-infrared (NIR), or short-wave infrared (SWIR), to mid-infrared (MIR, or MWIR) and long-infrared, or long-wave-infrared (LWIR) radiation [41]. Aside from phenotypic properties, bovine milk spectra have also been analyzed to identify the possible genetic parameters of wavelength absorbance [42–44] and to obtain genomic information [45,46].

Given that, unlike bovine milk, there is little knowledge of the FTIR spectrum of goat milk, and that this knowledge is needed for the correct use of goat spectra to predict and interpret chemical contents and metabolic properties, the aims of this research were as follows: (1) to study the absorbance values and their phenotypic variances of each wavenumber of the goat milk spectrum and to compare them with the characteristics of bovine milk spectrum from the literature; (2) to estimate the major components of phenotypic variance; and (3) to estimate the repeatability of 1060 wavenumbers in the FTIR region from 5000 to 930 $\times$ cm^{-1} of goat milk samples.

2. Materials and Methods

2.1. Experimental Design

This study is part of a research project (the Good-Milk project), which mainly aims to study the qualitative properties of goat milk compared to milk from other dairy species, with particular emphasis on milk protein fractions and genetic variants. The work package dedicated to the study and use of FTIR spectra to predict the qualitative traits of goat milk focuses on sampling milk from many goats representing different farming systems, breeds, parities, and lactation stages.

The 657 goats sampled for the present study were of 6 different breeds and were reared on 20 farms in Sardinia (Italy). The farms were classified into 3 dairy systems (traditional, intermediate, and modern) according to the feeding system, farm management, and conditions in the facilities. Information on the farming systems, animals, and sampling procedure are reported in a previous study [47]. Information on the qualitative and technological traits of the milk samples is provided in other studies carried out on the same database [48–50]. The goats belonged to the following breeds: Saanen (41 goats); Camosciata delle Alpi (164); Murciano-Granadina (143); Maltese (122); Sarda (44); and Sarda Primitiva (143). Parities on the day of sampling ranged from 1 to 15, and days in milk (DIM) ranged from 10 to 224.

A 50 mL milk sample was collected from each goat and stored immediately at 4 °C. Within 24 h of sampling, FTIR spectra (2 replicates per milk sample) were obtained with a MilkoScan FT6000 milk analyzer (Foss A/S, Hillerød, Denmark). Absorbance values of 1060 spectral wavenumbers from 5000 to 930 $\times$ cm^{-1} were recorded. Parity was recoded from a count variable to a factor with six levels (1–2, 3, 4, 5, 6, 7+), while DIM was recoded from a count variable to a factor with five levels (1–60, 61–90, 91–120, 121–150, 151–240).

2.2. Statistical Model

A database of 1,392,840 absorbance values (657 samples/goats $\times$ 2 replicates $\times$ 1060 wavenumbers) was compiled. Outlier spectra were checked on the basis of the Mahalanobis distance. A linear mixed model was fitted to estimate the variance components and repeatability of the milk absorbance at each individual FTIR wavenumber. The model was as follows:

$$y_{ijklm} = \mu + G_i + B_j + F_k + P_l + D_m + e_{ijklm}, \quad (1)$$

where y_{ijklm} represents the 1314 absorbance values (A) recorded for a particular wavenumber (657 samples/goats in duplicate), μ is the overall absorbance mean or intercept (fixed) for a particular wavenumber, $G_i \sim iid\ N(0, \sigma_G^2)$ is the random effect of the ith sample/goat, $B_j \sim iid\ N(0, \sigma_B^2)$ is the random effect of the jth breed, $F_k \sim iid\ N(0, \sigma_F^2)$ is the random effect of the kth flock, $P_l \sim iid\ N(0, \sigma_P^2)$ is the random effect of the lth parity, $D_m \sim iid\ N(0, \sigma_D^2)$ is the random effect of the mth days in milk, and $e_{ijklm} \sim iid\ N(0, \sigma_e^2)$ is the model residual. Here, $N(\cdot, \cdot)$ stands for a normally distributed random variable, and 'iid' stands for one that is independent and identically distributed. Strictly speaking, model (1) is a linear mixed model where μ is the only fixed effect and the other components are treated as random. All random components were assumed to be independent of each other. This model was fitted 1060 times (one per wavenumber) using the BGLR-R package [51] in the R programming language [52].

From the 1060 individual mixed models as per Equation (1) fitted to each wavenumber of the FTIR spectrum, the variance components of the random effects, i.e., $\hat{\sigma}_G^2$, $\hat{\sigma}_B^2$, $\hat{\sigma}_F^2$, $\hat{\sigma}_P^2$, $\hat{\sigma}_D^2$, the residual term $\hat{\sigma}_e^2$, and the computed proportion of variance explained by each term were estimated. Under additivity, the estimate of total (phenotypic) variance is calculated as the sum of the variance components, $\hat{\sigma}_{Ph}^2 = \hat{\sigma}_G^2 + \hat{\sigma}_B^2 + \hat{\sigma}_F^2 + \hat{\sigma}_P^2 + \hat{\sigma}_D^2 + \hat{\sigma}_e^2$, and sample repeatability is expressed as $R_{sample} = \frac{\hat{\sigma}_G^2 + \hat{\sigma}_B^2 + \hat{\sigma}_F^2 + \hat{\sigma}_P^2 + \hat{\sigma}_D^2}{\hat{\sigma}_{Ph}^2}$. Sample repeatability was calculated as the sum of the variances due to the random effects included in the model as a percentage of the phenotypic variance (i.e., the sum of the variance of the random factors plus the residual variance).

3. Results

3.1. Descriptive Statistics of the Goat Milk Spectra

Figure 1 depicts the mean phenotypic absorbance values and the 0.025 and 0.975 quantiles of the milk samples for the entire FTIR spectrum (1060 individual wavelengths) obtained from the milk of the 657 goats included in the study (1314 milk spectra). The two regions with very high phenotypic variability in absorbance values in the goat milk dataset, which are characterized by the wavelengths approximately between 3669 and 3052 and between 1698 and 1586, are of note. These regions are known to be water zones in cow milk samples and are indicated by peaks of water absorption that can mask the effects of other milk components. Goat milk spectra are therefore similar to cow milk spectra, and we will refer to these regions henceforth as the water zones or water regions.

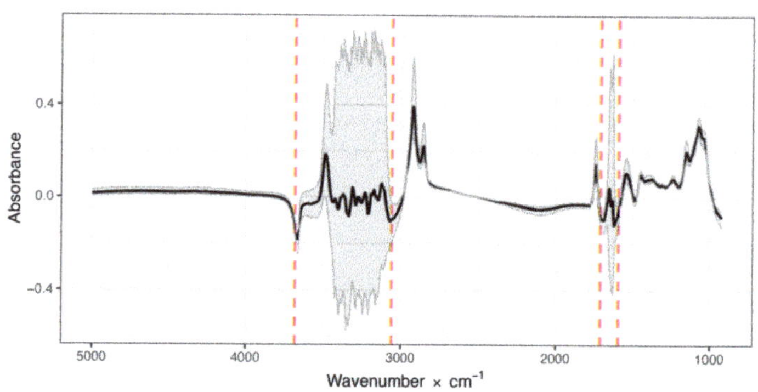

Figure 1. Mean absorbance values (dark solid line) and the 0.025 and 0.975 quantiles (gray lines) of the 1060 individual infrared wavelengths (5000 to 930 × cm^{-1}), measured from 1314 spectra of milk samples from 657 goats. Red dashed lines indicate regions with high variability.

3.2. Phenotypic Analysis of the FTIR Spectra

Figure 2 shows the sample/goat repeatability for the 1060 analyses carried out on the entire goat milk spectrum, i.e., the infrared region between wavelengths 2.0μm (wavenumber 5000 × cm^{-1}) and 10.8μm (wavenumber 930 × cm^{-1}), and the percentage variance explained by each random effect included in model (1) (Goat = G, Flock = F, Herd = H, Parity = P, and Days in milk = D) and the residual. As can be seen, the sample/goat repeatability approaches one for almost the entire spectrum, with the exception of the two regions with very high variability (Figure 1), where it is significantly lower because the residual variance often increases to values of 50% of phenotypic variance. In the other regions, the sample/goat random effect explains an average of 39% of the phenotypic variance, versus 9% in the aforementioned two regions. The random effect of flock explained, on average, 23% in the group of regions with low variability, and 11.8% in the group with high variability, while the random effect of breed explained, on average, 17% and 11%, respectively, of phenotypic variability. There was little variation in the contributions of the random effects of Parity (9%) and Days in Milk (11%) to the phenotypic variability in the entire spectrum. Finally, the unexplained phenotypic variation (residual) was an average of 1% in the low and 50% in the high variability regions of the spectrum.

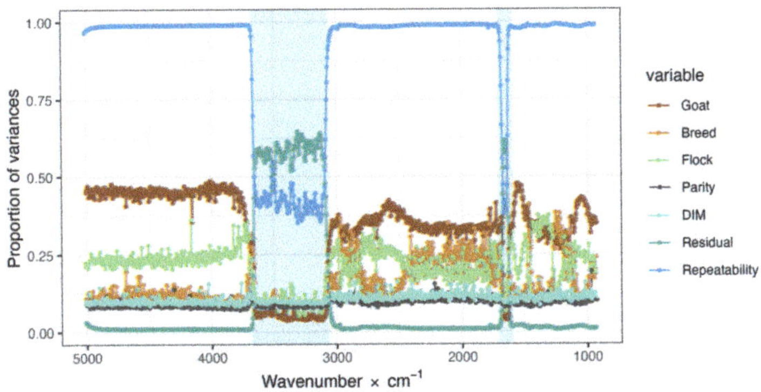

Figure 2. Sample/goat repeatability and proportions of variances for Goat, Breed, Flock, Parity, Days in Milk (DIM), and the residual term for the absorbance values of each of the 1060 wavenumbers analyzed.

4. Discussion

The discussion deals separately with the three specific aims of this study as follows: the patterns and phenotypic variances in the absorbance values of the goat milk spectrum; the estimation of the variance components of the major sources of variation of infrared absorbance; and the repeatability of 1060 wavenumbers in the FTIR spectrum of goat milk samples.

4.1. The Patterns and Phenotypic Variances in the Absorbance Values of the Goat Milk Spectrum

The infrared region analyzed in this study (wavelengths 2.0 to 10.8 μm, or wavenumbers 5000 to 930 $\times$ cm^{-1}) is a section of the near-, mid-, and long-infrared regions of the electromagnetic spectrum. The spectrometer used in this study is the one most commonly used to predict the composition of milk samples [53] in many countries of the world, especially within milk recording systems for the genetic improvement of dairy populations. It is widely used to analyze not just cow milk samples, but also buffalo [33], sheep [39,54,55], and goat [34,39,56] milk samples with specific calibrations [3,56].

Given that milk spectra can be expressed in different ways (as transmittance or absorbance, as the entire spectrum or specific regions, etc.), it is worth noting that the average spectrum obtained here for goat milk is very similar to that frequently obtained for cow milk.

In our previous study on the variability of FTIR spectra of bovine milk samples [41], we analyzed milk in the same wavenumber interval using the same type of spectrometer, but the spectra were expressed as transmittance, not absorbance, values. This explains why the spectra are centered on zero in this study, whereas in the previous study they were centered on a value of one, and why the pattern is reversed in the sign with respect to the center. Other studies obtained very similar patterns with transmittance spectra [43]. The absorbances and patterns reported for bovine milk in a subsequent study [57] were very similar to those observed here for caprine milk (Figure 1). Other studies on the absorbance FTIR spectra of bovine milk have also reported a similar pattern to that of caprine milk found here [58].

The phenotypic variation of absorbance, shown in Figure 1, was very different at every wavenumber analyzed. The two spectral regions with a much greater variation in absorbance than the rest of the spectrum were identified as the areas of absorbance of the O-H chemical bond, and therefore highly influenced by the presence of water. A large proportion of milk is constituted by water, so the transmittance (and consequently, the absorbance) spectrum of milk is very similar to that of water [59]. The milk spectrum is, as in the present study, frequently expressed as the ratio between the values measured in milk and those measured in pure water, taken as a reference. Based on the average values and standard deviations observed along the milk spectrum, and taking into account the heritability coefficients estimated for every wavenumber measured, in our previous study we proposed subdividing the spectrum of bovine milk into five sections [41]. Given the close similarity observed here, we decided to use the same subdivisions for goat milk, represented by the vertical dashed red lines in Figure 1 and the different background color of the areas in Figure 2 (white or light blue). The first (SWIR-MWIR) of the two "water" spectral regions is identified in the area of transition between the near- and mid-infrared (NIR and MIR) radiations (wavelength 2.73 to 3.27 μm) and the second (MWIR2) region in the central area of mid-infrared radiation (5.89 to 6.31 μm).

The two "water" spectral regions are often excluded when milk spectra are used to predict milk traits with appropriate chemometric procedures, as they are considered sources of "noise" and inexplicable variations [42]. However, the O-H bond is also present in many other chemical compounds that are important for defining milk quality, while other chemical bonds have been shown to correspond to the absorbance of electromagnetic radiation in these sections, and, lastly, the absorbance of several wavelengths in these sections has been found to be, in part, genetically controlled [43,46].

A better understanding and discussion of the role and importance of different spectral regions of the goat milk spectrum could be had by quantifying the major sources of variation in milk absorbance.

4.2. Variance Components of the Major Sources of Variation of Infrared Absorbance

In light of the results (see Figure 2), the phenotypic variability ($\hat{\sigma}^2_{Ph}$) of milk absorbance at each wavenumber was divided into their major sources of variation, and the variability due to individual sample/goat ($\hat{\sigma}^2_G$), breed of goat ($\hat{\sigma}^2_B$), flock ($\hat{\sigma}^2_F$), parity ($\hat{\sigma}^2_P$), stage of lactation ($\hat{\sigma}^2_D$), and the residual variation ($\hat{\sigma}^2_e$), and computed the sample/goat repeatability was treated as random effects. Bear in mind that in this study, only one milk sample (with two spectral replicates) per goat was taken, so the effects of individual goat and sample are combined, whereas the residual variation expresses the differences between the two spectral replicates obtained from each milk sample.

To facilitate discussion, these estimates were averaged according to the five spectral regions proposed in our previous study; these are summarized in Table 1.

The first of these five regions is the near-infrared or short-wavelength region (SWIR, 2–2.72 μm), followed by the first "water" region (SWIR-MWIR, 2.73–3.27 μm), the mid-infrared 1 region (MWIR-1, 3.28–5.88 μm), the second "water" region (MWIR-2, 5.89–6.31 μm), and lastly, the mid- to long-infrared (MWIR-LWIR, 6.32–10.76 μm) region. To facilitate comparison with other studies, equivalences in standard ISO, wavenumber (cycles per inverse centimeter, waves × cm^{-1} and frequencies (cycles per second, Hertz) are also listed in Table 1.

It is worth noting that the average absorbance of the wavenumbers in the two "water" regions is negative, whereas it is positive in the other three spectral regions. Moreover, almost one third of the wavenumbers had an average absorbance of less than 1 standard deviation from the overall mean. At the same time, in these two regions, 19% and 7% of the wavelengths had an average absorbance greater than 1 standard deviation from the overall mean, and the average phenotypic variability was 6 to 22 times larger than in the other three spectral regions (Table 1). Regarding the variance components, it is clear from the same table that these large differences in phenotypic variability are reflected in the variability in all the major sources of variation. It is therefore expected that, for each variance component, the "water" regions will exhibit the largest proportion of wavenumbers characterized by very high variance, while the other three regions will be characterized by very low variance.

Table 1. Characteristics of the five regions of the FTIR spectrum of goat milk.

Item [2]	Entire Spectrum	SWIR [1]	SWIR-MWIR [1]	MWIR-1 [1]	MWIR-2 [1]	MWIR-LWIR [1]
ISO	NIR-MIR	NIR	NIR-MIR	MIR	MIR	MIR
Wavenumber, cm^{-1}	5000–930	5000–3673	3669–3052	3048–1701	1698–1586	1582–930
Wavelength, μm	2.00–10.76	2.00–2.72	2.73–3.27	3.28–5.88	5.89–6.31	6.32–10.76
Frequency, THz	149.9–27.9	149.9–110.1	110.0–91.5	91.4–51.0	50.9–47.5	47.4–27.9
Waves tested, no.	1060	347	161	350	31	171
Absorbance:	medium	medium	low	medium	low	high
Average absorbance	0.0186	0.0109	−0.0109	0.0109	−0.0506	0.0910
Waves > 0.130 [a], %	10	0	19	10	7	25
Waves <−0.093 [b], %	7	2	31	1	31	1
Phenotypic variability:	medium	very low	very high	low	high	low
Mean of $\hat{\sigma}_{Ph}$	0.043	0.008	0.206	0.011	0.105	0.015
Waves $\hat{\sigma}_{Ph}$ > 0.09 [c], %	12	0	69	1	36	0
Waves $\hat{\sigma}_{Ph}$ <0.02 [d], %	74	99	0	86	4	77
Animal (Goat) variability:						
Mean of $\hat{\sigma}_G$	0.013	0.006	0.046	0.007	0.031	0.009
Proportion of $\hat{\sigma}^2_{Ph}$	0.33	0.45	0.07	0.35	0.22	0.36
Breed variability:						
Mean of $\hat{\sigma}_B$	0.014	0.003	0.063	0.005	0.034	0.006
Proportion of $\hat{\sigma}^2_{Ph}$	0.15	0.12	0.10	0.21	0.13	0.18
Flock variability:						
Mean of $\hat{\sigma}_F$	0.015	0.004	0.063	0.005	0.038	0.007
Proportion of $\hat{\sigma}^2_{Ph}$	0.21	0.24	0.10	0.22	0.20	0.25
Parity variability:						
Mean of $\hat{\sigma}_P$	0.013	0.002	0.059	0.004	0.031	0.005
Proportion of $\hat{\sigma}^2_{Ph}$	0.09	0.09	0.08	0.10	0.09	0.10
Lactation stage variability:						
Mean of $\hat{\sigma}_D$	0.014	0.003	0.064	0.004	0.033	0.005
Proportion of $\hat{\sigma}^2_{Ph}$	0.11	0.10	0.10	0.12	0.10	0.11
Repeatability:						
Mean Repeatibility	0.90	0.99	0.45	0.99	0.75	0.99

[1]: SWIR = short-wave infrared; MWIR = mid-wave infrared; LWIR = long-wave infrared; [2]: $\hat{\sigma}^2_{Ph}$ is the phenotypic variance calculated as the sum of the variance components of the random effects included in model 1 (goat, breed, flock, parity, lactation stage, and residual) [$\hat{\sigma}^2_{Ph} = \hat{\sigma}^2_G + \hat{\sigma}^2_B + \hat{\sigma}^2_F + \hat{\sigma}^2_P + \hat{\sigma}^2_D + \hat{\sigma}^2_e$]; the mean of $\hat{\sigma}_i$ is the mean of the standard deviation for the ith random effect of model 1 for the waves tested in the whole spectrum and in each of the five regions; the proportion of $\hat{\sigma}^2_{Ph}$ is the proportion of the phenotypic variance explained by the ith random effect of the model 1 [$\hat{\sigma}^2_i / \hat{\sigma}^2_{Ph}$]; repeatability is the proportion of the sum of the variances explained by the random effects (without the residual one) on the $\hat{\sigma}^2_{Ph}$ [$R_{sample} = \frac{\hat{\sigma}^2_G + \hat{\sigma}^2_B + \hat{\sigma}^2_F + \hat{\sigma}^2_P + \hat{\sigma}^2_D}{\hat{\sigma}^2_{Ph}}$]; [a]: Proportion of waves in the region (% of the total waves in the region) with a value higher than twice the average of the entire spectrum; [b]: number of waves in the region (% of the total waves in the region) with a value higher than half the average of the whole spectrum; [c]: number of waves in the region (% of the total waves in the region) with a value higher than the average + 1 SD of the entire spectrum; [d]: number of waves in the region (% of the total waves in the region) with a value lower than the average − 1 SD of the entire spectrum.

As the absorbance values measured at each wavenumber are centered and standardized before being used to predict milk traits, it is of interest to analyze the relative importance of different sources of variation in the five spectral regions. The sample/goat variance, expressed as an average of all the wavenumbers in the entire FTIR caprine milk spectrum (Table 1), accounts for one third of the phenotypic variance, and is, on average, greater in the three "non-water" regions and lower in the "water regions" (7% in the SWIR-MWIR and 22% in the MWIR2). It is worth noting that there were smaller differences in the effects of breed of goat among the five spectral regions, with average proportions ranging from 10% to 21%, and the effects of flock were similar in importance and variability, with values ranging from 10% to 25%. The individual goat factors that change with time (parity and stage of lactation) had a smaller, but significant, influence on infrared radiation absorbance, and presented similar values in the five spectral regions (8% to 10% for the effect of parity, 10% to 12% for stage of lactation).

No studies in the literature that analyze the major sources of variation in the absorbance of electromagnetic radiation at the level of individual wavenumbers or spectral regions for milk was found, but there are some studies, described later, that quantified the genetic and environmental components of the phenotype. As seen here, both components are much greater in the regions with a very large phenotypic variability ("water" regions) than in the rest of the spectrum [41]. In this case, too, it might be more informative to analyze the relative proportions of the genetic and environmental variance components, which are expressed as the heritability of absorbance. It is worth noting that our previous studies, among others, found heritabilities, albeit variable, for several wavenumbers also in the "water" regions [42,44]. Obviously, the genetic variance represents the major part of the variation due to the breed and a part of that due to the individual animal. The herd–year–season component has also been found to have a strong influence on the phenotypic variance of absorbance in bovine milk [44], and to affect the prediction of milk traits from FTIR caprine milk spectra [60]. Interactions between the cow's genetics, parity, and stage of lactation and spectral region have also been found [44].

We can conclude that the two "water" regions are affected by individual sample/goat, breed, herd, parity, and stage of lactation, albeit to a lesser extent than the other regions, and that they therefore probably contain valuable information that could be used in the prediction of milk traits or the authentication of breed and feeding and production systems, provided that it is combined with suitable chemometric methods.

It is worth noting that breed of cow had very little effect on the patterns and variability in the FTIR spectra of bovine milk [61]; however, when calibration equations developed on one breed for predicting milk traits were then applied to other breeds, validation accuracy tended to be slightly lower than when the calibration equations were developed on multi-breed training sets. Similar results were obtained in a study predicting milk coagulation traits in four goat breeds [62]. As the spectra were not compared in either of the two studies, it is unclear whether the different results with different breeds are due to inherent differences in the predictors (FTIR spectra) or to the different values and characteristics of the predicted traits.

4.3. *Animal/Sample Repeatability of the Absorbance of 1060 Wavelengths of FTIR Goat Milk Spectra*

The variability not captured by the random effects included in the model is captured by $\hat{\sigma}_e^2$. As can be seen from Figure 2, the relative contribution of the error term to the total phenotypic variance of absorbance is close to zero for the SWIR, MWIR-1, and MWIR-LWIR regions, while for the two "water" regions (SWIR-MWIR and MWIR2), it rose dramatically to more than 50%. In fact, the average proportion of phenotypic variance not explained by the random term is 10% over the entire spectrum, but in the SWIR, MWIR-1, and MWIR-LWIR regions, it is 1.1%, 1.3%, and 1.4%, respectively, whereas in the "water" regions (SWIR-MWIR and MWIR-2), it is 55% and 25%, respectively. Conversely, as defined in this study, the repeatability of the absorbance measures is the complement of the residual proportion of phenotypic variance and is almost 99% in the three "non-water" spectral regions and about 45% and 75% in the two "water" regions. Given the different structures of the sources of variation of the absorbance spectrum, when using the entire spectrum (including the two "water" regions), it seems advisable to use the average of two–three replicates per milk sample instead of a single spectrum, or to use chemometric procedures that can select the most informative wavenumbers.

There are very few if any data in the literature regarding the repeatability of FTIR absorbance measured at each wavenumber. In the case of bovine meat, we also found large differences in repeatability along the NIR spectrum [63]: it was highly variable in the region of the electromagnetic spectrum corresponding to ultraviolet and visible light (0.35 to 0.74 μm), relatively high (about 80%) for wavenumbers in the IR-A region (NIR: 0.74 to 1.40 μm), and very low (10% to 30%) in the IR-B interval (SWIR: 1.40 to 1.85 μm). The heritability of the individual wavelengths were, correspondingly, very different [64].

5. Conclusions

Our study on a representative goat population (different farming systems, breeds, parities, and lactation stages) shows that the FTIR spectrum of caprine milk has many similarities with that of bovine milk reported in the literature. The major sources of variation were as follows: sample/goat (33% of the total); flock (21%); goat breed (15%); lactation stage (11%); parity (9%); and the residual unexplained variation (10%). As in cattle species, the spectrum is highly heterogeneous, and it was possible to distinguish five regions, two of which ("water" regions) presented much larger variability than the others, not only in terms of the residual variation, but also in terms of the effects of the major sources of variation. The similarity with the bovine milk spectrum, as well as the high repeatability (90% for the entire spectrum, 99% in the non-water regions), leads us to expect that caprine milk spectra could also be a valuable tool for predicting many milk properties.

Author Contributions: Conceptualization, G.B.; methodology, B.d.J.V.-H. and N.A.; software, B.d.J.V.-H.; validation, B.d.J.V.-H., N.A., G.B. and A.C.; formal analysis, B.d.J.V.-H.; investigation, M.P. and G.M.V.; resources, M.P., G.M.V. and G.B.; data curation, M.P., G.M.V. and N.A.; writing—original draft preparation, B.d.J.V.-H. and G.B.; writing—review and editing, N.A., A.C., M.P., G.M.V. and G.B.; visualization, B.d.J.V.-H., N.A. and G.B; supervision, M.P., G.M.V. and G.B.; project administration, G.B.; funding acquisition, G.B and M.P. All authors have read and agreed to the published version of the manuscript.

Funding: This research was supported by the Italian Ministry of Agriculture, Forestry and Food Policies (project GOOD-MILK, D.M. 9367185—09/12/2020, CUP C29C20000450001) and the Regional Government of Sardinia (Legge Regionale 7/2007; CUP J72I15000030007).

Data Availability Statement: Data can be obtained from the authors.

Acknowledgments: The authors thank the farmers and the Sardinian Provincial Farmers Associations (Cagliari, Italy) for their support in sample collection; and Regional Farmers Association of Sardinia (Cagliari, Italy) for support in milk analysis.

Conflicts of Interest: The authors declare no conflict of interest.

References

1. Yakubu, H.G.; Kovacs, Z.; Toth, T.; Bazar, G. The Recent Advances of Near-Infrared Spectroscopy in Dairy Production—A Review. *Crit. Rev. Food Sci. Nutr.* **2022**, *62*, 810–831. [CrossRef] [PubMed]
2. Smith, B.C. *Fundamental of Fourier—Infrared Spectroscopy*; CRC Press: Boca Raton, FL, USA, 2011.
3. *ISO 9622 | IDF 141*; 2013-Milk and Liquid Milk Products—Guidelines for the Application of Mid-Infrared Spectrometry—FIL-IDF. International Organisation for Standardisation: Geneva, Switzerland, 2013.
4. Niemöller, A.; Holroyd, S. Applications of near Infrared Spectrometry for the Analysis of Milk and Milk Products. *Bull. Int. Dairy Fed.* **2019**, *497*, 1–19.
5. *ISO 21543:2020*; Milk and Milk Products—Guidelines for the Application of near Infrared Spectrometry. International Organisation for Standardisation: Geneva, Switzerland, 2020.
6. Coppa, M.; Revello-Chion, A.; Giaccone, D.; Tabacco, E.; Borreani, G. Could Predicting Fatty Acid Profile by Mid-Infrared Reflectance Spectroscopy Be Used as a Method to Increase the Value Added by Milk Production Chains? *J. Dairy Sci.* **2017**, *100*, 8705–8721. [CrossRef]
7. Fleming, A.; Schenkel, F.S.; Chen, J.; Malchiodi, F.; Bonfatti, V.; Ali, R.A.; Mallard, B.; Corredig, M.; Miglior, F. Prediction of Milk Fatty Acid Content with Mid-Infrared Spectroscopy in Canadian Dairy Cattle Using Differently Distributed Model Development Sets. *J. Dairy Sci.* **2017**, *100*, 5073–5081. [CrossRef]
8. Rovere, G.; de los Campos, G.; Lock, A.L.; Worden, L.; Vazquez, A.I.; Lee, K.; Tempelman, R.J. Prediction of Fatty Acid Composition Using Milk Spectral Data and Its Associations with Various Mid-Infrared Spectral Regions in Michigan Holsteins. *J. Dairy Sci.* **2021**, *104*, 11242–11258. [CrossRef]
9. Rutten, M.J.M.; Bovenhuis, H.; Heck, J.M.L.; van Arendonk, J.A.M. Predicting Bovine Milk Protein Composition Based on Fourier Transform Infrared Spectra. *J. Dairy Sci.* **2011**, *94*, 5683–5690. [CrossRef]

10. Bonfatti, V.; Cecchinato, A.; Carnier, P. Short Communication: Predictive Ability of Fourier-Transform Mid-Infrared Spectroscopy to Assess CSN Genotypes and Detailed Protein Composition of Buffalo Milk. *J. Dairy Sci.* **2015**, *98*, 6583–6587. [CrossRef]
11. Sanchez, M.P.; Ferrand, M.; Gelé, M.; Pourchet, D.; Miranda, G.; Martin, P.; Brochard, M.; Boichard, D. Short Communication: Genetic Parameters for Milk Protein Composition Predicted Using Mid-Infrared Spectroscopy in the French Montbéliarde, Normande, and Holstein Dairy Cattle Breeds. *J. Dairy Sci.* **2017**, *100*, 6371–6375. [CrossRef]
12. Eskildsen, C.E.; Næs, T.; Skou, P.B.; Solberg, L.E.; Dankel, K.R.; Basmoen, S.A.; Wold, J.P.; Horn, S.S.; Hillestad, B.; Poulsen, N.A.; et al. Cage of Covariance in Calibration Modeling: Regressing Multiple and Strongly Correlated Response Variables onto a Low Rank Subspace of Explanatory Variables. *Chemom. Intell. Lab. Syst.* **2021**, *213*, 104311. [CrossRef]
13. Soyeurt, H.; Bruwier, D.; Romnee, J.M.; Gengler, N.; Bertozzi, C.; Veselko, D.; Dardenne, P. Potential Estimation of Major Mineral Contents in Cow Milk Using Mid-Infrared Spectrometry. *J. Dairy Sci.* **2009**, *92*, 2444–2454. [CrossRef]
14. Wu, D.; He, Y.; Shi, J.; Shuijuan, F. Exploring near and Midinfrared Spectroscopy to Predict Trace Iron and Zinc Contents in Powdered Milk. *J. Agric. Food Chem.* **2009**, *57*, 1697–1704. [CrossRef]
15. Patel, N.; Toledo-Alvarado, H.; Cecchinato, A.; Bittante, G. Predicting the Content of 20 Minerals in Beef by Different Portable Near-Infrared (NIR) Spectrometers. *Foods* **2020**, *9*, 1389. [CrossRef]
16. Calamari, L.; Gobbi, L.; Bani, P. Improving the Prediction Ability of FT-MIR Spectroscopy to Assess Titratable Acidity in Cow's Milk. *Food Chem.* **2016**, *192*, 477–484. [CrossRef] [PubMed]
17. El Jabri, M.; Sanchez, M.P.; Trossat, P.; Laithier, C.; Wolf, V.; Grosperrin, P.; Beuvier, E.; Rolet-Répécaud, O.; Gavoye, S.; Gaüzère, Y.; et al. Comparison of Bayesian and Partial Least Squares Regression Methods for Mid-Infrared Prediction of Cheese-Making Properties in Montbéliarde Cows. *J. Dairy Sci.* **2019**, *102*, 6943–6958. [CrossRef] [PubMed]
18. Mota, L.F.M.; Pegolo, S.; Baba, T.; Peñagaricano, F.; Morota, G.; Bittante, G.; Cecchinato, A. Evaluating the Performance of Machine Learning Methods and Variable Selection Methods for Predicting Difficult-to-Measure Traits in Holstein Dairy Cattle Using Milk Infrared Spectral Data. *J. Dairy Sci.* **2021**, *104*, 8107–8121. [CrossRef] [PubMed]
19. Pralle, R.S.; Weigel, K.W.; White, H.M. Predicting Blood β-Hydroxybutyrate Using Milk Fourier Transform Infrared Spectrum, Milk Composition, and Producer-Reported Variables with Multiple Linear Regression, Partial Least Squares Regression, and Artificial Neural Network. *J. Dairy Sci.* **2018**, *101*, 4378–4387. [CrossRef]
20. Ho, P.N.; Luke, T.D.W.; Pryce, J.E. Validation of Milk Mid-Infrared Spectroscopy for Predicting the Metabolic Status of Lactating Dairy Cows in Australia. *J. Dairy Sci.* **2021**, *104*, 4467–4477. [CrossRef] [PubMed]
21. Bach, K.D.; Barbano, D.M.; McArt, J.A.A. Association of Mid-Infrared-Predicted Milk and Blood Constituents with Early-Lactation Disease, Removal, and Production Outcomes in Holstein Cows. *J. Dairy Sci.* **2019**, *102*, 10129–10139. [CrossRef]
22. Wallén, S.E.; Prestløkken, E.; Meuwissen, T.H.E.; McParland, S.; Berry, D.P. Milk Mid-Infrared Spectral Data as a Tool to Predict Feed Intake in Lactating Norwegian Red Dairy Cows. *J. Dairy Sci.* **2018**, *101*, 6232–6243. [CrossRef]
23. Grelet, C.; Froidmont, E.; Foldager, L.; Salavati, M.; Hostens, M.; Ferris, C.P.; Ingvartsen, K.L.; Crowe, M.A.; Sorensen, M.T.; Fernandez Pierna, J.A.; et al. Potential of Milk Mid-Infrared Spectra to Predict Nitrogen Use Efficiency of Individual Dairy Cows in Early Lactation. *J. Dairy Sci.* **2020**, *103*, 4435–4445. [CrossRef]
24. Shetty, N.; Løvendahl, P.; Lund, M.S.; Buitenhuis, A.J. Prediction and Validation of Residual Feed Intake and Dry Matter Intake in Danish Lactating Dairy Cows Using Mid-Infrared Spectroscopy of Milk. *J. Dairy Sci.* **2017**, *100*, 253–264. [CrossRef]
25. Smith, S.L.; Denholm, S.J.; Coffey, M.P.; Wall, E. Energy Profiling of Dairy Cows from Routine Milk Mid-Infrared Analysis. *J. Dairy Sci.* **2019**, *102*, 11169–11179. [CrossRef] [PubMed]
26. Vanlierde, A.; Soyeurt, H.; Gengler, N.; Colinet, F.G.; Froidmont, E.; Kreuzer, M.; Grandl, F.; Bell, M.; Lund, P.; Olijhoek, D.W.; et al. Short Communication: Development of an Equation for Estimating Methane Emissions of Dairy Cows from Milk Fourier Transform Mid-Infrared Spectra by Using Reference Data Obtained Exclusively from Respiration Chambers. *J. Dairy Sci.* **2018**, *101*, 7618–7624. [CrossRef]
27. Bittante, G.; Cipolat-Gotet, C. Direct and Indirect Predictions of Enteric Methane Daily Production, Yield, and Intensity per Unit of Milk and Cheese, from Fatty Acids and Milk Fourier-Transform Infrared Spectra. *J. Dairy Sci.* **2018**, *101*, 7219–7235. [CrossRef] [PubMed]
28. Wang, Q.; Bovenhuis, H. Validation Strategy Can Result in an Overoptimistic View of the Ability of Milk Infrared Spectra to Predict Methane Emission of Dairy Cattle. *J. Dairy Sci.* **2019**, *102*, 6288–6295. [CrossRef] [PubMed]
29. Bastin, C.; Théron, L.; Lainé, A.; Gengler, N. On the Role of Mid-Infrared Predicted Phenotypes in Fertility and Health Dairy Breeding Programs. *J. Dairy Sci.* **2016**, *99*, 4080–4094. [CrossRef]
30. Ho, P.N.; Bonfatti, V.; Luke, T.D.W.; Pryce, J.E. Classifying the Fertility of Dairy Cows Using Milk Mid-Infrared Spectroscopy. *J. Dairy Sci.* **2019**, *102*, 10460–10470. [CrossRef]
31. Toledo-Alvarado, H.; Pérez-Cabal, M.A.; Tempelman, R.J.; Cecchinato, A.; Bittante, G.; de los Campos, G.; Vazquez, A.I. Association between Days Open and Milk Spectral Data in Dairy Cows. *J. Dairy Sci.* **2021**, *104*, 3665–3675. [CrossRef]
32. Kamal, M.; Karoui, R. Analytical Methods Coupled with Chemometric Tools for Determining the Authenticity and Detecting the Adulteration of Dairy Products: A Review. *Trends Food Sci. Technol.* **2015**, *46*, 27–48. [CrossRef]
33. Sen, S.; Dundar, Z.; Uncu, O.; Ozen, B. Potential of Fourier-Transform Infrared Spectroscopy in Adulteration Detection and Quality Assessment in Buffalo and Goat Milks. *Microchem. J.* **2021**, *166*, 106207. [CrossRef]
34. Galvan, D.; Lelis, C.A.; Effting, L.; Melquiades, F.L.; Bona, E.; Conte-Junior, C.A. Low-Cost Spectroscopic Devices with Multivariate Analysis Applied to Milk Authenticity. *Microchem. J.* **2022**, *181*, 107746. [CrossRef]

35. Scampicchio, M.; Eisenstecken, D.; De Benedictis, L.; Capici, C.; Ballabio, D.; Mimmo, T.; Robatscher, P.; Kerschbaumer, L.; Oberhuber, M.; Kaser, A.; et al. Multi-Method Approach to Trace the Geographical Origin of Alpine Milk: A Case Study of Tyrol Region. *Food Anal. Methods* **2016**, *9*, 1262–1273. [CrossRef]
36. Coppa, M.; Martin, B.; Agabriel, C.; Chassaing, C.; Sibra, C.; Constant, I.; Graulet, B.; Andueza, D. Authentication of Cow Feeding and Geographic Origin on Milk Using Visible and Near-Infrared Spectroscopy. *J. Dairy Sci.* **2012**, *95*, 5544–5551. [CrossRef]
37. Bergamaschi, M.; Cipolat-Gotet, C.; Cecchinato, A.; Schiavon, S.; Bittante, G. Chemometric Authentication of Farming Systems of Origin of Food (Milk and Ripened Cheese) Using Infrared Spectra, Fatty Acid Profiles, Flavor Fingerprints, and Sensory Descriptions. *Food Chem.* **2020**, *305*, 125480. [CrossRef]
38. Dagnachew, B.S.; Meuwissen, T.H.E.; Ådnøy, T. Genetic Components of Milk Fourier-Transform Infrared Spectra Used to Predict Breeding Values for Milk Composition and Quality Traits in Dairy Goats. *J. Dairy Sci.* **2013**, *96*, 5933–5942. [CrossRef]
39. Ferrand-Calmels, M.; Palhière, I.; Brochard, M.; Leray, O.; Astruc, J.M.; Aurel, M.R.; Barbey, S.; Bouvier, F.; Brunschwig, P.; Caillat, H.; et al. Prediction of Fatty Acid Profiles in Cow, Ewe, and Goat Milk by Mid-Infrared Spectrometry. *J. Dairy Sci.* **2014**, *97*, 17–35. [CrossRef]
40. Núñez-Sánchez, N.; Martínez-Marín, A.L.; Polvillo, O.; Fernández-Cabanás, V.M.; Carrizosa, J.; Urrutia, B.; Serradilla, J.M. Near Infrared Spectroscopy (NIRS) for the Determination of the Milk Fat Fatty Acid Profile of Goats. *Food Chem.* **2016**, *190*, 244–252. [CrossRef] [PubMed]
41. Bittante, G.; Cecchinato, A. Genetic Analysis of the Fourier-Transform Infrared Spectra of Bovine Milk with Emphasis on Individual Wavelengths Related to Specific Chemical Bonds. *J. Dairy Sci.* **2013**, *96*, 5991–6006. [CrossRef] [PubMed]
42. Soyeurt, H.; Misztal, I.; Gengler, N. Genetic Variability of Milk Components Based on Mid-Infrared Spectral Data. *J. Dairy Sci.* **2010**, *93*, 1722–1728. [CrossRef]
43. Wang, Q.; Hulzebosch, A.; Bovenhuis, H. Genetic and Environmental Variation in Bovine Milk Infrared Spectra. *J. Dairy Sci.* **2016**, *99*, 6793–6803. [CrossRef]
44. Rovere, G.; de los Campos, G.; Tempelman, R.J.; Vazquez, A.I.; Miglior, F.; Schenkel, F.; Cecchinato, A.; Bittante, G.; Toledo-Alvarado, H.; Fleming, A. A Landscape of the Heritability of Fourier-Transform Infrared Spectral Wavelengths of Milk Samples by Parity and Lactation Stage in Holstein Cows. *J. Dairy Sci.* **2019**, *102*, 1354–1363. [CrossRef]
45. Wang, Q.; Bovenhuis, H. Genome-Wide Association Study for Milk Infrared Wavenumbers. *J. Dairy Sci.* **2018**, *101*, 2260–2272. [CrossRef]
46. Tiplady, K.M.; Lopdell, T.J.; Reynolds, E.; Sherlock, R.G.; Keehan, M.; Johnson, T.J.; Pryce, J.E.; Davis, S.R.; Spelman, R.J.; Harris, B.L.; et al. Sequence-Based Genome-Wide Association Study of Individual Milk Mid-Infrared Wavenumbers in Mixed-Breed Dairy Cattle. *Genet. Sel. Evol.* **2021**, *53*, 1–24. [CrossRef] [PubMed]
47. Vacca, G.M.; Stocco, G.; Dettori, M.L.; Pira, E.; Bittante, G.; Pazzola, M. Milk Yield, Quality, and Coagulation Properties of 6 Breeds of Goats: Environmental and Individual Variability. *J. Dairy Sci.* **2018**, *101*, 7236–7247. [CrossRef]
48. Vacca, G.M.; Stocco, G.; Dettori, M.L.; Bittante, G.; Pazzola, M. Goat Cheese Yield and Recovery of Fat, Protein, and Total Solids in Curd Are Affected by Milk Coagulation Properties. *J. Dairy Sci.* **2020**, *103*, 1352–1365. [CrossRef] [PubMed]
49. Vacca, G.M.; Stocco, G.; Dettori, M.L.; Summer, A.; Cipolat-Gotet, C.; Bittante, G.; Pazzola, M. Cheese Yield, Cheesemaking Efficiency, and Daily Production of 6 Breeds of Goats. *J. Dairy Sci.* **2018**, *101*, 7817–7832. [CrossRef] [PubMed]
50. Stocco, G.; Pazzola, M.; Dettori, M.L.; Paschino, P.; Bittante, G.; Vacca, G.M. Effect of Composition on Coagulation, Curd Firming, and Syneresis of Goat Milk. *J. Dairy Sci.* **2018**, *101*, 9693–9702. [CrossRef]
51. De Los Campos, G.; Pataki, A.; Pérez, P. The BGLR (Bayesian Generalized Linear Regression) R-Package; 2022 v.1.1.0. Available online: http://cran.r-project.org/web/packages/BGLR/BGLR.pdf (accessed on 18 January 2023).
52. Team, R.C. *R: A Language and Environment for Statistical Computing*; R Foundation for Statistical Computing: Vienna, Austria, 2021.
53. Orlandini, S. ICAR Certification Report MilkoScan 7RM Foss Analytical A/S. 2020. Available online: https://www.icar.org/wp-content/uploads/2020/03/Certification-Report-Milk-Analyser-MilkoScan%E2%84%A2-7-RM-by-FOSS.pdf (accessed on 18 January 2023).
54. Bittante, G.; Pellattiero, E.; Malchiodi, F.; Cipolat-Gotet, C.; Pazzola, M.; Vacca, G.M.; Schiavon, S.; Cecchinato, A. Quality Traits and Modeling of Coagulation, Curd Firming, and Syneresis of Sheep Milk of Alpine Breeds Fed Diets Supplemented with Rumen-Protected Conjugated Fatty Acid. *J. Dairy Sci.* **2014**, *97*, 4018–4028. [CrossRef] [PubMed]
55. Pazzola, M.; Dettori, M.L.; Cipolat-Gotet, C.; Cecchinato, A.; Bittante, G.; Vacca, G.M. Phenotypic Factors Affecting Coagulation Properties of Milk from Sarda Ewes. *J. Dairy Sci.* **2014**, *97*, 7247–7257. [CrossRef]
56. Bittante, G.; Amalfitano, N.; Bergamaschi, M.; Patel, N.; Haddi, M.L.; Benabid, H.; Pazzola, M.; Vacca, G.M.; Tagliapietra, F.; Schiavon, S. Composition and Aptitude for Cheese-Making of Milk from Cows, Buffaloes, Goats, Sheep, Dromedary Camels, and Donkeys. *J. Dairy Sci.* **2022**, *105*, 2132–2152. [CrossRef]
57. Ferragina, A.; Cipolat-Gotet, C.; Cecchinato, A.; Bittante, G. The Use of Fourier-Transform Infrared Spectroscopy to Predict Cheese Yield and Nutrient Recovery or Whey Loss Traits from Unprocessed Bovine Milk Samples. *J. Dairy Sci.* **2013**, *96*, 7980–7990. [CrossRef] [PubMed]
58. Eskildsen, C.E.; Rasmussen, M.A.; Engelsen, S.B.; Larsen, L.B.; Poulsen, N.A.; Skov, T. Quantification of Individual Fatty Acids in Bovine Milk by Infrared Spectroscopy and Chemometrics: Understanding Predictions of Highly Collinear Reference Variables. *J. Dairy Sci.* **2014**, *97*, 7940–7951. [CrossRef]

59. Kaylegian, K.E.; Lynch, J.M.; Fleming, J.R.; Barbano, D.M. Influence of Fatty Acid Chain Length and Unsaturation on Mid-Infrared Milk Analysis. *J. Dairy Sci.* **2009**, *92*, 2485–2501. [CrossRef]
60. Dadousis, C.; Cipolat-Gotet, C.; Stocco, G.; Ferragina, A.; Dettori, M.L.; Pazzola, M.; do Nascimento Rangel, A.H.; Vacca, G.M. Goat Farm Variability Affects Milk Fourier-Transform Infrared Spectra Used for Predicting Coagulation Properties. *J. Dairy Sci.* **2021**, *104*, 3927–3935. [CrossRef] [PubMed]
61. Mota, L.F.M.; Pegolo, S.; Baba, T.; Morota, G.; Peñagaricano, F.; Bittante, G.; Cecchinato, A. Comparison of Single-breed and Multi-breed Training Populations for Infrared Predictions of Novel Phenotypes in Holstein Cows. *Animals* **2021**, *11*, 1993. [CrossRef] [PubMed]
62. Stocco, G.; Dadousis, C.; Vacca, G.M.; Pazzola, M.; Paschino, P.; Dettori, M.L.; Ferragina, A.; Cipolat-Gotet, C. Breed of Goat Affects the Prediction Accuracy of Milk Coagulation Properties Using Fourier-Transform Infrared Spectroscopy. *J. Dairy Sci.* **2021**, *104*, 3956–3969. [CrossRef] [PubMed]
63. Savoia, S.; Albera, A.; Brugiapaglia, A.; Di Stasio, L.; Ferragina, A.; Cecchinato, A.; Bittante, G. Prediction of Meat Quality Traits in the Abattoir Using Portable and Hand-Held near-Infrared Spectrometers. *Meat Sci.* **2020**, *161*, 108017. [CrossRef]
64. Bittante, G.; Savoia, S.; Cecchinato, A.; Pegolo, S.; Albera, A. Phenotypic and Genetic Variation of Ultraviolet–Visible-Infrared Spectral Wavelengths of Bovine Meat. *Sci. Rep.* **2021**, *11*, 13946 . [CrossRef]

foods

Article

Extracting Tissue Optical Properties and Detecting Bruised Tissue in Pears Quickly and Accurately Based on Spatial Frequency Domain Imaging and Machine Learning

Shengqiang Xing [1], Jiaming Zhang [1], Yifeng Luo [1], Yang Yang [1] and Xiaping Fu [1,2,*]

1 School of Information Science and Engineering, Zhejiang Sci-Tech University, Hangzhou 310018, China
2 Key Laboratory of Transplanting Equipment and Technology of Zhejiang Province, Hangzhou 310018, China
* Correspondence: fuxp@zstu.edu.cn

Abstract: Recently, Spatial Frequency Domain Imaging (SFDI) has gradually become an alternative method to extract tissue optical properties (OPs), as it provides a wide-field, no-contact acquisition. SFDI extracts OPs by least-square fitting (LSF) based on the diffuse approximation equation, but there are shortcomings in the speed and accuracy of extracting OPs. This study proposed a Long Short-term Memory Regressor (LSTMR) solution to extract tissue OPs. This method allows for fast and accurate extraction of tissue OPs. Firstly, the imaging system was developed, which is more compact and portable than conventional SFDI systems. Next, numerical simulation was performed using the Monte Carlo forward model to obtain the dataset, and then the mapping model was established using the dataset. Finally, the model was applied to detect the bruised tissue of 'crown' pears. The results show that the mean absolute errors of the absorption coefficient and the reduced scattering coefficient are no more than 0.32% and 0.21%, and the bruised tissue of 'crown' pears can be highlighted by the change of OPs. Compared with the LSF, the speed of extracting tissue OPs is improved by two orders of magnitude, and the accuracy is greatly improved. The study contributes to the rapid and accurate extraction of tissue OPs based on SFDI and has great potential in food safety assessment.

Keywords: spatial frequency domain imaging (SFDI); optical properties; absorption; reduced scattering; long short-term memory (LSTM)

Citation: Xing, S.; Zhang, J.; Luo, Y.; Yang, Y.; Fu, X. Extracting Tissue Optical Properties and Detecting Bruised Tissue in Pears Quickly and Accurately Based on Spatial Frequency Domain Imaging and Machine Learning. *Foods* **2023**, *12*, 238. https://doi.org/10.3390/foods12020238

Academic Editors: Mourad Kharbach and Samuli Urpelainen

Received: 20 November 2022
Revised: 28 December 2022
Accepted: 29 December 2022
Published: 4 January 2023

1. Introduction

The study of the propagation process of light in biological tissue has been a hot issue. It has been found that tissue optical properties (OPs) show great potential in biomedical detection [1,2], OPs' detection of fruit [3,4] and OPs' detection of milk [5]. The propagation behavior of light in biological tissue consists mainly of absorption and scattering, which are generally quantitatively described by the absorption coefficient (μ_a) and the reduced scattering coefficient (μ'_s). The μ_a reflects the chemical composition of biological tissue, whereas the μ'_s reflects the physical structural properties of the tissues [6]. Therefore, obtaining μ_a and μ'_s of biological tissue is important for assessing the physicochemical properties of biological tissue. There are various methods to obtain tissue OPs, such as the temporally resolved [7], spatially resolved [8], and integrating sphere methods [9]. As a new method to obtain tissue OPs, Spatial Frequency Domain Imaging (SFDI) is widely used in burned tissue assessment [10], meat classification [11], and bruised fruit detection [12,13]. The SFDI technique is commonly used in the biomedical field, but it is rarely used in food safety evaluation and agricultural product quality assessment.

There are two homogeneous forward models of mapping from OPs to diffuse reflectance in Spatial Frequency Domain Imaging. One model is an analytic approach based on the diffusion approximation equation and another model is based on transport using Monte Carlo (MC) simulations [14,15]. The main task of extracting tissue OPs by transport models is to deal with an inverse process of mapping tissue OPs to spatial frequency diffuse reflectance. There are two ways to implement the inversion process, one is the error minimization, and the other is the search method. For the first approach, the error minimization problem ($\min \sum \left(R_{d,model}(f_x) - R_{d,sample}(f_x) \right)$) is solved by inputting a guess value of the optical properties into the model to obtain the diffuse reflectance ($R_{d,model}$) closest to the actual value ($R_{d,sample}$). The second approach is a search problem, which first generates a large amount of data using a forward model and then compares the diffuse reflectance of the sample and dataset to find the optical properties values. Regardless of which forward model is used, there are two common methods used for inversion so far. One is the least-square fitting (LSF) method, and the other is the look-up table (LUT) method [16,17]. Generally, to obtain accurate and stable results, diffuse reflectance at multiple spatial frequencies is used [18,19]. However, whether using analytic approach based on the diffusion approximation equation or MC simulations based on transport, the LSF is computationally slow and unsuitable for fitting large numbers of pixels, which is an inherent drawback of the fitting method. The LUT method generates a diffuse reflectance dataset from a forward model and then builds a mapping table from diffuse reflectance to OPs, and the inversion process usually uses interpolation to estimate the OPs. In theory, if the interval of the LUT is small enough, extremely high inversion accuracy can be obtained. However, with the decrease of LUT interval and the increase of frequency number, the inversion time increases exponentially. The LUT therefore requires a compromise between accuracy and speed. In conclusion, traditional inversion methods are not good to balance accuracy and speed at the same time. Therefore, it is necessary to improve the speed and accuracy of SFDI inversion to quantify the OPs of tissue quickly and accurately.

Machine learning is widely used in visual inspection [20], quality assessment of agricultural products [21], and metal material research [22]. Since machine learning techniques have great advantages in dealing with regression problems with large amounts of data, they are used to replace the time-consuming model-based inversion process in diffuse reflectance optics [23,24]. The mapping between OPs and diffuse reflectance is strongly nonlinear in SFDI. Meanwhile, machine learning and regression techniques were found to be highly advantageous in solving nonlinear problems; for example, an artificial neural network (ANN) implementation for extraction of tissue OPs [25], and extraction of tissue OPs based on random forest regressor (RFR) [26]. According to the literature [27,28], machine learning-based extraction of OPs can be two orders of magnitude faster than conventional methods, without degrading the accuracy of OPs, based on the SFDI technique. Although these methods are based on machine learning, which greatly improves the prediction speed, the prediction accuracy is still lacking.

The analysis of OPs allows for the assessment of physiological indicators such as firmness and Soluble Solids Content (SSC) [6], which helps in the evaluation and classification of fruits. Fruits are prone to receive crushing and bruising during the picking, transportation, and marketing process. Over time, the bruised tissues of pears will decay and spread to the surrounding tissue, which eventually leads to a decrease in the economic efficiency of pears. Furthermore, it is a good mean to detect the bruised tissue of fruits by Ops. Therefore fast, accurate, and portable extraction of tissue OPs is of great importance in agricultural production and food safety.

Researchers have been looking for fast and accurate inversion methods, aiming to achieve real-time, accurate, and portable extraction of tissue OPs based on the SFDI technique. Common mapping models based on machine learning methods are used to extract OPs, which greatly improve the prediction speed and prediction accuracy. However, accuracy is still lacking. In this study, a mapping method based on Long Short-term Memory (LSTM) [29] was proposed to extract OPs, which is not only fast, but also improves accuracy. This work lays a foundation for solving the problem of real-time, accurate, and portable extraction of tissue OPs based on the SFDI technique. The purpose of this study was to look for an alternative approach to extract OPs quickly and accurately from diffuse reflectance images for bruised tissue detection in 'crown' pears. Therefore, the main objectives of this research are as follows: (1) build a compact and portable system; (2) obtain data through

Monte Carlo simulation; (3) establish a mapping model; and (4) detect the change of tissue OPs after a 'crown' pear has been bruised.

2. Materials and Methods

2.1. Spatial Frequency Domain Imaging Instrumentation

The Spatial Frequency Domain Imaging system is shown in Figure 1a. The grayscale illumination pattern is generated by a miniature projection module. In this study, we used a digital projector (M1, Lenovo, Beijing, China), based on a digital micromirror-based digital light processing (DLP) light engine (Texas Instruments, Dallas, TX, USA) and an LED light source. A filter (λ = 525 nm, $\Delta\lambda$ = 10 nm, Beijing Optical Century Instrument co. LTD (BOCIC), Beijing, China) was used in front of the lens of the projector to filter out light with a wavelength of 525 nm. The diffuse light reflected from the sample surface was captured by an 8-bit CCD camera (MV-CA060-11GM, Hikvision, Hangzhou, China). The generation of the illumination pattern, the projection, and the acquisition of the diffuse reflectance image of the sample were implemented by two ARM boards (Jetson Nano, Nvidia Corporation, Santa Clara, CA, USA). The synchronization of the sinusoidal illumination pattern projection and the sample diffuse reflectance acquisition was ensured by network communication. Compared to conventional systems, this study abandoned the strategy of using a personal computer as the control core and used miniature components, making the system more portable and compact.

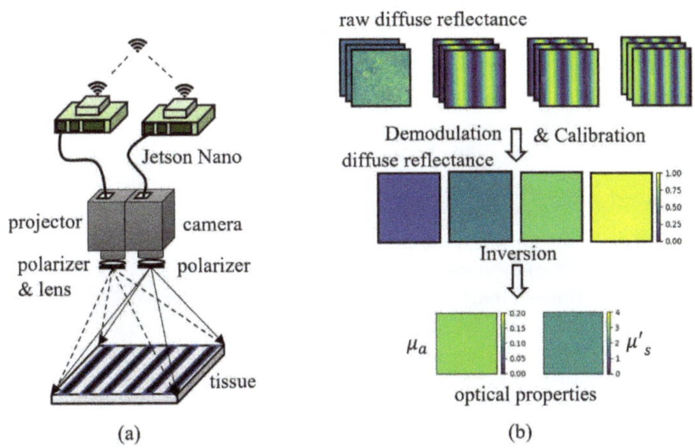

Figure 1. (**a**) for SFDI instrumentation and (**b**) for data processing.

2.2. SFDI Processing

For the Spatial Frequency Domain Imaging technique, the sinusoidally modulated light is projected onto the surface of the scattering medium first, and then the raw diffuse reflectance image is captured with a camera. Sinusoidally modulated light at each frequency needs to be projected three times with the phase 0π, $2\pi/3$, and $4\pi/3$. At least two frequencies are required to map the optical properties (OPs) using the optical transport model [30]. The data processing is shown in Figure 1b.

After obtaining the raw diffuse reflectance image, the modulation amplitude ($M(f_x)$) needs to be obtained by three-phase demodulation, as in Equation (1).

$$M(f_x) = \frac{\sqrt{2}}{3}\left\{(I_1(f_x) - I_2(f_x))^2 + (I_2(f_x) - I_3(f_x))^2 + (I_3(f_x) - I_1(f_x))^2\right\}^{1/2} \quad (1)$$

where $I_i(f_x)$ is the raw diffuse reflectance with different phases at the same frequency, and $M(f_x)$ is the modulation amplitude. The correction is then performed using a reference whiteboard with known diffuse reflectance, as in Equation (2),

$$R_d(f_x)_{measured} = \frac{M(f_x)_{measured}}{M(f_x)_{reference}} R_d(f_x)_{reference} \quad (2)$$

where $M(f_x)_{measured}$ is the modulated amplitude of the sample, $M(f_x)_{reference}$ is the modulated amplitude of the reference whiteboard, $R_d(f_x)_{measured}$ is the diffuse reflectance of the sample, and $R_d(f_x)_{reference}$ is the diffuse reflectance of the reference whiteboard. According to the diffuse approximation theory, the optical properties (μ_a, μ'_s) can be solved using curve fitting based on the diffuse equation [30] after the diffuse reflectance is obtained. This is shown in Equation (3),

$$R_d(f_x) = \frac{3A\mu'_s/\mu_{tr}}{(\mu'_{eff}/\mu_{tr}+1)(\mu'_{eff}/\mu_{tr}+3A)} \quad (3)$$

where $\mu_{tr} = \mu_a + \mu'_s$ is the transport coefficient, $\mu'_{eff} = \sqrt{3\mu_a\mu_{tr} + 2\pi f_x{}^2}$ represents the scalar attenuation coefficient in the spatial frequency domain, n is the refractive index of sample, $R_{eff} = 0.63n + 0.668 + 0.71/n - 1.44/n^2$ is effective reflection coefficient, and $A = (1 - R_{eff})/2(1 + R_{eff})$ is a proportionality constant.

2.3. Monte Carlo Simulations

Unlike the diffuse approximation equation, Monte Carlo (MC) simulation is a stochastic statistical method that simulates the transport of photons through tissue. After the photons enter the tissue, they constantly interact, and some of the photons are absorbed and disappear. Photons emitted from the upper surface of the tissue form diffuse light, and photons emitted from the lower surface of the tissue form transmitted light. Given the optical properties (OPs) parameters, the purpose of MC simulations is to simulate the photons transport process and then accurately calculate the corresponding diffuse reflectance. Many researchers have implemented MC simulations programs for different purposes, some for time-domain MC simulations [31], some for single-layer tissue MC simulations, and some for multi-layer tissue MC simulations [32]. This study uses a GPU-accelerated simulations program developed by Eric [33]. A large amount of mapping data (from OPs to diffuse reflectance) was obtained through MC simulations, which was used to construct the Long Short-term Memory Regressor model.

Given the value of the OPs, the diffuse reflectance $R_d(r)$ can be obtained using the MC simulations program. However, this spatially distributed diffuse reflectance $R_d(r)$ obtained by MC simulations is independent of the frequency of the structured light. The diffuse reflectance $R_d(f_x)$ in the spatial frequency domain (SFD) can be derived by Fourier transform [30]. As shown in Equation (4),

$$R_d(f_x) = 2\pi \sum_{i=1}^{n} r_i J_0(2\pi f_x r_i) R_d(r_i) \Delta r_i \quad (4)$$

where $R_d(f_x)$ is diffuse reflectance in SFD, r_i is the radial distance of the ith photon from the incident point of the light source in the MC simulation, f_x is the frequency of the sinusoidally modulated light, $R_d(r_i)$ is the reflection weight of the photon at the point r_i, Δr_i is the distance between radially adjacent photons, and J_0 is the zeroth-order Bessel function of the first kind. The initialization parameters of MC simulations are shown in Table 1.

Table 1. Parameter settings for the Monte Carlo simulations.

Parameters	Symbols [1]	Value
Number of photons	N	3,000,000
Resolution	dz [2], dr [3]	0.01 cm
Anisotropy	g	0.7
The thickness of tissue	d	5 cm
Refractive index air	n_0	1
Refractive index of tissue	n_1	1.34

[1] The symbols corresponding to the parameters. [2] The thickness of each tissue layer in the Monte Carlo simulations. [3] The thickness of each tissue layer in the radial direction of the light source.

2.4. Long Short-Term Memory Regressor Method

The Long Short-term Memory (LSTM) network model has great potential to solve problems where input sequences have context relations. Meanwhile, the LSTM network model performs very well in solving complex nonlinear problems. Therefore, this study decided to build a mapping model based on a LSTM network model for mapping tissue optical properties (OPs) from diffuse reflectance.

The Long Short-term Memory Regressor (LSTMR) model takes the n-dimensional diffuse reflectance vector and maps it to a 2-dimensional OPs vector. The input of the LSTM network is generally an n-dimensional vector with dimensions from 1 to n representing n moments. The LSTM has a memory cell that records the memory of each moment. Furthermore, the operations at each moment include adding memory and deleting memory to extract the relevant details of the context. The structure of the LSTMR model is shown in Figure 2, where the n-dimensional vector is the diffuse reflectance at different frequencies. The model of the network could be a deep neural network, and only one layer of the neural network is shown in the figure. The basic structure of the model is shown in the upper of Figure 2, and equations are shown in Equations (5)–(10),

$$F_t = \sigma(W_f[Y_{t-1}, X_t] + z_f) \tag{5}$$

$$I_t = \sigma(W_i[Y_{t-1}, X_t] + z_i) \tag{6}$$

$$\widetilde{C}_t = \tanh(W_c[Y_{t-1}, X_t] + z_c) \tag{7}$$

$$O_t = \sigma(W_o[Y_{t-1}, X_t] + z_o) \tag{8}$$

$$C_t = F_t \odot C_{t-1} + I_t \odot \widetilde{C}_t \tag{9}$$

$$Y_t = \tanh(F_t \odot C_{t-1} + I_t \odot \widetilde{C}_t) \odot O_t \tag{10}$$

where $\odot$ is the pointwise multiplication operation, σ is the sigmoid function, W is the weight matrix of the network layer, z is the bias term of the network layer, F_t is the forget gate, $=I_t$ is the input gate, $\widetilde{C}_t$ is the current memory, O_t is the output gate, C_t is the memory cell at moment t, h_t is the output at moment t, and X_t is the input at moment t. LSTM modifies the content of the memory cell through all the forget and input gates to extract context-related information. The final output of the model can be written as Equation (11),

$$[\mu_a, \mu'_s] = \sum w_t Y_t \tag{11}$$

where w_t is the weight of the output corresponding to each component of the input vector, and μ_a and μ'_s are the OPs.

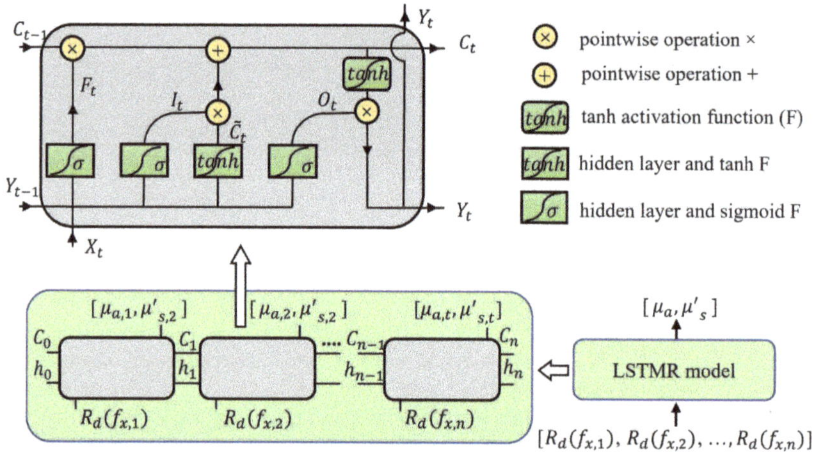

Figure 2. The Long Short-term Memory Regressor Model.

The Long Short-term Memory Regressor model was constructed based on the PyTorch framework (PyTorch, version 1.10.0+cu113, Meta, Menlo Park, CA, USA), which is a mainstream framework for building machine learning models. To obtain reliable and stable models, five-fold cross-validation was used in the optimization of model parameters. Theoretically, if there are enough nodes, an artificial neural network with one hidden layer can fit any complex function. This study obtained the best results when the number of nodes was 2^5 and the number of hidden layers was 5. Too small a batch size will lead to model oscillation and difficult convergence. Setting the batch size to the training set size is a good choice due to the small dataset, and choosing the Resilient Propagation optimizer, which is preferred when the batch size is equal to the training set size, has proven to be a smart choice. When the initial learning rate of the model was 0.0001, the model had a good convergence effect, and the model converged quickly before 200 epochs. Finally, the dropout algorithm was used to prevent overfitting, and the final model training took 30 min.

2.5. Model Testing

2.5.1. Simulations Experiments

The Long Short-term Memory Regressor (LSTMR) model was tested using a simulated dataset, and the tested dataset never appeared in the training dataset. To accelerate the inversion speed, the two-frequency inversion strategy is usually adopted. In this study, there were six alternative frequencies (f_x = 0.167, 0.180, 0.200, 0.220, 0.250, 0.300 mm^{-1}). Different mapping models were built with different high frequencies, and a five-fold cross-validation was used in the model-building process. Different models were used to map the optical properties (OPs), and then the mean absolute error of OPs was used as the basis for the preference.

After determining the optimal frequency, the full set of training dataset was used to train the Long Short-term Memory Regressor mapping model. To highlight the advantages of the model, least-square fitting (LSF), artificial neural network (ANN), random forest regressor (RFR), and recurrent neural networks (RNN) mapping methods were implemented in the experiments, respectively. The strengths and weaknesses of the models were evaluated by the normalized mean absolute error (NMAE), the determination coefficient (R^2), the root mean square error (RMSE), and the mean absolute error (MAE). Look-up table inversion was not chosen because it required a tradeoff in time and accuracy.

2.5.2. Phantoms Experiments

By using Indian ink (Royal Talens, Apeldoorn, The Netherlands) as an absorbing agent and titanium dioxide (T104950-500 g, Aladdin Biochemical Technology Corporation., Shanghai, China) as a scattering agent, deionized-water-based optical phantoms were prepared. Five gradients were set for absorption and reduced scattering, respectively. The volume fraction of India ink is 0.006–0.014% with a 0.002% interval and the volume fraction of TiO_2 is 0.04–0.12% with a 0.02% interval. Twenty-five small liquid phantoms with various optical properties (OPs) were fabricated. The absorption range of these phantoms was from 0.0945 mm^{-1} to 0.1905 mm^{-1}, and the scattering range was from 1.3689 mm^{-1} and 4.1063 mm^{-1}. According to Lambert's law, the standard value of the absorption coefficient was derived using the collimated transmittance T, and the collimated transmittance T is obtained by spectrometer acquisition (QE65pro, Ocean Insight Corporation, Orlando, FL, USA), as shown in Equation (12),

$$\mu_a = \frac{-\ln(T)}{D} = \frac{-\ln(I/I_0)}{D} \quad (12)$$

where D is the length of the optical path passed in the liquid during collimated transmission, and I_0 and I are the transmitted light intensity of water and the transmitted light intensity of the absorber, respectively. The standard value of the reduced scattering coefficient can be calculated by using the Mie program [34] given the parameters of TiO_2, the refractive index of water, and the wavelength of light. The parameters of TiO_2 include diameter, volume fraction, and refractive index.

The phantom experiments were performed using a two-frequency strategy (high frequency is the optimal frequency 0.25 mm^{-1}) and an inversion was performed using the Long Short-term Memory Regressor (LSTMR), the least-square fitting (LSF), the artificial neural network (ANN), the random forest regressor (RFR), recurrent neural network (RNN), respectively. A 300 × 300-pixel area near the center pixel of each phantom was selected as the target area and the OPs images were computed.

2.5.3. Pear Experiments

'Crown' pears were selected as experimental objects, and all crown pears came from fruit supermarkets. The surface of these pears was not damaged, and they were very fresh. The experiments were conducted in March at an ambient temperature of 20 degrees Celsius. During the bruising treatment, the pendulum motion was simulated by using a small iron ball to hit the pear around the equator, thus inducing the formation of bruised tissue. During the experiments, the experimental subjects were consistent before and after bruising, and the images of normal pears were collected first, and then the images of bruised pears were collected after bruising treatment. All experimental procedures used the same system to acquire images and the same program to extract optical properties.

3. Results

3.1. Simulation Experiment Results

To obtain the best high frequency, the mapping models with different high frequencies were built based on the training dataset, and the most suitable high frequencies were determined in the range of 0.167–0.300 mm^{-1}. Figure 3 illustrates the mean absolute error (MAE) of the optical properties (OPs), where the horizontal axis is the mapping model for different high frequencies. The results show that the model has the best accuracy when the frequency is chosen to be 0.25 mm^{-1}, and when the MAE of the absorption coefficient (μ_a) and the reduced scattering coefficient (μ'_s) are 0.6240% and 0.5939%, respectively. The optimal frequency of 0.25 mm^{-1} is very close to the commonly used optimal frequency of 0.2 mm^{-1}, which is consistent with the experimental results of Luo's frequency preference [5].

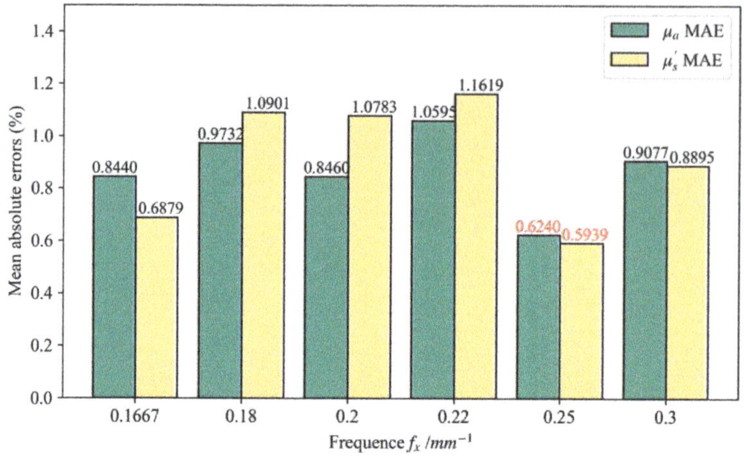

Figure 3. The Mean absolute errors of different models in predicting optical properties. The horizontal coordinate is the mapping model for different high frequencies. The numbers marked in red are the minimum mean absolute errors.

The prediction results of different models are shown in Table 2. The prediction results of the Long Short-term Memory (LSTMR) are optimal in terms of normalized mean absolute error (NMAE), MAE, root mean square error (RMSE), and determined coefficient (R^2). Except for the LSTMR model, the RFR model has the best prediction results. For LSTMR, the MAE of the μ_a and the μ'_s are 0.32% and 0.21%, respectively. This is an order-of-magnitude improvement compared to the prediction accuracy of the LSF. As shown in Figure 4, there is an extremely high linearity between the predicted and target values of the LSTMR, with the R^2 approaching 1. As can be seen in Figure 4, the target and predicted values almost exactly coincide and overlap in a straight line, both for the μ_a and the μ'_s. This indicates that the model is an excellent fitting and that the model fits well as a function of the diffuse reflectance and OPs. The experiments illustrate that LSTMR is an ideal model for accurately mapping OPs.

Table 2. Predictive performance of different mapping models in simulation experiments.

OPs	Metric	LSTMR	ANN	RNN	RFR	LSF
μ_a	NMAE	0.0012	0.0069	0.0077	0.0049	0.0506
	MAE	0.0032	0.0151	0.0181	0.0171	0.0597
	RMSE	0.0002	0.0010	0.0012	0.0007	0.0207
	R^2	1.0000	0.9999	0.9999	0.9999	0.9999
μ'_s	NMAE	0.0009	0.0046	0.0053	0.0036	0.0598
	MAE	0.0021	0.0127	0.0139	0.0023	0.0770
	RMSE	0.0060	0.0298	0.0350	0.0226	0.3323
	R^2	1.0000	0.9971	0.9966	0.9978	0.9996

3.2. Phantoms Experiments Results

To verify that the proposed Long Short-term Memory Regressor (LSTMR) mapping model can be used to extract optical properties (OPs) accurately and quickly, 25 optical phantoms with known OPs were produced. As shown in Table 3, LSTMR mapped OPs at a speed of 253 ms for a 300 × 300-pixel image (CPU, Intel-I7-11800H). However, for the least-square fitting (LSF) method, extracting the OPs of a 300 × 300-pixel image took 57,970 ms. The results show that the LSTMR inversion speed is improved by 2 to 3 orders of magnitude compared to LSF. The speed of predicting tissue OPs based on machine learning methods depends on the complexity of the model (number of nodes and number of network layers), so this study only implemented a speed comparison between LSTMR and LSF.

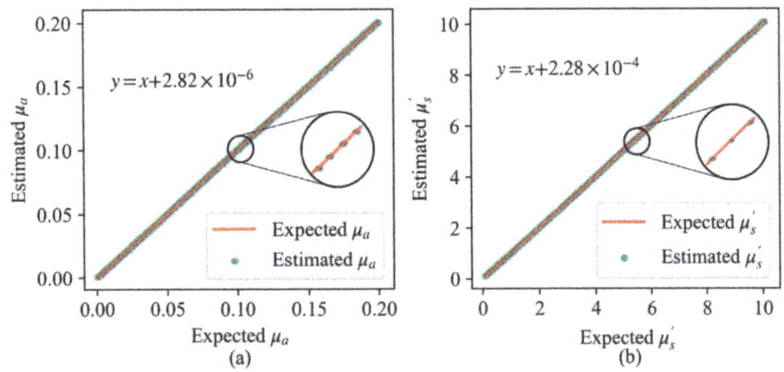

Figure 4. Linearity between the measured and calculated values of the optical properties in the simulation experiments, (**a**) is for the absorption coefficient and (**b**) is for the reduced scattering coefficient.

Table 3. The speed of mapping from diffuse reflectance to optical properties.

Methods	Resolution/Pixel	Speed ms/Pixel
LSF	300 × 300	0.6441
LSTMR	300 × 300	0.0028

The results of different inversion methods for the phantoms experiments are shown in Figures 5 and 6, and the mean absolute error (MAE) of μ_a and μ'_s are 0.0211 and 0.0674 using the LSTMR method, respectively. Furthermore, the R^2 of μ_a and μ'_s are 0.9916 and 1.0, respectively, which indicates that the predicted results of LSTMR have a good linear relationship with the expected values. It confirms that LSTMR is an ideal choice for inversion in Spatial Frequency Domain Imaging. Due to the inevitable experimental error, the actual value of the phantom is different from the reference value, so the prediction result of the phantom would be slightly worse than the simulation result. The relative error of μ'_s is larger than that of μ_a because the scattering agent is easily precipitated and is more influenced by whether the liquid surface is stationary or not, resulting in a larger error in the prediction of μ'_s. Obviously, the mapping results of the LSTMR inversion model are better than other models in the experiments.

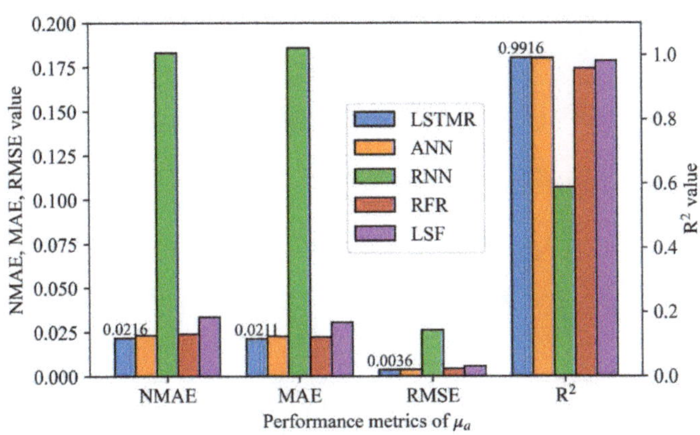

Figure 5. Performance metrics for different models in the phantom experiment for absorption coefficient.

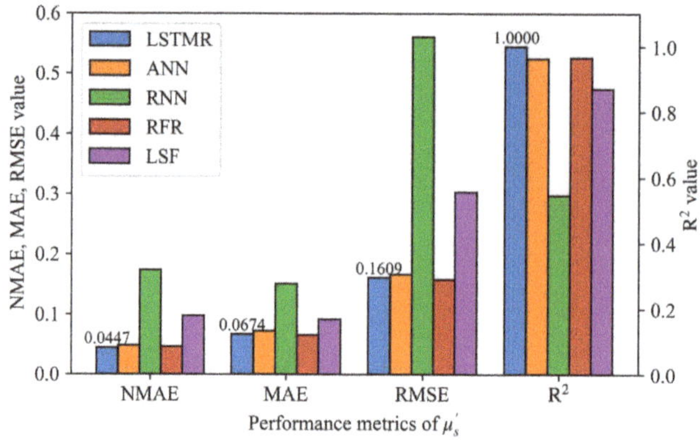

Figure 6. Performance metrics for different models in the phantom experiment for reduced scattering coefficient.

3.3. Pear Experiment Results

The results of the bruised tissue detection experiment for pears are shown in Figures 7 and 8. Pears will form bruised tissue at an early stage after being slightly crushed. In this study, it was demonstrated that bruised tissue forms on the surface of the pear after a slight impact. The absorption coefficient of the tissue increases during the formation of bruised tissue. The opposite is true for the reduced scattering coefficient, which is consistent with the experimental results of Sun [25] and Luo [35]. Moreover, both absorption and reduced scattering images could highlight the areas of bruised tissue. Therefore, using the Spatial Frequency Domain Imaging technique, early bruised detection of fruits can be performed. Furthermore, it can effectively control the bruised of fruits during transportation, thus controlling the cost of the fruit industry. The optical properties (OPs) of apple tissues can be used for nondestructive quality or ripeness prediction of apples [4], and the Long Short-term Memory method proposed in this study can obtain prediction results more accurately and quickly. Therefore, the rapid acquisition of OPs in tissues is of particular importance. This further illustrates the need to improve the speed and accuracy of extracting the OPs of tissues. It also lays the foundation for the real-time, portable acquisition of tissue OPs.

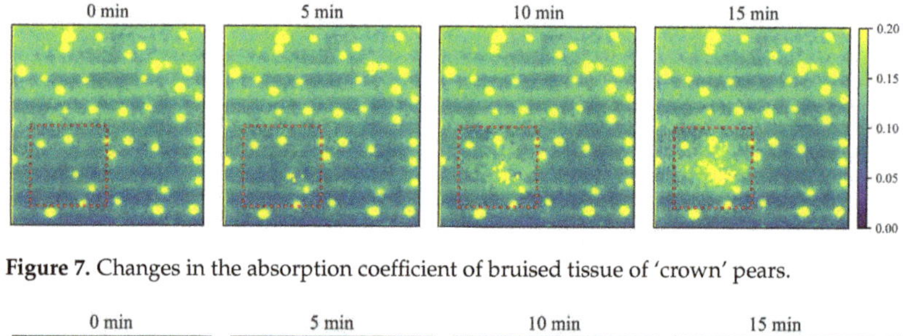

Figure 7. Changes in the absorption coefficient of bruised tissue of 'crown' pears.

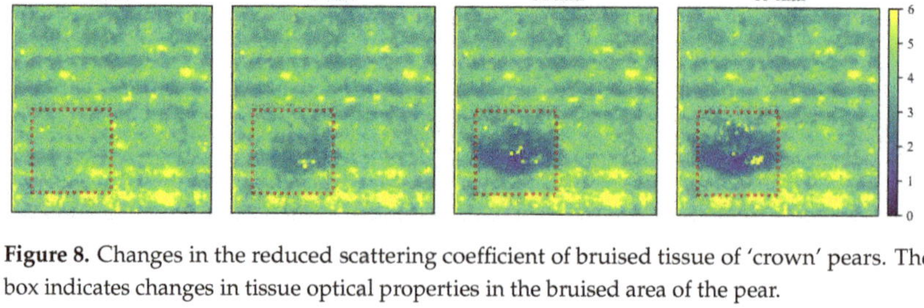

Figure 8. Changes in the reduced scattering coefficient of bruised tissue of 'crown' pears. The red box indicates changes in tissue optical properties in the bruised area of the pear.

4. Discussion

Absorption and scattering have different sensitivities to frequency. Absorption is mainly sensitive to low frequency, whereas scattering is mainly sensitive to high frequency. The non-zero frequency should not be too large or too small for the two-frequency inversion. The simulations experimental results showed that $f_x = 0.25\ mm^{-1}$ was the most suitable frequency under these experimental conditions, which was also closer to the non-zero frequency used in the existing literature [5]. The simulations experimental results show that the mapping accuracy of the Long Short-term Memory Regressor (LSTMR) model could be substantially improved, and the mean absolute error (MAE) of μ_a and μ'_s could reach 0.32% and 0.21%, respectively. The mapping accuracy of LSTMR is much better than that of the traditional LSF method, and it also performs better than the other machine learning methods in the experiments. Compared with the LSF method, the phantoms experiment not only shows that LSTMR has an advantage in mapping accuracy, but also has a huge performance advantage in inversion speed.

Jäger et al. [23] combined spatial resolution technology with multiple artificial neural network to extract optical properties (OPs). According to their reports, the normalized mean absolute errors (NMAEs) of μ_a and μ'_s are 6.1% and 2.9%, respectively. The MAE of the deep neural network mapping model proposed by Stier [1] for μ'_s is 6.8%. Song [28] developed an OPs mapping model based on the deep neural network, and according to their study, the mean and standard deviation of the percentage errors of μ_a and μ'_s were 0.0 ± 1.4 and $0.0 \pm 0.28\%$, respectively. Sun proposed an artificial neural network method [25] for the inversion of OPs based on multi-frequency inversion, where the NMAEs of μ_a and μ'_s are 0.18% and 0.027%, respectively. Sun used seven frequencies for inversion, whereas we used only two frequencies to achieve comparable accuracy. Panigrahi [26] demonstrated that the random forest regressor (RFR) method was a highly accurate and fast inversion method, and the MAE of OPs could be reduced to 0.556% and 0.126%, respectively. Comparing the MAE, it could be found that the μ_a of LSTMR was more accurate, whereas the μ'_s of RFR was more accurate. The μ'_s of LSTMR is slightly less accurate than RFR due to the large gradient (the interval of μ'_s is $0.126\ mm^{-1}$) of the μ'_s of the dataset.

The phantom experimental results showed that the LSTMR not only has better inversion accuracy than other methods, but also had a dramatic improvement in inversion speed, with a speed improvement of 2 to 3 orders of magnitude compared to the LSF. The LSF method requires continuous iterations for optimization until the error is within an acceptable range, which consumes a lot of time during the iterations, and which is evident as the number of frequency increases. The look-up table uses a search strategy in which the time taken for the search process increases exponentially as the number of frequencies increases. However, using multiple spatial frequencies for inversion can improve the robustness of the model [18]. The machine learning method can solve the slow speed problem in the process of multi-frequency inversion, and the mapping accuracy can be improved at the same time. As can be seen from Table 3, LSTMR is more than 100 times faster than LSF. This is also consistent with the results of Zhao's study [27] and Song's study [28].

5. Conclusions

The proposed Long Short-term Memory Regressor (LSTMR) method is an ideal mapping model to replace the inversion method based on the optical transport model. It can quickly extract optical properties (OPs), but without loss of estimation accuracy. This study not only compared the LSTMR method to the traditional LSF method, but also to other machine learning methods that appeared in journals, and it turns out that the LSTMR method is indeed a good choice. The experimental results show that the accuracy of LSTMR inversion is comparable to or even better than that of the previous literature. Furthermore, the speed of LSTMR is improved by 2~3 orders of magnitude compared with LSF. These pear experiments proved that LSTMR can accurately distinguish bruised tissue, which provides a feasible solution for the quality assessment of pears. All experiments are based on our developed miniaturized Spatial Frequency Domain Imaging system. This study laid the hardware foundation and method foundation for real-time and portable OPs acquisition of pears, and further applied it to pear quality evaluation.

Author Contributions: Conceptualization, S.X. and X.F.; methodology, S.X.; software, S.X.; validation, S.X., Y.L., Y.Y. and J.Z.; formal analysis, S.X.; investigation, S.X.; resources, X.F.; data curation, S.X.; writing—original draft preparation, S.X.; writing—review and editing, S.X. and X.F.; visualization, S.X.; supervision, X.F.; project administration, X.F.; funding acquisition, X.F. All authors have read and agreed to the published version of the manuscript.

Funding: This research was supported by the National Natural Science Fund of China (32071904), the Natural Science Fund of Zhejiang Province (LY20C130008), and the Science Foundation of Zhejiang Sci-Tech Univ. (ZSTU) (Grand No. 16022177-Y).

Data Availability Statement: The data that support the findings of this study are available from the corresponding author upon reasonable request.

Acknowledgments: The authors thank anonymous reviewers for providing helpful suggestions for improving the quality of this manuscript.

Conflicts of Interest: The authors declare no conflict of interest.

References

1. Stier, A.C.; Goth, W.; Hurley, A.; Brown, T.; Feng, X.; Zhang, Y.; Lopes, F.C.P.S.; Sebastian, K.R.; Ren, P.; Fox, M.C.; et al. Imaging sub-diffuse optical properties of cancerous and normal skin tissue using machine learning-aided spatial frequency domain imaging. *J. Biomed. Opt.* **2021**, *26*, 096007. [CrossRef] [PubMed]
2. Yafi, A.; Vetter, T.S.; Scholz, T.; Patel, S.; Saager, R.B.; Cuccia, D.J.; Evans, G.R.; Durkin, A.J. Postoperative quantitative assessment of reconstructive tissue status in a cutaneous flap model using spatial frequency domain imaging. *Plast. Reconstr. Surg.* **2011**, *127*, 117–130. [CrossRef] [PubMed]
3. Hu, D.; Fu, X.; He, X.; Ying, Y. Noncontact and wide-field characterization of the absorption and scattering properties of apple fruit using spatial-frequency domain imaging. *Sci. Rep.* **2016**, *6*, 37920. [CrossRef] [PubMed]
4. Lohner, S.A.; Biegert, K.; Nothelfer, S.; Hohmann, A.; McCormick, R.; Kienle, A. Determining the optical properties of apple tissue and their dependence on physiological and morphological characteristics during maturation. Part 1: Spatial Frequency Domain Imaging. *Postharvest Biol. Technol.* **2021**, *181*, 111647. [CrossRef]
5. Luo, Y.; Dai, L.; Jiang, X.; Fu, X. Measurement of absorption and scattering properties of milk using a hyperspectral spatial frequency domain imaging system. *J. Food Meas. Charact.* **2022**, *16*, 753–761. [CrossRef]
6. He, X.; Fu, X.; Rao, X.; Fang, Z. Assessing firmness and SSC of pears based on absorption and scattering properties using an automatic integrating sphere system from 400 to 1150 nm. *Postharvest Biol. Technol.* **2016**, *121*, 62–70. [CrossRef]
7. Lurie, S.; Vanoli, M.; Dagar, A.; Weksler, A.; Lovati, F.; Zerbini, P.E.; Spinelli, L.; Torricelli, A.; Feng, J.; Rizzolo, A. Chilling injury in stored nectarines and its detection by time-resolved reflectance spectroscopy. *Postharvest Biol. Technol.* **2011**, *59*, 211–218. [CrossRef]
8. Chen, Y.W.; Chen, C.C.; Huang, P.J.; Tseng, S.H. Artificial neural networks for retrieving absorption and reduced scattering spectra from frequency-domain diffuse reflectance spectroscopy at short source-detector separation. *Biomed. Opt. Express* **2016**, *7*, 1496–1510. [CrossRef]
9. Hu, D.; Lu, R.; Huang, Y.; Ying, Y.; Fu, X. Effects of optical variables in a single integrating sphere system on estimation of scattering properties of turbid media. *Biosyst. Eng.* **2020**, *194*, 82–98. [CrossRef]
10. Rowland, R.A.; Ponticorvo, A.; Baldado, A.L.; Kennedy, G.T.; Burmeister, D.M.; Christy, R.J.J.; Bernal, N.P.; Durkin, A.J. Burn wound classification model using spatial frequency-domain imaging and machine learning. *J. Biomed. Opt.* **2019**, *24*, 056007.
11. Fu, X.; Jiang, X.; Dai, L.; Luo, Y. Study of spatial frequency domain imaging technique for turbid media optical property estimation and application. In *Optical Metrology and Inspection for Industrial Applications VII (OMIIA 2020)*; SPIE: Bellingham, WA, USA, 2020; Volume 11552. [CrossRef]
12. He, X.; Fu, X.; Li, T.; Rao, X. Spatial frequency domain imaging for detecting bruises of pears. *J. Food Meas. Charact.* **2018**, *12*, 1266–1273. [CrossRef]
13. He, X.; Hu, D.; Fu, X.; Rao, X. Spatial frequency domain imaging for determining absorption and scattering properties of bruised pears based on profile corrected diffused reflectance. *Postharvest Biol. Technol.* **2021**, *179*, 111570. [CrossRef]
14. Jacques, S.L.; Wang, L. *Monte Carlo Modeling of Light Transport in Tissues*; Welch, A.J., Van Gemert, M.J.C., Eds.; Optical-Thermal Response of Laser-Irradiated Tissue, Lasers, Photonics, and Electro-Optics; Springer: Boston, MA, USA, 1995; pp. 73–100.
15. Marquet, P.; Bevilacqua, F.P.; Depeursinge, C.D.; de Haller, E.B. Determination of reduced scattering and absorption coefficients by a single charge-coupled-device array measurement, Part I: Comparison between experiments and simulations. *Opt. Eng.* **1995**, *34*, 2055–2063. [CrossRef]
16. Erickson, T.A.; Mazhar, A.; Cuccia, D.J.; Durkin, A.J.; Tunnell, J.W. Lookup-table method for imaging optical properties with structured illumination beyond the diffusion theory regime. *J. Biomed. Opt.* **2010**, *15*, 036013. [CrossRef] [PubMed]

17. Angelo, J.; Vargas, C.R.; Lee, B.T.; Bigio, I.J.; Gioux, S. Ultrafast optical property map generation using lookup tables. *J. Biomed. Opt.* **2016**, *21*, 110501. [CrossRef]
18. Pera, V.; Karrobi, K.; Tabassum, S.; Teng, F.; Roblyer, D. Optical property uncertainty estimates for spatial frequency domain imaging. biomed. *Opt. Express* **2018**, *9*, 661–678. [CrossRef] [PubMed]
19. Lin, A.J.; Ponticorvo, A.; Konecky, S.D.; Cui, H.; Rice, T.B.; Choi, B.; Durkin, A.J.; Tromberg, B.J. Visible spatial frequency domain imaging with a digital light microprojector. *J. Biomed. Opt.* **2013**, *18*, 096007. [CrossRef]
20. Mahadevkar, S.V.; Khemani, B.; Patil, S.; Kotecha, K.; Vora, D.R.; Abraham, A.; Gabralla, L.A. A review on machine learning styles in computer vision—Techniques and future directions. *IEEE Access* **2022**, *10*, 107293–107329. [CrossRef]
21. Kaya, F.; Keshavarzi, A.; Francaviglia, R.; Kaplan, G.; Başayiğit, L.; Dedeoğlu, M. Assessing machine learning-based prediction under different agricultural practices for digital mapping of soil organic carbon and available phosphorus. *Agriculture* **2022**, *12*, 1062. [CrossRef]
22. Geng, X.; Mao, X.; Wu, H.; Wang, S.; Xue, W.; Zhang, G.; Ullah, A.; Wang, H. A hybrid machine learning model for predicting continuous cooling transformation diagrams in welding heat-affected zone of low alloy steels. *J. Mater. Sci. Technol.* **2022**, *107*, 207–215. [CrossRef]
23. Jäger, M.; Foschum, F.; Kienle, A. Application of multiple artificial neural networks for the determination of the optical properties of turbid media. *J. Biomed. Opt.* **2013**, *18*, 057005. [CrossRef] [PubMed]
24. He, X.; Jiang, X.; Fu, X.; Gao, Y.; Rao, X. Least squares support vector machine regression combined with Monte Carlo simulation based on the spatial frequency domain imaging for the detection of optical properties of pear. *Postharvest Biol. Technol.* **2018**, *145*, 1–9. [CrossRef]
25. Sun, Z.; Xie, L.; Hu, D.; Ying, Y. An artificial neural network model for accurate and efficient optical property mapping from spatial-frequency domain images. *Comput. Electron. Agric.* **2021**, *188*, 106340. [CrossRef]
26. Panigrahi, S.; Gioux, S. Machine learning approach for rapid and accurate estimation of optical properties using spatial frequency domain imaging. *J. Biomed. Opt.* **2018**, *24*, 071606. [CrossRef] [PubMed]
27. Zhao, Y.; Deng, Y.; Bao, F.; Peterson, H.; Istfan, R.; Roblyer, D. Deep learning model for ultrafast multifrequency optical property extractions for spatial frequency domain imaging. *Opt. Lett.* **2018**, *43*, 5669–5672. [CrossRef] [PubMed]
28. Song, B.; Jia, W.; Zhao, Y.; Huang, H.; Fan, Y. Ultracompact deep neural network for ultrafast optical property extraction in Spatial Frequency Domain Imaging (SFDI). *Photonics* **2022**, *9*, 327. [CrossRef]
29. Hochreiter, S.; Schmidhuber, J. Long Short-term Memory. *Neural Comput.* **1997**, *9*, 1735–1780. [CrossRef]
30. Cuccia, D.J.; Bevilacqua, F.P.; Durkin, A.J.; Ayers, F.R.; Tromberg, B.J. Quantitation and mapping of tissue optical properties using modulated imaging. *J. Biomed. Opt.* **2009**, *14*, 024012. [CrossRef]
31. Jacques, S.L. Time-resolved reflectance spectroscopy in turbid tissues. *IEEE. Trans. Biomed. Eng.* **1989**, *36*, 1155–1161. [CrossRef]
32. Wang, L.; Jacques, S.L.; Zheng, L. MCML-Monte Carlo modeling of light transport in multi-layered tissues. *Comput. Methods Programs Biomed.* **1995**, *47*, 131–146. [CrossRef]
33. Alerstam, E.; Svensson, T.; Andersson-Engels, T. Parallel computing with graphics processing units for high-speed monte carlo simulation of photon migration. *J. Biomed. Opt.* **2008**, *13*, 060504. [CrossRef] [PubMed]
34. Mätzler, C. MATLAB Functions for Mie Scattering and Absorption. 2002. Available online: https://omlc.org/software/mie/maetzlermie/Maetzler2002.pdf (accessed on 17 June 2021).
35. Luo, Y.; Jiang, X.; Fu, X. Spatial frequency domain imaging system calibration, correction and application for pear surface damage detection. *Foods* **2021**, *10*, 2151. [CrossRef] [PubMed]

foods

Article

Discriminant Analysis of Brazilian Stingless Bee Honey Reveals an Iron-Based Biogeographical Origin

Flavia C. Lavinas [1], Brendo A. Gomes [2], Marcos V. T. Silva [1], Renata M. Nunes [1], Suzana G. Leitão [3], Mirian R. L. Moura [3], Rosineide C. Simas [4,5], Carla S. Carneiro [3] and Igor A. Rodrigues [3,*]

1 Programa de Pós-Graduação em Ciências Farmacêuticas, Faculdade de Farmácia, Universidade Federal do Rio de Janeiro, Rio de Janeiro 21941-902, Brazil
2 Programa de Pós-Graduação em Biotecnologia Vegetal, Universidade Federal do Rio de Janeiro, Rio de Janeiro 21941-902, Brazil
3 Departamento de Produtos Naturais e Alimentos, Faculdade de Farmácia, Universidade Federal do Rio de Janeiro, Rio de Janeiro 21941-902, Brazil
4 Laboratorio de Cromatografia e Espectrometria de Massas (LaCEM), Universidade Federal de Goiás, Goiania 74690-900, Brazil
5 Escola de Engenharia, Universidade Presbiteriana Mackenzie, São Paulo 01302-907, Brazil
* Correspondence: igor@pharma.ufrj.br

Abstract: Stingless bee honey (SBH) is gaining attention due to its nutritional, sensorial, and medicinal characteristics. This study focuses on the combination of physicochemical properties, antioxidant capacity, mineral profile, and mass spectrometry-based fingerprints, using a chemometric approach to differentiate SBH ($n = 18$) from three different Brazilian biogeographical zones (Caatinga, Cerrado, and Atlantic Forest). The physicochemical properties of SBH varied, resulting in a wide range of water activity, moisture, total soluble solids, pH, and total and free acidity. The Caatinga honey showed the highest and the lowest contents of phenolics and flavonoids, respectively. The antioxidant free-radical scavenging assays demonstrated that the Brazilian SBH has a high antioxidant potential. The mineral profile of honey samples from the Atlantic Forest revealed higher contents of Ca and Fe while the Cerrado and Caatinga honey showed the highest P contents. Partial Least-Squares Discriminant Analysis (PLS-DA) analysis separated the samples into three groups based on the biogeographical zones of harvest. The main separation factors between groups were the m/z 326 ion and the Fe content. Univariate analysis confirmed that Fe content is important for SBH discrimination. The present results indicate that the origin of SBH can be determined on the basis of mineral profile, especially Fe content.

Keywords: meliponine honey; physicochemical properties; biomes; antioxidant potential; mineral profile; mass spectrometry analysis; chemometrics

Citation: Lavinas, F.C.; Gomes, B.A.; Silva, M.V.T.; Nunes, R.M.; Leitão, S.G.; Moura, M.R.L.; Simas, R.C.; Carneiro, C.S.; Rodrigues, I.A. Discriminant Analysis of Brazilian Stingless Bee Honey Reveals an Iron-Based Biogeographical Origin. *Foods* **2023**, *12*, 180. https://doi.org/10.3390/foods12010180

Academic Editors: Samuli Urpelainen and Mourad Kharbach

Received: 14 November 2022
Revised: 9 December 2022
Accepted: 12 December 2022
Published: 1 January 2023

1. Introduction

Brazil is a country of continental dimensions (about 8.5 million km^2), with marked climatic variations over its territory. Climate zones such as tropical, semi-arid, and temperate areas occur from the north to the south of the country. These climatic differences, allied to the various soil types, lead to major ecological variations with the consequent formation of distinct biomes or biogeographical zones. Brazilian biogeographical zones can be mainly divided into six distinct areas: the Amazon Rainforest; Pantanal (wetlands), Cerrado (Brazilian savannah), Caatinga (semi-arid scrub forest), Pampas (grasslands), and the Atlantic Forest [1]. Unsurprisingly, 70% of the world's cataloged animal and plant species can be found in this country [2].

Brazilian native meliponines (Apidae, subfamily Meliponinae), or simply stingless bees, are insects of great economic and socioenvironmental importance. These bees play an important role as pollinators of wild botanical individuals, including endemic species, as well as several crops such as passion fruit, Brazilian nut, watermelon, guava, tomato, and açaí [3]. In addition, stingless bees (SB) can produce a distinguished honey with unique compositional and sensory characteristics. Stingless bee honey (SBH) is a high sugar-containing solution often showing higher moisture and free acidity compared to *Apis mellifera* honey [4]. Notably, the physicochemical characteristic of honey is influenced by its complex chemical composition, which may vary according to environmental factors, bee species, botanical resources, storage conditions, and processing methods [5].

Substances bearing phenolic groups are identified as the main factors responsible for the antioxidant capacity of honey. Phenolic substances commonly reported in SBH include phenolic acids (gallic acid, caffeic acid, coumaric acid, ellagic acid, hydroxycinnamic acid) and flavonoids (taxifolin, naringenin, luteolin, quercetin, catechin, apigenin) [6,7]. Other components that may contribute to the antioxidant property of honey are amino acids, proteins, enzymes, carotenoids, and organic acids [8]. Here, we combined physicochemical properties, antioxidant capacity, mineral profile, and mass spectrometry-based fingerprints using a chemometric approach to provide new insights into the influence of biogeographical zones on the compositional singularities of SBH. In addition, the identification of the main factors responsible for honey discrimination would permit the use of these analytes for the investigation of honey origin.

2. Materials and Methods

2.1. Chemicals

High-Performance Liquid Chromatography (HPLC) grade chloroform, methanol and formic acid were obtained from Tedia Brazil® (RJ, Brazil). Folin-Ciocalteu, ABTS (2,2′-azinobis-(3-ethylbenzothiazoline-6-sulfonic acid), TPTZ (2,4,6-tripyridyl-s-Triazine) and nitric acid were obtained from Sigma-Aldrich (St. Louis, MO, USA). Acetic acid, chloride acid, ferric chloride heptahydrate, iron (II) sulfate heptahydrate, and aluminum chloride hexahydrate were obtained from Dinâmica® (Sao Paulo, Brazil).

2.2. Honey Samples

SBH samples (n = 18) were provided by the Rio de Janeiro Stingless Beekeepers Association (AME-Rio, Rio de Janeiro, Brazil). The samples were originally collected in different Brazilian biogeographical zones between 2017 and 2018 and stored at low temperature (4 °C) in the dark until the time for the analytical procedures. Table 1 summarizes the SB species and honey origin.

Table 1. Details of stingless bee species providing samples.

Biogeographical Zones	Bee Species	Popular Name	Year (Season)	Sample Identifier
Atlantic Forest	*Melipona mondury*	Uruçu Amarela	2017 (autumn)	MM1
	Melipona mondury	Uruçu Amarela	2017 (autumn)	MM2
	Melipona mondury	Uruçu Amarela	2017 (autumn)	MM3
	Melipona mondury	Uruçu Amarela	2018 (summer)	MM4
	Melipona quadrifasciata	Mandaçaia	2018 (spring)	MQ1
Caatinga	*Melipona subnitida*	Jandaíra	2018 (autumn)	MS1
	Melipona subnitida	Jandaira	2018 (winter)	MS2
	Melipona scutellaris	Uruçu Nordestina	2018 (winter)	MSC1
	Melipona scutellaris	Uruçu Nordestina	2018 (summer)	MSC2
	Melipona scutellaris	Uruçu Nordestina	2018 (summer)	MSC3
	Melipona mondury	Uruçu Amarela	2018 (summer)	MM5 (6)
	Scaptotrigona aff. postica	Tubi	2018 (winter)	SCA1
Cerrado	*Melipona fasciculata*	Tiúba	2018 (winter)	MF1
	Melipona fasciculata	Tiúba	2018 (winter)	MF2
	Melipona fasciculata	Tiúba	2018 (winter)	MF3
	Melipona fasciculata	Tiúba	2018 (spring)	MF4

Table 1. *Cont.*

Biogeographical Zones	Bee Species	Popular Name	Year (Season)	Sample Identifier
	Scaptotrigona polysticta	Benjoi	2018 (winter)	SCP1
	Melipona mondury	Uruçu Amarela	2018 (summer)	MM6 (5)

2.3. Physicochemical Analysis

The physicochemical properties of SBH, including water activity (Aw), moisture, total soluble solids (TSS), pH, total and free acidity, and hydroxymethylfurfural (HMF), were determined as described by the Association of Official Analytical Chemists [9]. The color features of SBH were determined spectrophotometrically. Initially, the absorbance (Abs) of honey sample solutions (0.5 g honey in 1 mL distilled water) was measured at 635 nm [10]. The results were converted to the Pfund scale (mm) using the equation below. Brown pigment formation was determined by measuring the absorbance (420 nm) of honey samples previously diluted to 4° Brix with distilled water [11].

$$\text{Pfund} = -38.70 + 371.39 \times \text{Abs} \quad (1)$$

2.4. Total Phenolic (TPC) and Total Flavonoid (TFC) Contents

TPC was determined spectrophotometrically by the Folin-Ciocalteu method [12]. First, 1 g of honey was diluted to 10 mL with distilled water and then filtered through a Whatman® (Maidstone, UK) Grade 1 qualitative filter paper (11 µm). Next, 0.1 mL of this solution was added to 0.5 mL of Folin-Ciocalteau reagent (10% *w*/*v*). After 5 min in the dark, 0.4 mL of a 7.5% sodium bicarbonate solution was added and the mixture was kept in the dark at room temperature for 2 h. The absorbance of the reaction mixture was measured at 760 nm (SpectraMax M5, Molecular Devices, Sunnyvale, CA, USA). Gallic acid was used to produce a calibration curve (7–200 µg/mL), and TPC was expressed as mg gallic acid equivalent per 100 g fresh honey weight (mg GAE/100 g FW).

TFC was determined by a spectrophotometric assay based on flavonoid-aluminum complex formation [10]. Initially, a honey solution was prepared by mixing 0.25 mL of the sample with 1.25 mL of distilled water and 75 µL of 5% $NaNO_2$. Then, 0.15 mL of a 10% $AlCl_3$ solution was added and the mixture was allowed to stand for 5 min. After adding 0.5 mL of a 1 M NaOH solution, the final volume was adjusted to 2.5 mL with distilled water. The absorbance of the sample was measured at 510 nm (SpectraMax M5, Molecular Devices, Sunnyvale, CA, USA). Quercetin calibration curves were prepared (50–550 µg/mL) and TFC was expressed as mg quercetin equivalent per 100 g fresh honey weight (mg QE/100 g FW).

2.5. Antioxidant Capacity Assays

The radical scavenging capacity of SBH samples was determined by the ABTS method [5]. After the $ABTS^{\bullet+}$ generation step, 2.7 mL aliquots of the radical solution were transferred to test tubes containing samples previously diluted in ethanol (100–500 µg/mL, final concentrations). After 10 min, sample absorbances were measured at 734 nm (SpectraMax M5, Molecular Devices, Sunnyvale, CA, USA). The $ABTS^{\bullet+}$ scavenging capacity percentage was determined as described below.

$$ABTS^{\bullet+} \text{ scavenging capacity } (\%) = (Abs_{cnt} - A_s) \times 100/A_{cnt} \quad (2)$$

where Abs_{cnt} = absorbance obtained from the $ABTS^{\bullet+}$ solution; A_s = absorbance obtained from samples in the presence of the $ABTS^{\bullet+}$ solution.

The ferric reducing antioxidant power (FRAP) of honey samples was determined as reported by Benzie & Strain (1996) with some modifications [13]. Initially, the FRAP reagent (2 mL of 10 mM TPTZ solution in 6 N HCl, 2 mL of 20 mM $FeCl_3$ solution, and 20 mL of 300 mM acetate buffer, pH 3.6) was warmed to 37 °C prior to analysis. A 20 µL aliquot of honey solution (100 mg.mL^{-1}) was added to 180 μL of freshly prepared FRAP reagent, the reaction mixture was incubated at 37 °C for 4 min and absorbance was then measured at 595 nm. A calibration curve was prepared with a ferrous sulfate solution (50–800 μM). FRAP values were expressed as micromoles of ferrous equivalent per 100 g fresh honey weight (μmol Fe^{2+}/100 g FW).

2.6. Mineral Profile Analysis

The mineral and trace contents of stingless bee honey were determined by Total Reflection X-ray Fluorescence (TXRF). Sample preparation was based on acid digestion (65% HNO_3) and heat [14]. Blank samples were also prepared to assess possible contamination. Eight replicates were prepared for each sample. Standard solutions (CertiPUR Reference Material, Merck®, Rahway, NJ, USA) with different concentrations of well-known elements were prepared and Ga (5.0 μg/g, final concentration) was added as an internal standard. To check the accuracy of the procedure used for quantitative analysis, standard reference material (Bovine liver, NIST®, SRM 1577, Gaithersburg, MD, USA) was analyzed, and the calculated data were compared to certified values. An X-ray fluorescence system operating under total reflection conditions was used for sample excitation. For characteristic X-ray detection, a portable automated total reflection X-ray fluorescence system using a low-power X-ray tube and a compact Si-PIN detector was employed [15]. The quantitative analysis of X-ray spectra was performed using the QXAS software 1.2 (International Atomic Energy Agency – IAEA, Seibersdorf, Austria). The mineral content of honey samples was expressed as μg/g fresh weight (FW).

2.7. Mass Spectrometry Analysis

Honey samples were prepared from liquid-liquid partitions using ultra-purified water and chloroform (1:1). The chloroform portion was dried and solubilized in methanol at a concentration of 20 mg/mL. Next, 1 μL of each sample was analyzed with an LCQ Fleet mass spectrometer (MS) with an electrospray source (ESI), operated in positive ionization mode. Samples were analyzed by automated direct injection in 0.1 mL.min−1 flow for 5 min. High purity nitrogen (N2) was used as sheath gas (35 arbitrary units) and auxiliary gas (10 arbitrary units). High purity helium (He) was used as collision gas. The MS parameters were as follows: source voltage 5 kV, source current 100 μA, source temperature 450 °C, capillary voltage 7 V, tube lens voltage 65 V, and capillary temperature 400 °C. MS spectra were acquired with a range of *m/z* 50–1000.

2.8. Statistical Analyses

Assays were performed in triplicate, and the results were expressed as mean values with standard deviations (SD). The Shapiro-Wilks test was used to determine the normality of physicochemical, antioxidant, and mineral data ($p > 0.05$). Next, analysis of variance (one-way ANOVA) followed by Tukey's multiple comparison post hoc test was performed using XLSTAT® software (version 2014, Addinsoft, Paris, France). Mass spectrometry data from each sample were analyzed in XCalibur® 2.2 (ThermoScientific, Waltham, MA, USA), where the one thousand most intense peaks were exported in centroid mode. Variables present in less than 7 samples were discarded to avoid biases. The matrix data consisted of 277 variables that corresponded to mass peak intensities; 13 were physicochemical parameters and 12 were mineral concentrations. Multivariate (PLS-DA) and univariate (One-Way ANOVA and Tukey HSD post hoc test) analyses were carried out using the MetaboAnalyst 5.0 webserver. The Shapiro-Wilks test was used to determine the normality of data. Finally, correlations were established using Pearson's correlation coefficient (r). The correlations were determined using XLSTAT®. A p-value < 0.05 was considered significant.

3. Results and Discussion

3.1. Physicochemical Characterization of SBH

The physicochemical properties of SBH from different biogeographical zones of Brazil are summarized in Table 2. We observed that most samples differed significantly ($p < 0.05$) from one another, even those belonging to the same species and collected in the same region. The variation in the physicochemical properties of SBH has been frequently reported [16], demonstrating that quality standards may be extremely hard to establish.

Water content is considered to be one of the most important features of honey since it affects several properties such as viscosity, specific weight, maturity, flavor, crystallization [17], and microbial growth [18]. Here, SBH displayed Aw values ranging from 0.65 to 0.75. There was a significant difference ($p < 0.05$) between honey samples from the Atlantic Forest and Cerrado zones and between those from the Caatinga and Cerrado zones. Ávila et al., (2019), reported Aw values ranging from 0.77 to 0.91 in thirty-two honey samples of two south Brazilian SB species (*Scaptotrigona* spp. and *Melipona* spp.) [19]. In addition, the Aw values of SBH collected in several cities in the state of Santa Catarina (SC, south Brazil) ranged from 0.67 to 0.78 [20]. The *Tetragonisca angustula* honey collected in the municipality of Piracicaba (SP, Southeast Brazil) displayed Aw values ranging from 0.59 to 0.82 [21]. In the present study, all honey samples showed high water activity (greater than 0.6), which means that stingless bee honey is susceptible to microbial fermentation. Indeed, osmophilic yeasts can grow at Aw values above 0.6 [22].

SBH is more fluid and displays higher moisture content than *A. mellifera* honey [8]. Here, the moisture content of SBH ranged from 19.47% to 36.31% (w/w). There was a significant difference ($p < 0.05$) in moisture content between the samples from the Atlantic Forest and the Cerrado zones and between the samples from the Caatinga and Cerrado zones. The honey samples MF2 (Cerrado) and MM2 (Atlantic Forest) showed the lowest and highest moisture contents, respectively. The MF2 result may be due to the influence of the climate condition of the Cerrado. During the dry season, relative air humidity can reach low levels (about 9–11%) similar to those of desert regions. Even during the rainy season, high temperatures and a decrease in air relative humidity may eventually occur [23]. SBH produced in different regions of the semi-arid northeastern region of Brazil (Caatinga zone) displayed moisture content values ranging from 23.17% to 28.9% (w/w) [24–26]. Biluca et al., (2016), reported that SBH from the Atlantic Forest zone showed high moisture content (23.1% to 43.5%, w/w) [27]. The climate of this region is classified as tropical humid [28], which could explain the moisture levels of SBH collected in this biogeographical zone. Similarly, SBH collected in different Ecuadorian regions showed moisture contents ranging from 22% to 30% [29]. The mean annual humidity in Ecuador can range from 65% to 85% [30]. Taken together, these findings suggest that relative air humidity can influence the moisture content of SBH.

Among the SBH samples investigated in the present study, MM2 and MF2 displayed TSS values of 61.33 and 75.75° Brix, the lowest and the highest values, respectively. In addition, there was a significant difference between samples obtained from the Atlantic Forest and Cerrado zones and those obtained from the Caatinga and Cerrado zones. Previous studies conducted by Brazilian [25,31] and Malaysian [32] research groups demonstrated TSS values similar to those determined here. TSS is closely related to the moisture and sugar content of honey, representing an important indicator of adulteration. In addition, honey produced by stingless bees exhibits lower TSS values than honey produced by *A. mellifera* [33].

Table 2. Physicochemical parameters of honey produced by stingless bees from different Brazilian biogeographical zones.

Biogeogr. Zone	Samples	Aw	Moisture (%)	TSS (°Brix)	pH	Total Acidity (meq/Kg)	Free Acidity (meq/Kg)	Brown Pigment	HMF (mg/Kg)	Pfund (mm)	Color Name
Atlantic Forest	MM1	0.75 ± 0.00 ^{c,d}	30.43 ± 0.37 ^{c,d}	67.65 ± 0.35 ^{g}	3.24 ± 0.10 ^{f}	71.04 ± 0.55 ^{k}	59.74 ± 0.99 ^{f}	0.05 ± 0.00 ^{d}	0.38 ± 0.08 ^{d}	23.82 ± 0.00 ^{d}	White
	MM2	0.75 ± 0.00 ^{c}	36.31 ± 0.09 ^{a}	61.33 ± 0.13 ^{i}	3.57 ± 0.02 ^{b}	81.79 ± 1.65 ^{j}	67.11 ± 1.87 ^{e}	0.12 ± 0.00 ^{a}	5.16 ± 1.17 ^{b}	46.72 ± 0.00 ^{a}	Extra light amber
	MM3	0.73 ± 0.00 ^{f,g}	28.55 ± 0.70 ^{e,f}	69.33 ± 0.45 ^{e,f}	3.26 ± 0.03 ^{e,f}	143.86 ± 2.77 ^{i}	83.80 ± 1.86 ^{c}	0.12 ± 0.00 ^{a}	nd	28.77 ± 0.00 ^{c}	White
	MM4	0.74 ± 0.01 ^{d,e}	30.79 ± 0.65 ^{c}	69.05 ± 1.13 ^{e,f,g}	3.40 ± 0.00 ^{c,d}	240.67 ± 0.00 ^{c}	75.26 ± 0.00 ^{d}	0.07 ± 0.00 ^{b,c,d}	nd	≤8.00	Water white
	MQ1	0.81 ± 0.00 ^{a}	35.79 ± 0.10 ^{a}	61.92 ± 0.13 ^{i}	-	-	-	0.07 ± 0.01 ^{b,c,d}	nd	0.00	Water white
Caatinga	MS1	0.72 ± 0.00 ^{g,h}	27.96 ± 0.10 ^{e,f,g}	70.08 ± 0.13 ^{e}	3.37 ± 0.00 ^{c,d,e}	204.91 ± 0.00 ^{g}	50.80 ± 0.00 ^{h}	0.07 ± 0.02 ^{b,c,d}	nd	≤8.00	Water white
	MS2	0.73 ± 0.00 ^{e,f}	28.83 ± 0.00 ^{d,e,f}	69.25 ± 0.00 ^{e,f}	3.26 ± 0.05 ^{e,f}	211.09 ± 0.00 ^{f}	42.50 ± 0.00 ^{i}	0.06 ± 0.02 ^{c,d}	1.05 ± 0.00 ^{d}	≤8.00	Water white
	MSC1	0.77 ± 0.00 ^{b}	32.90 ± 0.20 ^{b}	65.00 ± 0.23	3.26 ± 0.11 ^{e,f}	299.42 ± 0.00 ^{a}	130.79 ± 0.00 ^{a}	0.12 ± 0.00 ^{a}	0.15 ± 0.00 ^{d}	15.39 ± 0.00 ^{e}	Extra white
	MSC2	0.74 ± 0.00 ^{e,f}	28.13 ± 0.18 ^{e,f,g}	70.08 ± 0.13 ^{e}	3.47 ± 0.00 ^{b,c}	201.97 ± 0.00 ^{g}	33.35 ± 0.00 ^{j}	0.09 ± 0.01 ^{a,b,c,d}	nd	15.27 ± 0.00 ^{f}	Extra white
	MSC3	0.73 ± 0.00 ^{e,f}	29.16 ± 1.36 ^{c,d,e,f}	67.67 ± 1.32 ^{g}	4.78 ± 0.00 ^{a}	179.11 ± 0.00 ^{h}	17.26 ± 0.00 ^{k}	0.09 ± 0.01 ^{a,b,c,d}	1.03 ± 0.00 ^{d}	≤8.00	Water white
	MM5 (6)	0.75 ± 0.00 ^{c}	29.52 ± 0.95 ^{c,d,e}	67.92 ± 0.13 ^{f,g}	3.54 ± 0.00 ^{b}	220.17 ± 0.00 ^{e}	58.13 ± 0.00 ^{f}	0.10 ± 0.03 ^{a,b,c}	3.56 ± 0.00 ^{c}	≤8.00 ^{h}	Water white
	SCA1	0.71 ± 0.00 ^{i}	25.92 ± 0.00 ^{h}	72.25 ± 0.00 ^{c}	3.06 ± 0.00 ^{g}	288.71 ± 0.00 ^{b}	120.52 ± 0.00 ^{b}	0.12 ± 0.02 ^{a}	3.43 ± 0.00 ^{c}	40.78 ± 0.00 ^{b}	Extra light amber
Cerrado	MF1	0.68 ± 0.00 ^{j}	23.87 ± 0.10 ^{i}	74.17 ± 0.13 ^{b}	3.46 ± 0.00 ^{b,c}	227.45 ± 0.00 ^{d}	53.51 ± 0.00 ^{g}	0.07 ± 0.01 ^{b,c,d}	nd	≤8.00	Water white
	MF2	0.65 ± 0.00 ^{k}	19.47 ± 0.63 ^{j}	75.75 ± 0.22 ^{a}	3.30 ± 0.00 ^{d,e,f}	218.32 ± 0.00 ^{e}	51.69 ± 0.00 ^{g,h}	0.07 ± 0.00 ^{b,c,d}	nd	≤8.00	Water white
	MF3	0.67 ± 0.00 ^{j}	23.13 ± 0.21 ^{i}	75.08 ± 0.13 ^{a,b}	-	-	-	0.07 ± 0.01 ^{b,c,d}	nd	≤8.00	Water white
	MF4	0.67 ± 0.00 ^{j}	23.40 ± 0.00 ^{i}	74.92 ± 0.13 ^{a,b}	-	-	-	0.07 ± 0.01 ^{b,c,d}	nd	≤8.00	Water white
	SCP1	0.71 ± 0.00 ^{i}	26.61 ± 0.10 ^{g,h}	71.75 ± 0.23 ^{c,d}	-	-	-	0.11 ± 0.01 ^{a,b}	12.64 ± 0.00 ^{a}	13.11 ± 0.00 ^{g}	Extra White
	MM6 (5)	0.72 ± 0.00 ^{h}	27.72 ± 0.00 ^{f,g}	70.50 ± 0.00 ^{d,e}	-	-	-	0.08 ± 0.03 ^{a,b,c,d}	0.45 ± 0.00 ^{d}	≤8.00	Water white

- = not analyzed; nd = not detected. Values are expressed as mean ± standard deviation. $^{a–k}$ = different superscript letters in the same column denote significant differences (ANOVA and Tukey, $p < 0.05$).

Low pH and high acidity values contribute to honey stability and shelf life due to their impact on microbial growth. In addition, these parameters have a strong impact on honey taste diversity. In the present study, all SBH samples showed an acidic character typical of honey, with pH values ranging from 3.06 to 4.78, and free and total acidity values ranging from 17.26 to 130.79 mEq/kg and 71.04 to 299.42 mEq/kg, respectively (Table 2). In addition, we observed that most samples differed significantly ($p < 0.05$) from one another in their free and total acidity content. However, the range of variation was similar to those previously reported for Brazilian SBH [5,19,27]. SBH produced in other countries also has a higher acidic character. Thai SBH exhibited total acidity values ranging from 440 to 592 mEq/kg [8]. Moreover, high free acidity values were reported for SBH collected in the Ecuadorian Amazon (mean value of 318 mEq/kg) [33].

Brown pigments and HMF are generated by honey heating or prolonged storage (Maillard reaction) which may lead to important color changes [34]. The present SBH samples showed low brown pigment and HMF content ranging from 0.05 to 0.12 and "not detected" to 12.64 $mg.kg^{-1}$, respectively (Table 2). The low amounts of HMF in SBH are expected and have been attributed to the high moisture and acidity content of this honey in addition to the predominance of fructose. These properties contribute to the inhibition of HMF formation in honey [33]. Color evaluation with the Pfund scale showed that the SBH samples ranged from $\leq$ 8.00 mm (water white) to 46.72 mm (extra light amber). The predominance of water-like colors in SBH samples collected in distinct biogeographical zones draws attention. Darker honey samples (amber-like colors) were reported by Sant'ana et al. (2020) [31] and de Sousa et al. (2016) [26] for SBH collected in the Brazilian semiarid region. Honey color is influenced by several factors, including climate conditions, botanical origin, harvest time, and degree of honey maturation [32].

3.2. Phenolic Content and Antioxidant Capacity of SBH

Table 3 shows that TPC and TFC ranged from 16.3 to 62.33 mg GAE/100 g and 5.39 to 27.22 mg QE/100 g, respectively. The SCA1 and MS2 samples, both collected in the Caatinga zone, had the highest and the lowest TPC and TFC values, respectively. The antioxidant capacity of SBH was evaluated by FRAP and ABTS•+ scavenging assays. As shown in Table 3, SBH displayed FRAP values ranging from 79.53 to 215.07 μmol Fe^{+2}/100 g. In the ABTS assay, SBH samples displayed values ranging from 37.84% to 54.95%. Recently, Carina Biluca et al., (2021), reported TPC values of 14 honey samples of six different Brazilian SB ranging from 11.01 to 38.92 mg GAE/100 g [35]. In a previous report, the same research group demonstrated the antioxidant capacity of Brazilian *T. angustula* honey with a FRAP value of 734.5 μmol Fe^{+2}/100 g [36]. Few studies have demonstrated the radical scavenging ability of SBH using ABTS•+. Majid et al., (2020), showed that methanol and aqueous extracts of Malaysian SBH displayed ABTS•+ scavenging values ranging from 18.77% to 65.02% and 15.61 to 65.77%, respectively [37]. These values are similar to those obtained in the present study.

Table 3. Contents of phenolic compounds, total flavonoids, and antioxidant activity of stingless bee honey of different species from different biogeographical zones of Brazil. TPC and TFC results are expressed as mean concentration (mg/100 g FW) ± standard deviation. The FRAP result is expressed as the mean value of μmol Fe^{2+} ± standard deviation of fresh honey samples. The ABTS result is expressed as the mean percentage of free radical scavenging power ± standard deviation of fresh honey samples.

Biogeographical Zone	Samples	TPC (mg GAE/100 g FW)	TFC (mg QE/100 g FW)	Antioxidant Activity	
				FRAP (μmol Fe^{2+}/100 g FW)	ABTS (%)
Atlantic Forest	MM1	47.11 ± 1.39 ^{a,b,c}	16.70 ± 0.32 ^{b}	110.85 ± 4,42 ^{b,c}	38.22 ± 1.47 ^{b,c}
	MM2	44.46 ± 1.16 ^{a,b,c,d}	15.02 ± 0.18 ^{b,c}	114.49 ± 9.38 ^{b,c}	40.08 ± 2.07 ^{b,c}
	MM3	35.69 ± 2.45 ^{b,c,d,e}	13.53 ± 0.04 ^{c,d}	84.87 ± 8.53 ^{c}	51.61 ± 8.24 ^{a,b}
	MM4	46.74 ± 5.75 ^{a,b,c}	9.93 ± 0.38 ^{f,g}	212.16 ± 8.84 ^{a}	39.82 ± 1.70 ^{b,c}
	MQ1	29.80 ± 4.50 ^{c,d,e,f}	9.04 ± 0.38 ^{f,g,h,i}	-	-
	MS1	32.99 ± 9.44 ^{b,c,d,e,f}	6.99 ± 0.63 ^{j,k}	108.95 ± 1.18 ^{b,c}	39.82 ± 2.20 ^{b,c}
	MS2	16.30 ± 4.26 ^{f}	5.39 ± 1.07 ^{k}	-	-
	MSC1	25.26 ± 1.66 ^{d,e,f}	8.89 ± 1.22 ^{f,g,h,i}	-	39.20 ± 3.42 ^{b,c}
	MSC2	22.24 ± 18.18 ^{e,f}	10.55 ± 0.21 ^{e,f}	118.71 ± 23.28 ^{b,c}	48.74 ± 6.30 ^{a,b,c}
	MSC3	39.29 ± 7.23 ^{b,c,d,e}	8.33 ± 0.21 ^{g,h,i,j}	215.07 ± 7.27 ^{a}	54.95 ± 11.88 ^{a}
	MM5(6)	36.03 ± 6.17 ^{b,c,d,e}	7.91 ± 0.04 ^{h,i,j}	143.96 ± 15.80 ^{b}	43.71 ± 4.79 ^{a,b,c}
	SCA1	62.33 ± 4.33 ^{a}	27.22 ± 3.04 ^{a}	-	-
Cerrado	MF1	39.86 ± 6.88 ^{b,c,d,e}	9.26 ± 0.07 ^{f,g,h}	88.70 ± 0.19 ^{c}	44.20 ± 5.49 ^{a,b,c}
	MF2	37.98 ± 6.00 ^{b,c,d,e}	7.66 ± 0.10 ^{h,i,j}	84.53 ± 2.55 ^{c}	37.84 ± 3.24 ^{c}
	MF3	50.95 ± 4.59 ^{a,b}	7.39 ± 0.07 ^{i,j}	79.53 ± 2.55 ^{c}	40.08 ± 3.21 ^{b,c}
	MF4	26.68 ± 2.19 ^{d,e,f}	6.67 ± 0.60 ^{j,k}	88.56 ± 5.91 ^{c}	38.22 ± 1.30 ^{b,c}
	SCP1	34.30 ± 11.23 ^{b,c,d,e,f}	12.13 ± 0.07 ^{d,e}	140.03 ± 3.34 ^{b}	45.09 ± 6.68 ^{a,b,c}
	MM6 (5)	36.41 ± 9.89 ^{b,c,d,e}	7.91 ± 0.18 ^{h,i,j}	-	-

- = not analyzed. Values are expressed as mean ± standard deviation. $^{A-k}$ = different superscript letters in the same column denote significant differences (ANOVA and Tukey test, $p < 0.05$). TPC = total phenolic content; TFC = total flavonoid content.

3.3. Correlation Analysis of Physicochemical Properties and Antioxidant Activity

As expected, strong positive correlations were obtained between some physicochemical parameters such as moisture content and water activity (r = 0.934), and between free and total acidity (r = 0.810) (Table S1). A positive and significant linear correlation was observed between color intensity and TPC (r = 0.654) and between color and TFC (r = 0.866) content of SBH. Moniruzzaman et al., (2013), reported that the color intensity of honey was a consistent parameter that specified the existence of pigments with antioxidant activity, including carotenoids and flavonoids [38]. In the present study, the TPC of Brazilian SBH from different biogeographical zones was positively correlated with TFC (r = 0.705). No significant correlations were found between antioxidant activity and TPC and TFC in SHB samples collected in different biogeographical zones of Brazil (Table S1). These results indicate that the antioxidant activity is not solely dependent on the phenolic and flavonoid content, with other antioxidant compounds present in honey also possibly being involved [32]. The antioxidant capacity of honey could be the result of the combined activity of a narrow range of compounds, including phenolic compounds, peptides, organic acids, and possibly other minor components.

3.4. Mineral Profile of SBH

In the present study, we used the TXRF technique to investigate the mineral composition of SBH. The method was validated in terms of TXRF measurements, quantitative analysis, and accuracy (Table S2). TXRF analysis showed a good equivalence of the concentration values of the elements present in the certified sample. The relative errors ranged from 1% to 14%, attesting to the accuracy of the technique.

The results of total concentration of mineral elements in samples of SBH from different biomes are listed in Table 4. Statistically significant differences were found for most minerals investigated. The order of average mineral concentration was K > Ca > Fe > Cu > P > Cl > Mn > Zn > Ni > Cr > Sr in the SBH samples from the Atlantic Forest; K > P > Ca > Cl > Cu > Fe > Mn > Sr > Zn > Cr > Ni in the Caatinga samples, and K > Ca > P > Cu > Zn > Fe > Mn > Sr > Ni > Cu in the Cerrado samples. The mineral profile analysis of honey is important because of the presence of metallic species which can have a nutritional or toxic effect depending on the metal present and/or the amount ingested [39]. The difference in the concentrations of metals in honey can also be attributed to biological, chemical, and physical factors and the existence of industrial contamination, thus providing important information about the region where the honey is produced [40].

Table 4. Mineral profile of SBH from different Brazilian biogeographical zones. The results are expressed as mean concentration (μg/g FW) ± standard deviation. Minimum and maximum concentrations are shown in parenthesis.

Mineral	Atlantic Forest	*Caatinga*	*Cerrado*
P	3.54 ± 0.56 (3.01–4.06) ^{a}	49.20 ± 29.81 (4.60–100.19) ^{b}	20.19 ± 14.36 (2.59–40.75) ^{b}
Cl	1.71 ± 1.69 (0.69–0.16) ^{a}	4.20 ±1.46 (2.54–6.41) ^{a}	nd
K	411.58 ± 107.74 (249.01–679.50) ^{a}	234.22 ± 109.23 (79.73–419.23) ^{a}	127.03 ± 44.57 (67.35–177.99) ^{b}
Ca	91.65 ± 48.98 (44.28–165.00) ^{a}	39.18 ± 15.72 (21–47–57.28) ^{b}	31.84 ± 6.20 (24.45–39.75) ^{b}
Cr	0.27 ± 0.22 (0.04–0.55) ^{a}	0.26 ± 0.13 (0.15–0.48) ^{a}	0.19 ± 0.01 (0.18–0.19) ^{a}
Mn	1.24 ± 0.91 (0.47–2.66) ^{a}	0.90 ± 0.48 (0.37–1.75) ^{a}	0.56 ± 0.16 (0.36–0.77) ^{a}
Fe	8.71 ± 4.41 (1.68–12.92) ^{a}	1.32 ± 0.47 (0.96–2.32) ^{b}	0.84 ± 0.10 (0.73–0.97) ^{c}
Ni	0.61 ± 0.45 (0.01–0.99) ^{a}	0.22 ± 0.08 (0.13–0.30) ^{a}	0.21 ± 0.01 (0.20–0.21) ^{a}
Cu	3.76 ± 3.80 (0.34–9.82) ^{a}	3.86 ± 1.73 (1.70–6.72) ^{a}	2.31 ± 0.67 (1.66–3.33) ^{a}
Zn	0.87 ± 0.38 (0.44–1.36) ^{a}	0.36 ± 0.22 (0.10–0.73) ^{b}	0.88 ± 0.29 (0.49–1.23) ^{a}
Rb	1.22 ± 0.41 (0.85–1.59) ^{a}	0.54 ± 0.36 (0.20–1.08) ^{a,b}	0.52 ± 0.01 (0.51–0.52) ^{b}
Sr	0.10 ± 0.02 (0.08–0.11) ^{a}	0.38 ± 0.26 (0.15–0.81) ^{b}	0.26 ± 0.08 (0.19–0.33) ^{b}

Nd = not detected. Different superscript letters on the same line indicate significant differences (ANOVA and Tukey test, $p < 0.05$).

Quantitatively, K (67.35–679.50 μg/g) > Ca (21.47–165 μg/g) > P (2.59–100.19 μg/g) were the main minerals found in honey from stingless bees in the biomes studied. This is consistent with the findings reported by Biluca et al., (2016), which demonstrated that K (27.3 to 448 mg/100 g) and Ca (1.12 to 35.2 mg/100 g) were the most abundant minerals in samples of SBH [27]. These results also corroborate those described by Pucholobek et al., (2022), who also observed that Ca (4.3 to 363.8 μg/g) is one of the most abundant minerals in the honey of stingless bees from Brazil [39]. These same authors observed a strong correlation between the concentration of minerals in honey and the type of soil in the study region. Plants accumulate metallic elements and this accumulation is influenced by the mineral character of the soil. In the present study, the Atlantic Forest honey had a higher content of Ca and Fe than the honey from the Cerrado and Caatinga, especially considering Fe. On the other hand, the Cerrado and Caatinga honey had the highest P content. The biogeographical regions where the samples were collected have soil types with compositional characteristics that might influence the amounts of these metals in honey. Indeed, these characteristics may explain the differences between regions for the elements P, K, Ca, Fe, Zn, Rb, and Se. It is noteworthy that the Fe content detected was about eight times higher in honey samples from the Atlantic Forest region than in samples from the other regions. This result was not observed for any of the other minerals, which showed differences of much smaller orders of magnitude. The Atlantic Forest has a variety of soil types, but with a predominance of soil rich in iron oxide (Latosol/Nitosol), which may influence the amounts of this metal in honey [39].

3.5. Discriminant Analysis of SBH

The fingerprinting of SBH is shown as Supplementary Materials (Figures S1–S3). Ion masses within the *m/z* range of 50–1000, physicochemical features, and mineral content were analyzed statistically to discriminate honey samples using the producing stingless bee species or the biogeographical origin. Minor ions responsible for group separation are shown in Figures S1–S3.

Different species of stingless bees may have specific preferences for botanical individuals when exploiting food sources [41]. Thus, we initially evaluated the influence of stingless bee species on the chemical and physicochemical features of honey. Figure S4 shows that bee species has little impact on sample discrimination. Nonetheless, when samples were categorized by their biogeographical origin, better separation of the groups was achieved. PLS-DA analysis (Figure 1a) shows little overlap between the Cerrado and Caatinga ellipsoids, while the Atlantic Forest group did not overlap with any other group. These results demonstrate that the chemical and physicochemical features of SBH are mainly influenced by their respective biogeographical origin.

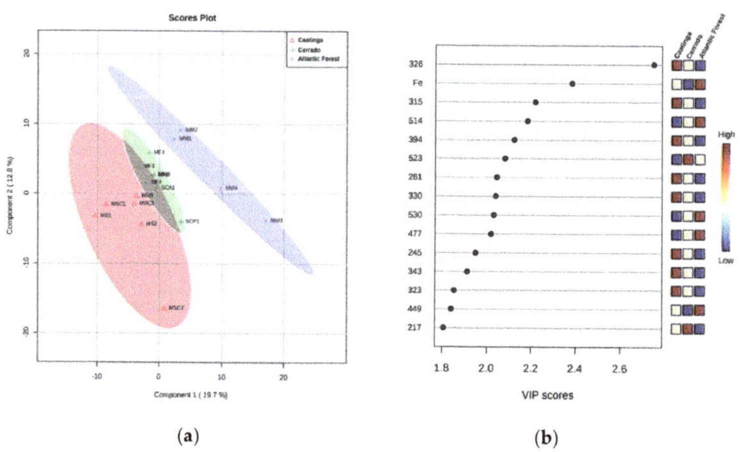

(a) (b)

Figure 1. Discriminant analysis of SBH based on Brazilian biogeographical zones. (**a**) PLS-DA score plots; and (**b**) VIP score of physicochemical parameters, mineral profile, antioxidant capacity, and mass fingerprint analysis of SBH.

The main separation factors between groups were the *m/z* 326, 315, 514, 394, 523, 261, 330, 530, 477 ions, and Fe content. Each of these factors showed VIP scores > 2, which confirmed their weight for group separation (Figure 1b). The *m/z* 326, 315, 394, 291, and 330 ions had a high weight in the Caatinga group and a low weight in the Atlantic Forest group. On the other hand, iron content had a high weight in the Atlantic Forest group and a low weight in the Cerrado group, while the *m/z* 514, 530, and 477 ions had a high weight in the Atlantic Forest group and a low weight in the Caatinga group. Univariate statistical analysis of iron content confirmed its influence on group separation, with a significant difference ($p < 0.05$) between samples from the Cerrado and Atlantic Forest, and between samples from the Caatinga and Atlantic Forest. These results suggest that iron content could be used to indicate the biogeographical origin of the sample. Interestingly, the iron content of 88 SBH was positively correlated with the soil types of areas where the honey samples were collected. High amounts of iron were detected in samples collected from areas of typical Atlantic Forest flora, where iron-rich soil is predominant [39]. Here, we did not observe a significant difference in iron content between samples from the Cerrado and Caatinga. The Cerrado soil is acidic, poor in nutrients, and with a low capacity to retain water [42], characteristics that are very close to those of crystalline and sedimentary caatingas (two subgroups of the Caatinga zone) [43].

The other variables identified as VIP in multivariate analysis did not show significant differences in univariate analyses. This can be explained by the difference between univariate and multivariate methods. A variable-by-variable (univariate) analysis, such as ANOVA with post hoc tests, uses few criteria to determine significant differences between samples, requiring these criteria to be extremely strong to determine separations into groups. An analysis model that uses variable information and correlations with other variables (multivariate), such as the PLS-DA, allows the identification of many more criteria to determine significant differences between samples and determine separations into groups.

4. Conclusions

This study demonstrated that Brazilian SBH has diverse physicochemical properties, antioxidant capacity, and chemical and mineral profiles. Despite the phenolic and flavonoid contents, no correlation was observed between them and the antioxidant capacity, suggesting that other compounds contribute to the antioxidant properties of SBH. The mineral profile displayed high amounts of K, Ca, and P in all samples. PLS-DA analysis revealed that the composition of SBH is influenced by the biogeographical zone of collection. In addition, Fe content was identified as the main factor responsible for SBH separation into the biogeographical zone groups. Taken together, the present results highlight the mineral content, especially the amounts of Fe, as an important factor for SBH discrimination.

Supplementary Materials: The following supporting information can be downloaded at: https://www.mdpi.com/article/10.3390/foods12010180/s1, Table S1: Correlation matrix between physicochemical and antioxidant activity of samples of stingless bee honey from different geographical regions of Brazil; Table S2: Results of the standard reference material 1577b; Figure S1: Mass spectra of stingless bee honey collected in the Atlantic Forest biogeographical zone; Figure S2: Mass spectra of stingless bee honey collected in the Caatinga biogeographical zone; Figure S3: Mass spectra of stingless bee honey collected in the Cerrado biogeographical zone; Figure S4: Discriminant analysis of SBH based on bee species.

Author Contributions: Conceptualization, R.C.S., C.S.C. and I.A.R.; formal analysis, F.C.L., B.A.G., M.V.T.S. and R.C.S.; investigation, F.C.L., R.M.N., B.A.G. and M.V.T.S.; resources, S.G.L., M.R.L.M., C.S.C. and I.A.R.; writing—original draft preparation, F.C.L., B.A.G., M.V.T.S., R.C.S., C.S.C. and I.A.R.; writing—review and editing, S.G.L. and M.R.L.M.; visualization, R.C.S., C.S.C. and I.A.R.; F.C.L., supervision, R.C.S., C.S.C. and I.A.R.; project administration, I.A.R.; funding acquisition, C.S.C. and I.A.R. All authors have read and agreed to the published version of the manuscript.

Funding: This research was funded by *Fundação de Amparo à Pesquisa do Estado do Rio de Janeiro* (FAPERJ), grant numbers E-26/202.752/2018, E-26/201.362/2022, and E-26/201.418/2022; Conselho Nacional Científico e Tecnológico (CNPq), and Coordenação de Aperfeiçoamento de Pessoal de Nível Superior (CAPES), (graduate fellowships).

Data Availability Statement: The data that support the findings of this study are available from the corresponding author upon reasonable request.

Acknowledgments: The authors thank Kamila Marques Sette for technical assistance. This study was registered with the Brazilian National System for the Management of Genetic Heritage and Associated Traditional Knowledge (SISGen) under the code A6DAB60.

Conflicts of Interest: The authors declare that they have no conflicts of interest. The funding agencies had no role in the design of the study, in the collection, analyses, or interpretation of data, in the writing of the manuscript, or in the decision to publish the results.

References

1. Brazil. Biodiversidade. 2022. Available online: https://www.gov.br/mma/pt-br/assuntos/biodiversidade (accessed on 1 August 2022).
2. CBD. Brazil—Biodiversity Facts. 2021. Available online: https://www.cbd.int/countries/profile/?country=br (accessed on 18 April 2021).
3. Giannini, T.C.; Alves, D.A.; Alves, R.; Cordeiro, G.D.; Campbell, A.J.; Awade, M.; Bento, J.M.S.; Saraiva, A.M.; Imperatriz-Fonseca, V.L. Unveiling the contribution of bee pollinators to Brazilian crops with implications for bee management. *Apidologie* **2020**, *51*, 406–421. [CrossRef]
4. Lage, L.G.A.; Coelho, L.L.; Resende, H.C.; Tavares, M.G.; Campos, L.A.O.; Fernandes-Salomão, T.M. Honey physicochemical properties of three species of the Brazilian *Melipona*. *An. Acad. Bras. Ciênc.* **2012**, *84*, 605–608. [CrossRef]
5. da Silva, I.A.; da Silva, T.M.; Camara, C.A.; Queiroz, N.; Magnani, M.; de Novais, J.S.; Soledade, L.E.; de Oliveira Lima, E.; de Souza, A.L.; de Souza, A.G. Phenolic profile, antioxidant activity and palynological analysis of stingless bee honey from Amazonas, Northern Brazil. *Food Chem.* **2013**, *141*, 3552–3558. [CrossRef]
6. Ávila, S.; Hornung, P.S.; Teixeira, G.L.; Malunga, L.N.; Apea-Bah, F.B.; Beux, M.R.; Beta, T.; Ribani, R.H. Bioactive compounds and biological properties of Brazilian stingless bee honey have a strong relationship with the pollen floral origin. *Food Res. Int.* **2019**, *123*, 1–10. [CrossRef]
7. Ranneh, Y.; Ali, F.; Zarei, M.; Akim, A.M.; Hamid, H.A.; Khazaai, H. Malaysian stingless bee and Tualang honeys: A comparative characterization of total antioxidant capacity and phenolic profile using liquid chromatography-mass spectrometry. *LWT* **2018**, *89*, 1–9. [CrossRef]
8. Chuttong, B.; Chanbang, Y.; Sringarm, K.; Burgett, M. Physicochemical profiles of stingless bee (Apidae: Meliponini) honey from South East Asia (Thailand). *Food Chem.* **2016**, *192*, 149–155. [CrossRef] [PubMed]
9. AOAC—Association of Official Analytical Chemists. *Official Methods of Analysis of AOAC International*; AOAC: Gaithersburg, MD, USA, 2012.
10. Ferreira, I.C.F.R.; Aires, E.; Barreira, J.C.M.; Estevinho, L.M. Antioxidant activity of Portuguese honey samples: Different contributions of the entire honey and phenolic extract. *Food Chem.* **2009**, *114*, 1438–1443. [CrossRef]
11. Turkmen, N.; Sari, F.; Poyrazoglu, E.S.; Velioglu, Y.S. Effects of prolonged heating on antioxidant activity and colour of honey. *Food Chem.* **2006**, *95*, 653–657. [CrossRef]
12. Meda, A.; Lamien, C.E.; Romito, M.; Millogo, J.; Nacoulma, O.G. Determination of the total phenolic, flavonoid and proline contents in Burkina Fasan honey, as well as their radical scavenging activity. *Food Chem.* **2005**, *91*, 571–577. [CrossRef]
13. Benzie, I.F.; Strain, J.J. The ferric reducing ability of plasma (FRAP) as a measure of "antioxidant power": The FRAP assay. *Anal. Biochem.* **1996**, *239*, 70–76. [CrossRef]
14. Ribeiro, R.O.R.; Mársico, E.T.; de Jesus, E.F.O.; Carneiro, C.S.; Júnior, C.A.C.; de Almeida, E.; Filho, V.F.N. Determination of trace elements in honey from different regions in Rio de Janeiro State (Brazil) by total reflection x-ray fluorescence. *J. Food Sci.* **2014**, *79*, T738–T742. [CrossRef] [PubMed]
15. De Araújo, U.B.; Costa, A.C.M.; Oliveira, D.F.; Jesus, E.F.O.; Anjos, M.J.; Marsico, E.T.; Carneiro, C.S.; Ribeiro, R.O.R.; Lopes, R.T. Analysis of milk trace elements with a home-made portable automated total reflection x-ray fluorescence system. *Radiat. Phys. Chem.* **2019**, *156*, 216–221. [CrossRef]
16. Bijlsma, L.; De Bruijn, L.L.M.; Martens, E.P.; Sommeijer, M.J. Water content of stingless bee honeys (Apidae, Meliponini): Interspecific variation and comparison with honey of *Apis mellifera*. *Apidologie* **2006**, *37*, 480–486. [CrossRef]
17. Nascimento, A.; Marchini, L.; Carvalho, C.; Araújo, D.; Olinda, R.; Silveira, T. Physical-chemical parameters of honey of stingless bee (Hymenoptera: Apidae). *Chem. Sci. Int. J.* **2015**, *7*, 139–149. [CrossRef]
18. Kek, S.P.; Chin, N.L.; Yusof, Y.A.; Tan, S.W.; Chua, L.S. Classification of entomological origin of honey based on its physicochemical and antioxidant properties. *Int. J. Food Prop.* **2017**, *20*, S2723–S2738. [CrossRef]
19. Ávila, S.; Lazzarotto, M.; Hornung, P.S.; Teixeira, G.L.; Ito, V.C.; Bellettini, M.B.; Beux, M.R.; Beta, T.; Ribani, R.H. Influence of stingless bee genus (*Scaptotrigona* and *Melipona*) on the mineral content, physicochemical and microbiological properties of honey. *J. Food Sci. Technol.* **2019**, *56*, 4742–4748. [CrossRef] [PubMed]
20. Batiston, T.F.T.P.; Frigo, A.; Stefani, L.M.; Silva, A.S.D.; Araujo, D.N. Physicochemical composition and antimicrobial potential of stingless honey: A food of differentiated quality. *Res. Soc. Dev.* **2020**, *10*, e709910822. [CrossRef]
21. Anacleto, D.A.; Souza, B.A.; Marchini, L.C.; Moreti, A.C.C.C. Composição de amostras de mel de abelha Jataí (*Tetragonisca angustula* latreille, 1811). *Cienc. Tecnol. Aliment.* **2009**, *29*, 535–541. [CrossRef]
22. Jimenez, M.; Beristain, C.I.; Azuara, E.; Mendoza, M.R.; Pascual, L.A. Physicochemical and antioxidant properties of honey from *Scaptotrigona mexicana* bee. *J. Apic. Res.* **2016**, *55*, 151–160. [CrossRef]
23. ISPN. Cerrado—The Heart of Brazil. 2022. Available online: https://ispn.org.br/en/biomes/cerrado/ (accessed on 5 July 2022).
24. Braga, D.C.; Liberato, M.C.T.C.; Lima, V.L.F.; de Araújo Neto, J.A.M. Analytical study of the physicochemical characteristics from *Melipona subnitida* d. Honey in adequation to Brazilian law. *Food Sci. Technol.* **2020**, *40*, 217–221. [CrossRef]
25. de Melo, F.H.C.; Menezes, F.N.D.D.; de Sousa, J.M.B.; dos Santos Lima, M.; da Silva Campelo Borges, G.; de Souza, E.L.; Magnani, M. Prebiotic activity of monofloral honeys produced by stingless bees in the semi-arid region of Brazilian Northeastern toward *Lactobacillus acidophilus* LA-05 and *Bifidobacterium lactis* BB-12. *Food Res. Int.* **2020**, *128*, 108809. [CrossRef] [PubMed]

26. de Sousa, J.M.; de Souza, E.L.; Marques, G.; de Toledo Benassi, M.; Gullón, B.; Pintado, M.M.; Magnani, M. Sugar profile, physicochemical and sensory aspects of monofloral honeys produced by different stingless bee species in Brazilian semi-arid region. *LWT* **2016**, *65*, 645–651. [CrossRef]
27. Biluca, F.C.; Braghini, F.; Gonzaga, L.V.; Costa, A.C.O.; Fett, R. Physicochemical profiles, minerals and bioactive compounds of stingless bee honey (Meliponinae). *J. Food Compos. Anal.* **2016**, *50*, 61–69. [CrossRef]
28. Francisquini, M.I.; Lorente, F.L.; Pessenda, L.C.; Junior, A.A.; Mayle, F.E.; Cohen, M.C.; França, M.C.; Bendassolli, J.A.; Giannini, P.C.; Schiavo, J.; et al. Cold and humid Atlantic Rainforest during the last glacial maximum, northern Espírito Santo State, southeastern Brazil. *Quat. Sci. Rev.* **2020**, *244*, 106489. [CrossRef]
29. Villacrés-Granda, I.; Coello, D.; Proaño, A.; Ballesteros, I.; Roubik, D.W.; Jijón, G.; Granda-Albuja, G.; Granda-Albuja, S.; Abreu-Naranjo, R.; Maza, F.; et al. Honey quality parameters, chemical composition and antimicrobial activity in twelve Ecuadorian stingless bees (Apidae: Apinae: Meliponini) tested against multiresistant human pathogens. *LWT* **2021**, *140*, 110737. [CrossRef]
30. Juna, C.F.; Cho, Y.H.; Ham, D.; Joung, H. Associations of relative humidity and lifestyles with metabolic syndrome among the Ecuadorian adult population: Ecuador National Health and Nutrition Survey (ENSANUT-ECU) 2012. *Int. J. Environ. Health Res.* **2020**, *17*, 9023. [CrossRef]
31. Sant'ana, R.S.; de Carvalho, C.A.L.; Oda-Souza, M.; Souza, B.D.A.; Dias, F.D.S. Characterization of honey of stingless bees from the Brazilian semi-arid region. *Food Chem.* **2020**, *327*, 127041. [CrossRef]
32. Shamsudin, S.; Selamat, J.; Sanny, M.; Abd Razak, S.B.; Jambari, N.N.; Mian, Z.; Khatib, A. Influence of origins and bee species on physicochemical, antioxidant properties and botanical discrimination of stingless bee honey. *Int. J. Food Prop.* **2019**, *22*, 238–263. [CrossRef]
33. Guerrini, A.; Bruni, R.; Maietti, S.; Poli, F.; Rossi, D.; Paganetto, G.; Muzzoli, M.; Scalvenzi, L.; Sacchetti, G. Ecuadorian stingless bee (Meliponinae) honey: A chemical and functional profile of an ancient health product. *Food Chem.* **2009**, *114*, 1413–1420. [CrossRef]
34. Bulut, L.; Kilic, M. Kinetics of hydroxymethylfurfural accumulation and color change in honey during storage in relation to moisture content. *J. Food Process. Preserv.* **2009**, *33*, 22–32. [CrossRef]
35. Carina Biluca, F.; Braghini, F.; de Campos Ferreira, G.; Costa dos Santos, A.; Helena Baggio Ribeiro, D.; Valdemiro Gonzaga, L.; Vitali, L.; Amadeu Micke, G.; Carolina Oliveira Costa, A.; Fett, R. Physicochemical parameters, bioactive compounds, and antibacterial potential of stingless bee honey. *J. Food Process Preserv.* **2021**, *45*, e15127. [CrossRef]
36. Biluca, F.C.; da Silva, B.; Caon, T.; Mohr, E.T.B.; Vieira, G.N.; Gonzaga, L.V.; Vitali, L.; Micke, G.; Fett, R.; Dalmarco, E.M.; et al. Investigation of phenolic compounds, antioxidant and anti-inflammatory activities in stingless bee honey (Meliponinae). *Food Res. Int.* **2020**, *129*, 108756. [CrossRef]
37. Majid, M.; Ellulu, M.S.; Abu Bakar, M.F. Melissopalynological study, phenolic compounds, and antioxidant properties of *Heterotrigona itama* honey from Johor, Malaysia. *Scientifica* **2020**, *2020*, 2529592. [CrossRef]
38. Moniruzzaman, M.; Sulaiman, S.A.; Khalil, M.I.; Gan, S.H. Evaluation of physicochemical and antioxidant properties of sourwood and other Malaysian honeys: A comparison with manuka honey. *Chem. Cent. J.* **2013**, *7*, 138. [CrossRef] [PubMed]
39. Pucholobek, G.; de Andrade, C.K.; Rigobello, E.S.; Wielewski, P.; de Toledo, V.D.; Quináia, S.P. Determination of the Ca, Mn, Mg and Fe in honey from multiple species of stingless bee produced in Brazil. *Food Chem.* **2022**, *367*, 130652. [CrossRef] [PubMed]
40. do Nascimento, A.S.; Chambó, E.D.; de Jesus Oliveira, D.; Andrade, B.R.; Bonsucesso, J.S.; de Carvalho, C.A.L. Honey from stingless bee as indicator of contamination with metals. *Sociobiology* **2018**, *65*, 727–736. [CrossRef]
41. Basari, N.; Ramli, S.N.; Abdul-Mutalid, N.A.; Shaipulah, N.F.; Hashim, N.A. Flowers morphology and nectar concentration determine the preferred food source of stingless bee, *Heterotrigona itama*. *J. Asia-Pac. Entomol.* **2021**, *24*, 232–236. [CrossRef]
42. Coutinho, L. *Biomas brasileiros*, 1st ed.; Oficina de Textos: São Paulo, Brazil, 2016; Available online: https://books.google.com.br/books?hl=pt-BR&lr=&id=KlLdDgAAQBAJ&oi=fnd&pg=PT4&dq=coutinho+and+biomas+brasileiros&ots=sj7bwkppna&sig=9s_q-KFYGslNzq2aASMv9sAAcY0#v=onepage&q=coutinho%20and%20biomas%20brasileiros&f=false (accessed on 14 November 2022).
43. Da Silva, J.M.C.; Lacher, T.E. Caatinga—South America. Reference module in earth systems and environmental sciences. In *Encyclopedia of the World's Biomes*; Goldstein, M.I., DellaSala, D.A., Eds.; Elsevier: Berlin/Heidelberg, Germany, 2020; pp. 554–561. [CrossRef]

foods

MDPI

Article

Quality Evaluation of Fair-Trade Cocoa Beans from Different Origins Using Portable Near-Infrared Spectroscopy (NIRS)

Matteo Forte [1], Sarah Currò [2,*], Davy Van de Walle [3], Koen Dewettinck [3], Massimo Mirisola [4], Luca Fasolato [2] and Paolo Carletti [1]

1 Dipartimento di Agronomia, Animali, Alimenti, Risorse Naturali e Ambiente, Università degli Studi di Padova, Viale dell'Università 16, 35020 Legnaro, PD, Italy
2 Dipartimento di Biomedicina Comparata e Alimentazione, Università degli Studi di Padova, Viale dell'Università 16, 35020 Legnaro, PD, Italy
3 Food Structure and Function Research Group, Department of Food Technology, Safety and Health, Faculty of Bioscience Engineering, Ghent University 653, 9000 Ghent, Belgium
4 Dipartimento di Medicina Animale, Produzioni e Salute, Università degli Studi di Padova, Viale dell'Università 16, 35020 Legnaro, PD, Italy
* Correspondence: sarah.curro@unipd.it

Abstract: Determining cocoa bean quality is crucial for many players in the international supply chain. However, actual methods rely on a cut test protocol, which is limited by its subjective nature, or on time-consuming, expensive and destructive wet-chemistry laboratory procedures. In this context, the application of near infrared (NIR) spectroscopy, particularly with the recent developments of portable NIR spectrometers, may represent a valuable solution for providing a cocoa beans' quality profile, in a rapid, non-destructive, and reliable way. Monitored parameters in this work were dry matter (DM), ash, shell, fat, protein, total polyphenols, fermentation index (FI), titratable acidity (TA) and pH. Different chemometric analyses were performed on the spectral data and calibration models were developed using modified partial least squares regression. Prediction equations were validated using a fivefold cross-validation and a comparison between the different prediction performances for the portable and benchtop NIR spectrometers was provided. The NIRS benchtop instrument provided better performance of quantification considering the whole than the portable device, showing excellent prediction capability in protein and DM quantification. On the other hand, the NIRS portable device, although showing lower but valuable performance of prediction, can represent an appealing alternative to benchtop instruments for food business operators, being applicable in the field.

Keywords: *Theobroma cacao* L.; dry matter; chemometrics; fermentation index; protein content

Citation: Forte, M.; Currò, S.; Van de Walle, D.; Dewettinck, K.; Mirisola, M.; Fasolato, L.; Carletti, P. Quality Evaluation of Fair-Trade Cocoa Beans from Different Origins Using Portable Near-Infrared Spectroscopy (NIRS). *Foods* **2023**, *12*, 4. https://doi.org/10.3390/foods12010004

Academic Editors: Mourad Kharbach and Samuli Urpelainen

Received: 23 November 2022
Revised: 15 December 2022
Accepted: 16 December 2022
Published: 20 December 2022

1. Introduction

With an ancient history starting in the Preclassic period (1200–400 B.C.) with consumption among the Olmec and other pre-Colombian populations of the Americas [1], cocoa is now a ubiquitous food. Cocoa appears today in many different forms, mainly in chocolate, with consumption averaging around 8 kg per person per annum in many European countries [2]. The top four countries, which account for nearly 65% of the total world chocolate production, are the USA, Germany, Switzerland, and Belgium, which in terms of retail sales reach respectively USD 20, USD 10, USD 14, and USD 12 billion per year [3]. The consumption of cocoa and cocoa-based products is of great interest both for the highly appreciated sensorial profile and for the possible beneficial health effects which are being studied in current times [4,5].

Cocoa beans represent the essential raw material for chocolate, and they are obtained from the *Theobroma cacao* L. tree, which is almost exclusively cultivated in tropical/developing countries, where it represents a source of export earnings both at the level of families, communities, and nations [6]. Africa covers more than 75% of the world's total cocoa beans production and Côte d'Ivoire is the largest cocoa bean exporter in the world, with more than USD four billion exported in 2020, followed by Ghana and Ecuador.

Once the cacao pods have been harvested, they must undergo post-harvest operations on farms and plantations before becoming the so-called cocoa beans, which are then traded in the international market and processed into final industrial products. The post-harvest processing comprise pod opening and removal of beans from the pod, bean-pulp mass fermentation and bean drying. In this sequence, the fermentation constitutes an essential critical step for the development of desired flavor attributes of the commercial cocoa beans. In the further processing, cocoa beans are roasted, cracked and ground to give a powdery mass from which fat is expressed [7] and the release of fat ultimately leads to a liquid-like ingredient, namely cocoa liquor. Additionally, many processes are implemented in the chocolate industry that originate several products with different forms and functionalities.

Fermentation and drying constitute key farm(er)-based unit operations with strong influences in the final quality of cocoa beans and subsequent products [8].

Since 2010, approximately 4 million tons of cocoa beans have been produced annually around the world [9], and the three biggest importers are the Netherlands, Germany, and USA, with product worth 2,375,923, 1,209,366 and 1,026,931 USD imported in 2020, respectively (Sources: ITC calculations based on UN COMTRADE and ITC statistics).

The socio-economic importance and international interest towards cocoa beans are intelligible as an estimated five million farming households depend on cacao as a cash crop, and 70 per cent of cocoa is produced by smallholders living on less than USD 2 per day and relying on cocoa for 60 to 90 per cent of their income [9].

Cocoa beans are divided worldwide between "fine or flavor" (mainly Criollo and Trinitario) and "bulk" (mainly Forastero) varieties, with a common belief that fine and flavor cacao varieties receive significant price premiums in international markets. However, recent studies show how post-harvest processing has a central role in causing heterogeneity in cocoa prices, independently of the variety grown [10]. This is because although the primary factors influencing the quality attributes of cocoa beans are the cocoa tree cultivar and genotype, it is well-established that the agronomic and environmental conditions together with the harvest and post-harvest steps are crucial elements in the determination of the final quality of commercial cocoa beans [11–15].

Partially fermented or unfermented beans are prone to bitterness and astringency with poor chocolate flavor and aroma [2]. Moreover, an appropriate drying process will reduce the beans' water content of 55% to around 7% [16], preventing fermentation from continuing uncontrolled, slowing the development of molds that could give rise to unwanted, unpleasant flavors and equilibrating the beans acidity that would otherwise be excessive in the final products [17,18].

Although fermentation is considered as the "core stage" of the cocoa transformation process from seeds to chocolate, it is currently performed mostly by small third-world producers in an empirical way, with little or no technification, without control in processing conditions, originating cocoa batches of low and heterogeneous quality [13].

Chocolate and chocolate-based products sell in a very competitive market, where quality is crucial, and value is enormous. It should be clear that if the quality of cocoa beans is poor, final products will suffer this deficiency as well, and the whole industry sector will be affected [19].

Most of the existing commercial standards for cocoa beans base their quality requirements on the results of the cut test, or on sensory estimation by trained panels (Aculey et al., 2010). The cut test consists of cutting cocoa beans lengthwise, observing the number of defective beans. The ISO 1114 states that both halves of each bean shall be visually examined, and the result for each kind of defect shall be expressed as a percentage of the 300 beans examined. The ISO also defines nine categories of defects: those related to poor fermentation (slaty and violet/purple beans) and those being indicators of high FFA levels, poor flavor and/or other contaminants (bean clusters, broken beans, smoky beans, moldy beans, germinated beans, flat beans, insect-damaged/infested beans) [20].

On the one hand, the cut test is limited by its subjective nature and does not represent a sufficiently reliable methodology for a comprehensive description of the main quality contributors [21]. On the other hand, laboratory methodologies are often demanding in terms of time and cost effectiveness, which can be critical factors, and they are mainly based on destructive determinations. Moreover, in cocoa producing countries the availability of laboratory infrastructures is poor [22].

In this context, the application of near infrared (NIR) spectroscopy, particularly with the recent developments of portable NIR spectrometers, may represent a valuable solution for providing a cocoa beans' quality profile, in a rapid, non-destructive, and reliable way. This analytical technique could be useful to both cocoa bean producers, mostly in the developing countries, and processors, mostly in the developed countries, alike.

Many researchers have already investigated quality parameters of cocoa beans through NIR spectroscopy quantifying, for instance, fat, sugars, proteins, moisture, pH and titratable acidity, polyphenols, and other volatile and non-volatile compounds [23]. Even the assessment of the authenticity of cocoa powder has been studied by identifying the country of origin of raw materials, varietal purity, or the presence of adulterants [24].

It must be noted, however, that most of the studies that successfully predicted cocoa beans' quality parameters through NIR spectroscopy mainly analyzed samples which had been purposely subjected to different degrees of fermentation, e.g., analyzing the beans at different days during the fermentation process. Doing so, the samples are not in the status in which importers/exporters normally trade them in the international market, fully fermented and dried, therefore this might not be a representative "working condition" for these actors in the cocoa supply chain. Moreover, only cocoa bean samples coming from one or few countries are most often utilized in the previous studies, hence limiting the variability that can be included in the NIR prediction model. Finally, most of the studies present in literature utilize benchtop NIR instruments, which have some notable practical disadvantages if compared with the more recent portable NIR spectrometers, which are recently being considered in the literature [25,26].

On this basis, our study aimed to predict some quality parameters of commercial cocoa bean samples using portable NIR spectrometers, also in comparison with a benchtop spectrometer, on both whole and ground samples. Cocoa beans were provided by an Italian fair-trade importer and all the samples have been produced to meet internationally accepted merchantable quality standards: well-fermented and dry. The ultimate purpose of the study was, therefore, to evaluate the possibility of using portable NIR spectrometers in commercial contexts, while also assessing the practicability of using NIR spectroscopy on whole fermented coca beans to rapidly predict main quality parameters.

2. Materials and Methods

2.1. Samples

Fifty-six samples from commercially available cocoa beans have been provided by Altromercato Impresa Sociale Soc. Coop. (Via 9. Crispi, Bolzano, Italy), a major importer of fair-trade products in Italy in 2021. Thirty-three of them came from Africa while the remaining twenty-three came from South America. All the samples have been produced to meet internationally accepted merchantable quality standards i.e., well-fermented and dry, free from smoky beans and abnormal or foreign odors, free from evidence of adulteration, reasonably free from living insects, virtually free from broken beans, pieces of shell, and foreign matter. African countries included Togo, Uganda, Madagascar, and Sierra Leone, while American beans came from Honduras, Ecuador, Perú, Dominican Republic, Nicaragua, and Venezuela (Supplementary Table S1). Although it was not possible to obtain accurate information on the beans' variety for each sample, documents reported that only Trinitario and Forastero varieties were utilized.

The beans were shipped in 25-ton containers which roughly corresponds to 360 bags (70 kg of cocoa beans/bag). Around 4 kg of cocoa beans from each container were sampled as representative for the 33% of the total bags' number. Of these, 200 g of cocoa beans were randomly taken for analysis and stored at −20 °C in plastic bags until assay in a LGPv 8420 MediLine refrigerator (Liebherr, Kirchdorf an der Iller, Germany).

2.2. Beans Peeling and Grinding

Three random aliquots of 20 g of dry and nitrogen frozen cocoa beans from each sample were de-husked by hand. The peels and nibs were carefully collected and weighted to determine average shell percentage on a 4 digits balance (Adventurer model ARRV70, OHAUS, Parsippany, NJ, USA). About 100 g of nibs (de-husked cocoa beans) were ground in a multi-purpose grinder for 45 s (3 intervals of 15 s with 10 s pause), then sifted on a 0.5 mm sieve. Before grinding, the beans were frozen with liquid nitrogen to make them brittle and avoid becoming a mash. The obtained cocoa powder for each sample was stored at −20 °C in the dark prior to the following analyses.

2.3. Spectral Data Acquisition

Spectral data acquisition was performed both on whole cocoa beans and on de-husked cocoa bean powder. About 100 g of randomly chosen whole cocoa beans from each sample were scanned with a portable instrument (PoliSPEC-NIR, ITPhotonics, Breganze, Italy) and with the benchtop instrument (FOSS DS−2500 scanning monochromator FossNIR-System, Hillerød, Denmark). Both NIR data acquisitions were performed in reflectance mode, with the following parameters:

- FOSS DS-2500: scanning monochromator covering a range of 850–2500 nm at 0.5 nm intervals. Scans were performed using a slurry cup with quartz window of about a 12.6 cm^2 area.
- PoliSPEC-NIR: covering a range of 900–1680 nm at 2 nm intervals. Spectral data measurements were performed through a round scanning window (3.2 cm^2) placed in direct contact with the sample surface. Each spectrum was obtained by averaging 3 data acquisitions.

2.4. Chemical Analyses

Unless otherwise specified, analyses were performed according to official methods of analysis (AOAC, 2016). All chemical analyses were performed in triplicate on peeled and ground cocoa beans.

2.4.1. Dry Matter

Dry matter is measured as subtraction of the moisture content measured using a gravimetric method based on AOAC method 931.04 [21]. Hereto, approximately 2 g of powder sample were dried at 101–103 °C to constant weight in a forced-air electric oven (UF55 Plus, Memmert, Schwabach, Germany). After the drying process was completed, the samples were immediately closed with glass lids to avoid exposure and stored in desiccators for one hour to equilibrate samples towards ambient temperature [27]. The moisture content was expressed as average percentage (%) based on loss in weight of three independent samples.

2.4.2. Ash

For the measurement of ashes, the sample was charred on a plate and placed in a muffle furnace (Gefran Model 1200; Gefran Spa, Brescia, Italy) at 550 °C (AOAC 972.15A). Ash content was expressed as weight percentage (%).

2.4.3. Fat Content

The fat content was measured by extraction with petroleum ether [21] in a TE-188 Soxhlet lipid extractor (model SOXTEC 255 Tecator-Foss Analytical, Hillerød, Denmark) with the following parameters: 60 min boiling, 50 min washing, 15 min drying. Fat content was expressed as weight percentage (%).

2.4.4. Total Protein Content

Protein determination was carried out by the Kjeldahl method, as described in AOAC 2016 (method 970.22) (model Kjeltec 2300-Foss Analytical). The protein content was calculated from the concentration of total nitrogen by applying a conversion factor of 6.25.

2.4.5. Total Phenolic Content

The total phenolic content was determined according to the colorimetric method of Folin–Ciocalteu [28]. Samples were defatted using the Soxhlet method (AOAC 963.15). Defatted powder (0.05 g) was added to 10 mL of a methanol-water (70:30 v/v) mixture at room temperature and stirred for 45 min. After centrifugation, 0.1 mL of solution was mixed with 3 mL of distilled water and 0.5 mL Folin–Ciocalteu reagent. The mixture was stored for 3 min after which 1 mL of aqueous Na_2CO_3 (200 g L^{-1}) was added. The mixture was allowed to stand for 20 min at 40 °C and the total polyphenols were determined by spectrophotometry at 765 nm (spectrophotometer model Cary 60 UV-Vis Agilent Technologies Stevens Creek Blvd. Santa Clara, CA, USA). The standard curve was prepared using 0, 50, 100, 150, 200 and 250 mg L^{-1} solutions of gallic acid in methanol. Total phenol values were expressed in terms of gallic acid equivalents (mg g^{-1} of dry fat-free mass) [29]. The analyses were performed in triplicate.

2.4.6. Fermentation Index

Fermentation index (FI) corresponds to the color change within the bean cotyledons during fermentation. This change is due to the decreasing anthocyanin content as beans progress through fermentation [30]. A 50 mg sample of previously prepared cocoa powder was weighed and mixed with 5 mL MeOH:HCl (97:3 v/v). Samples were extracted at 4 °C for 16–18 h, centrifuged for 5 min at 3500$\times$ g, and the clear supernatant was collected. Absorbance of the supernatant was read at wavelengths 460 nm and 530 nm using UV-VIS spectrophotometer (model Cary 60 UV-Vis Agilent Technologies Stevens Creek Blvd. Santa Clara, CA, USA). All the measurements were performed in triplicate.

The FI was obtained by calculating the ratio of the absorbance at 460 nm and 530 nm (FI = A_{460}/A_{530}). Values greater than 1 are considered as well-fermented, while less than 1 as under-fermented beans [31,32]. However, it must be noted that this accounts for the Forastero variety and with some precautions for the Trinitario variety (which can contain both purple and white beans). Criollo beans do not contain anthocyanin pigments, therefore FI cannot be used to describe the fermentation level for this variety. In our study, both Trinitario and Forastero beans were used, but white beans were always absent.

2.4.7. pH and Titratable Acidity

Cocoa powder (5 g) was mixed with 100 mL hot water (100 °C), stirred, and allowed to stand for 30 min. After 30 min, when the suspension was cooled up to 25 °C, it was centrifuged for 10 min at 5000 rpm and vacuum filtered through Whatman No. 4 paper filter according to AOAC 2006 methods 970.21 (pH) and 942.15 (potentiometric titration) [21], AOAC Section 42.104 (16th Ed. 1995) [21,32–35].

The pH of the filtered solution was measured with a pH-meter model PC 80 + DHS (XS Instruments, Carpi, Italy) and then 25 mL aliquots of the same solution were titrated to pH 8.1 with 0.05 M NaOH. All data were measured in triplicate. Titratable acidity results are expressed as mMol NaOH/100 g powder [34] or % acetic acid [21].

It is important to note that this procedure was not for quantifying the actual pH of the cocoa bean itself, but rather to measure the acidity derived from bean acids diffusing into water; it is useful for comparison between the pH of solutions produced by different beans [30].

2.5. Wavelenght Selection and Chemometric Analyses

Spectral chemometric analyses were performed using firstly the wavelength selection and secondly the full spectra collected. Wavelengths selection was carried out through the interval partial least-square (iPLS) [36] and through the principal component regression (PCR) [37] by using R software version 3.2.5 (R Core Team, Auckland, New Zealand, 2016) and WinISI software (Infrasoft International, Port Matilda, PA, USA), respectively. In particular, the iPLS was carried out applying the forward mode, in which the full spectrum was subdivided in 30 intervals that are successively included in the analysis: the first step calculated 30 models (one for each interval) that were tested using the cross-validation; the interval which provides the lowest model root-mean-square error of cross-validation were selected as most informative. The selected intervals were calculated per each parameter investigated and used for the following modelling. The PCR is based on the identification of the principal factors variance among spectral absorbance data through the principal component analysis [38]. Wavelengths selection was performed on the spectra acquired with FOSS DS-2500 on cocoa powder.

The second approach considered the use of the full spectrum and mathematical treatment as reported by several authors [39–41] in foods for chemical prediction purpose. This procedure takes advantage of the mathematical treatment as multiplicative correction (MSC) of the dispersion used to correct the problems of dispersed light in reflectance spectroscopy or the spectra normalization using standard normal variation (SNV) and first or second derivatives often used to remove the deviation and slope of the baseline in the spectrum [42]. This approach was applied to spectra acquired with both instruments, on both whole and ground cocoa beans.

The calibration models were performed using the Modified PLS (MPLS) regression on wavelength selected and on full spectra, whereas PCR was applied on full spectra (WinISI software, Infrasoft International, Port Matilda, PA, USA). Prediction equations were validated using a 5-fold cross-validation. Samples with a predicted value that differed more than 2.5 SD from the reference value (T-statistics) were considered outliers and removed from the dataset. Several combinations of scatter corrections (NONE, no correction; SNV_DT, standard normal variate and detrending; MSC, multiplicative scatter correction) and derivative mathematical treatments (0,0,1,1; 1,4,4,1; 2,5,5,1; where the first digit is the number of the derivative, the second is the gap over which the derivative is calculated, the third is the number of data points in the first smoothing and the fourth is the number of data points in the second smoothing) were tested. The performances of the prediction models were evaluated based on the number of the standard error of calibration (SEC), cross-validation (SECV), the coefficient of determination of cross-validation (R^2cv) and the ratio performance to deviation of cross-validation (RPDcv) calculated as the ratio between SD and SECV [43]. Predictions were considered excellent when R^2 was greater than 0.91, good when R^2 ranged from 0.82 to 0.90, approximate when R^2 was between 0.66 and 0.81, and poor when R^2 was less than 0.66 [44]. Prediction models with RPD greater than 2.5 were considered adequate for analytical purposes [45], whereas prediction models with RPD smaller than 1.5 were considered unsatisfactory [44].

3. Results and Discussion

3.1. Chemical Properties

Shell content was on average 13.25% (Table 1), with minimum and maximum values (11.13% and 18.34%, respectively) in line with those reported in the literature (12–20%) [46,47]. Although the shell provides protection to the nib from mold and insects infestations, the shell content should be as low as possible (10–14%) because it has very little commercial value for the cocoa processor: it is removed during cocoa bean processing and it mainly constitutes a waste material [48].

Table 1. Descriptive statistics of cocoa beans: SD = standard deviation; CV = coefficient of variation; TPC = total phenolic component; TA = titratable acidity; FI = fermentation index; DM = dry matter.

	Minimum	Maximum	Mean	SD	CV (%)
Shell (%)	11.13	18.34	13.25	1.54	11.59
Fat (%)	36.96	48.39	44.72	1.94	4.35
Protein (%)	8.32	15.43	13.85	1.13	8.14
TPC (mg/g dry defatted powder)	32.58	98.04	56.42	13.32	23.60
pH	4.84	6.47	5.58	0.36	6.50
TA (mMol NaOH/100 g powder)	8.20	26.81	17.19	4.22	24.52
FI (A460/A530)	0.57	2.24	1.29	0.49	38.43
DM (%)	93.30	95.76	94.51	0.59	0.62
Ash (%)	2.34	3.66	2.99	0.30	10.10

Dry matter was on average 94.51%, with a minimum of 93.30%. These values correspond to an average moisture content of 5.49% and maximum of 6.70%, which are mainly below the optimal commercial levels of 6.5–8.0% as reported in CAOBISCO/ECA/FCC [19] but are in line with data found in the literature [49]. Moisture is a parameter that depends on storage conditions: since storage conditions of the studied samples varied, this may have affected the final moisture levels.

The average ash content of 2.99% found in our samples was in line with data reported in the literature [48,50]. With regards to fat content, which is the most abundant macronutrient in cocoa beans, only one sample presented a value below 40 g/100 g (i.e., 36.96 g/100 g), while the average fat content was 44.72 g/100 g. These data are in line with other studies [21,50]. African cocoa beans have generally higher fat content than American beans [16], but this was not observable in our set of samples. However, according to literature, the fat content can vary greatly from values of about 40 g/100 g to values of 57–58 g/100 g depending on different factors such as: genotype, plant age, growing practices, fermentation, drying processes and environmental conditions [51,52].

FI is one of the most used parameters for determining the degree of fermentation of cocoa beans as an indirect measure of the anthocyanin content [29,35]. In our case study, 22 out of 56 samples had a FI slightly below 1, with a minimum value of 0.57, which would indicate a low fermentation degree. The maximum value was 2.24 and the average was 1.29. The coefficient of variation for this parameter was particularly high (38.43%). Since the FI is an indirect measurement of anthocyanin content, the high dispersion of data might be due to factors other than solely the fermentation degree. It has been reported that different hybrids or genotypes have different pigments and that phenolic compounds are quantitatively affected by cocoa growth conditions (microclimate and position of pods on the tree) [29].

The TPC in the dried fat-free mass of our samples exhibited a wide variation, ranging from 32.58 to 98.04 mg/g dry defatted powder. In fermented beans, TPC should be approximately 5% in the dried fat-free mass, and values above 10% are considered a sign of a bad fermentation [53]. The average value of TPC in our samples was 56.42 mg/g dry defatted powder (equals to 5.6%) that would indicate well-fermented beans. Moreover, few samples showed values close to 10%. Overall, the values are in line with those reported in Anyidoho, et al. [54] and Djikeng, et al. [55].

In dried cocoa beans, a high degree of acidity is usually associated with a pH of 5.0 or less [19]. Some studies report that beans of higher pH (5.5–5.8) are considered unfermented, with a low fermentation index, and result in chocolates with high astringency [32], while beans of lower pH (4.75–5.19) are considered as well-fermented. Other studies report that pH of 5–6 is considered good for flavor development, and cocoa beans with pH below 4.5 are not accepted by cocoa bean processers because they show low levels of flavor precursors, and high acidic-derived products [35]. The pH can still be considered as a good indicator of fermentation as higher pH correlates to a lower fermentation degree [16] and an "international acceptable range" of 5.00–5.55 for dried cocoa beans [56] can be considered as a valid reference. In our case study, cocoa beans had an average pH of 5.58 with a minimum of 4.84. This describes a situation of well fermented samples.

The titratable acidity value is often associated with the beans' pH. The present results confirm an overall good fermentation of the samples with an average titratable acidity of 17.19 mmol NaOH/100 g powder, in line with data reported in the literature [57,58].

Overall, this set of samples included many variation factors (e.g., genetic variety, crop, fermentation and drying conditions, transport, and storage) giving rise to high coefficients of variation in most of the studied parameters [59].

3.2. Spectral Characteristics of Cocoa Samples

Figure 1 is representative for average NIR spectra of cocoa beans samples obtained by FOSS DS 2500. The spectra show high similarity with spectra found in the literature [21,27,50,51]. Since cocoa beans contain about 50% of fat (Table 1), absorption spectra are dominated by signals derived from C=O and CH_2 groups [49]. The absorptions around 1930 nm are caused by the second overtone vibration of ester C=O and O–H asymmetric stretching [49,60]. Caporaso, et al. [61] reported that wavelength of 1919 nm has been attributed to the C=O stretching second overtone in the carbonyl groups (–CO_2H or CONH) but this absorption band is very close to 1923 nm, which is assigned to the O–H group of water and therefore it might be influenced by this group.

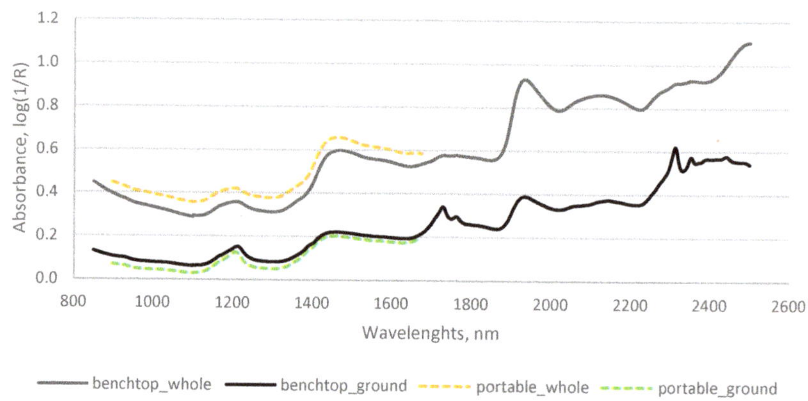

Figure 1. NIR spectra (mean) of whole (gray) and ground (black) cocoa beans acquired with benchtop spectrometer (FOSS DS 2500) and NIR spectra (mean) of whole (yellow) and ground (green) cocoa beans acquired with portable NIR spectrometer (PoliSPEC-NIR).

The combination vibrations of CH_2 stretch and CH_2 deformation appear around 2320 nm. Moreover, the absorption at 1744 nm has been previously assigned to C–H stretch first overtone (CH_2) of lipids, and the CH_2 group also absorbs at 1725 nm, due to the C–H stretch first overtone [61]. Similar wavelength values (i.e., 1750 nm and 1730 nm), associated with first overtones of symmetric and anti-symmetric C–H stretch vibration (CH_2-groups), are reported by Krahmer et al. [49].

Fat content is also related to the absorption bands visible around 1200 nm, as reported by Hayati et al. [27]. The authors also argued that the bands in the wavelength regions of 1460–1490 nm and 1920–1980 nm are most likely related to moisture content (O–H bonds). However, absorbances around 1450 nm have been attributed to carbonyl groups (e.g., ketones and aldehydes) as well as O–H polymeric groups, which can be due to complex carbohydrates, and the region between 1400 nm and 1440 nm has also been attributed to aliphatic alcohols and phenols [61].

Absorbance around 1490 nm has been attributed in the literature to several possible chemical bond vibrations, including N–H stretch first overtone and O–H stretch first overtone, thus indicating amides or compounds such as cellulose [61]. Accordingly,

Krahmer et al. [49] reported that first overtones of intermolecular H-bridges and stretch vibrations of amidic NH-groups can be observed in the region of 1400 to 1500 nm and the corresponding combination of two amides can be found around 2130 nm.

Barbin et al. [50] associated the broad peaks around 1190, 1460 and 1950 nm with O–H, C–H, N–H stretch first and second overtones and combination bands that can be attributed to water absorption and protein changes.

Peaks around 1215 nm are visible and are associated with –CH=CH second overtone [23] and even C–H stretching second overtone (–CH_3 or –CH_2) of carbohydrates is associated with this wavelength [61].

The absorbance at 2057 nm indicates an N–H stretch/amide 1st combination band, which has been attributed to protein, while the peaks at 2145 and 2313 nm have been tentatively attributed to C–H deformation and C–H deformation and C–H bend second overtones respectively, both indicating lipids [61].

3.3. Calibration Models for Cocoa Beans Quality

Variable selection is generally applied in the multivariate analysis to extract the most informative region, removing redundant information. However, among the approaches tested in this study, a lower prediction was observed for the PCR than the MPLS approach as observed in the study of Xie et al. [37]. In detail, in the present study, the PCR showed poor performance of prediction for all traits investigated (see Supplementary Table S2).

Comparing the performance of prediction using the MPLS between full and iPLS selected spectra, it was observed that among the eight parameters, the best prediction was achieved using the full spectra for seven of them (see Supplementary Table S3). The iPLS wavelength selection had a better performance in the fat prediction (R^2cv of 0.86 and RPD of 2.88) that did not differ substantially from the prediction obtained using the whole spectrum (900–1680 nm; R^2cv = 0.83 and RPD = 2.43).

The results of prediction performance for the benchtop (NIR FOSS DS 2500) and the portable (PoliSPEC-NIR) spectrometers are presented in Tables 2 and 3, which describe data obtained from whole cocoa beans and peeled-ground cocoa beans, respectively.

Table 2. Fitting statistics of prediction models for whole cocoa traits developed using cross-validation results for benchtop (NIR FOSS DS 2500) and portable (PoliSPEC-NIR) NIR-spectrometers.

Math Treatment		Constituent	N	Mean	SD	SEcal	R^2cal	SEcv	R^2cv	RPD
FOSS DS 2500										
NONE	1441	Fat	52	2.48	0.26	0.13	0.77	0.15	0.69	1.78
NONE	2551	Protein	51	0.77	0.09	0.04	0.81	0.06	0.66	1.71
SNV_DET	2551	TPC	54	3.04	0.66	0.61	0.15	0.67	0.03	0.98
NONE	0011	pH	54	5.60	0.36	0.21	0.65	0.24	0.58	1.52
MSC	0011	TA	53	16.81	3.99	2.73	0.53	2.98	0.46	1.34
MSC	2551	FI	55	0.07	0.03	0.02	0.25	0.03	0.07	1.03
NONE	1441	DM	55	94.49	0.57	0.26	0.80	0.31	0.72	1.86
MSC	2551	Ash	54	0.16	0.03	0.01	0.87	0.02	0.51	1.43
PoliSPEC-NIR										
Math Treatment		**Constituent**	**N**	**Mean**	**SD**	**SEcal**	**R^2cal**	**SEcv**	**R^2cv**	**RPD**
MSC	1441	Fat	54	2.45	0.27	0.18	0.57	0.21	0.38	1.28
SNV_DET	0011	Protein	50	0.78	0.10	0.04	0.83	0.07	0.56	1.49
MSC	2551	TPC	48	3.04	0.60	0.22	0.87	0.40	0.56	1.51
MSC	0011	pH	50	5.56	0.37	0.18	0.76	0.20	0.70	1.83
MSC	1441	TA	55	17.26	4.22	2.77	0.57	3.03	0.48	1.39
NONE	0011	FI	55	0.07	0.03	0.02	0.66	0.02	0.58	1.55
MSC	0011	DM	49	94.50	0.54	0.26	0.77	0.32	0.66	1.72
SNV_DET	0011	Ash	53	0.16	0.03	0.02	0.66	0.02	0.45	1.36

NONE = no correction; SNV_DET = SNV and detrend; MSC = multiplicative scatter correction; SD = standard deviation of reference data selected; SEcal = standard error in calibration; R^2cal = coefficient of determination of calibration; SEcv = standard error in cross-validation; R^2cv = coefficient of determination of cross-validation. TPC = total phenolic compound; TA = titratable acidity; FI = fermentation index; DM = dry matter.

Table 3. Fitting statistics of prediction models for ground cocoa traits developed using cross-validation results for benchtop (NIR FOSS DS 2500) and portable (PoliSpec NIR) NIR-spectrometers.

Math Treatment		Constituent	N	Mean	SD	SEcal	R^2cal	SEcv	R^2cv	RPD
FOSS DS 2500										
NONE	1441	Fat	56	2.45	0.27	0.11	0.84	0.13	0.76	2.11
SNV_DET	0011	Protein	54	0.77	0.11	0.02	0.95	0.03	0.91	3.40
NONE	2551	TPC	55	3.04	0.65	0.51	0.40	0.59	0.16	1.10
MSC	0011	pH	50	5.57	0.37	0.08	0.95	0.13	0.88	2.96
NONE	1441	TA	52	16.69	3.86	0.80	0.96	1.43	0.86	2.70
NONE	1441	FI	55	1.29	0.50	0.27	0.70	0.38	0.42	1.31
MSC	0011	DM	56	94.51	0.59	0.15	0.94	0.18	0.90	3.20
MSC	0011	Ash	50	0.16	0.03	0.01	0.90	0.01	0.89	2.98
PoliSPEC-NIR										
Math Treatment		**Constituent**	**N**	**Mean**	**SD**	**SEcal**	**R^2cal**	**SEcv**	**R^2cv**	**RPD**
NONE	2551	Fat	52	2.45	0.27	0.09	0.88	0.12	0.82	2.34
NONE	0011	Protein	53	0.77	0.11	0.04	0.84	0.05	0.79	2.17
MSC	0011	TPC	55	3.04	0.65	0.50	0.42	0.57	0.23	1.14
NONE	1441	pH	53	5.56	0.36	0.11	0.90	0.18	0.74	1.98
NONE	0011	TA	51	17.23	4.07	1.14	0.92	1.76	0.81	2.32
NONE	0011	FI	56	1.29	0.49	0.41	0.33	0.42	0.26	1.17
NONE	0011	DM	54	94.54	0.59	0.25	0.81	0.27	0.79	2.17
NONE	1441	Ash	54	0.16	0.03	0.01	0.89	0.01	0.76	2.08

NONE = no correction; SNV_DET = SNV and detrend; MSC = multiplicative scatter correction; SD = standard deviation of reference data selected; SEcal = standard error in calibration; R^2cal = coefficient of determination of calibration; SEcv = standard error in cross-validation; R^2cv = coefficient of determination of cross-validation. TPC = total phenolic compound; TA = titratable acidity; FI = fermentation index; DM = dry matter.

Generally, most of the cocoa studies were performed on ground cocoa to reduce the effects of the physical sample properties on spectra collection [24]. Indeed, for both NIRS devices, the best performances of prediction were observed on ground sample, probably due to the enhanced homogeneity of the samples characterized by a similarity in the particles size and in a more compacted powder that affects the scattering of light.

In this study, spectra corrections by mathematical treatments to remove irrelevant data such as noise and background information were evaluated. In particular, SNV and MSC were used as pre-processed methods to remove the influence of solid particle size and the surface scattering; moreover, the methods above are mainly recognized as the best mathematical treatment in the equation models developed for whole cocoa. The SNV_DT and MSC treatments improved the prediction accuracy for some quality parameters of both whole and ground cocoa bean samples, while for other parameters raw spectra gave the best results. This was in line with Barbin et al. [50] who found no considerable improvement of the predictive ability when comparing different pre-processing methods with the original raw data. Indeed, Barbin et al. [50] stated that since the complexity of the models was similar to that obtained with the original data, it is feasible to use the raw spectra to build prediction models for both whole beans and ground cocoa samples.

Moreover, to evaluate the performance of technologies on the market, the whole spectrum was considered to perform the prediction equations, although some researchers suggest that selection of spectral intervals could lead to higher prediction performances [35,62]. All the predictions performed against whole bean sample can be considered as approximate to poor [44] with the highest capability achieved for DM (R^2cv = 0.72; RPDcv = 1.86) for the benchtop and for pH (R^2cv = 0.70; RPDcv = 1.83) with portable device (Table 2). In general, the minor prediction capability in whole cocoa beans compared to the ground sample has been confirmed also in the study of Hernández-Hernández et al. [63], in which the poor performance of chemical predictions was attributed to the shell that reflects the incident light hindering the interaction with internal constituents. Although predictions on whole cocoa beans were not adequate for quantitative purposes, they could represent a fast approach for food business operators to sort cocoa beans towards a specific transformation according to high or low value. Moreover, at germplasm banks and breeding programs, a rapid whole cocoa analysis reduces the time required for the shell removing (usually carried out by hand in the laboratory), suggesting NIRS devices are capable to identify functional genotypes to improve qualitative aspects in cocoa products [63].

Excellent performance was obtained in ground cocoa for protein content ($R^2cv = 0.91$; RPDcv = 3.40) and very good prediction was achieved for DM ($R^2cv = 0.90$; RPDcv = 3.20), ash ($R^2cv = 0.89$; RPDcv = 2.98), pH ($R^2cv = 0.88$; RPDcv = 2.96) and TA ($R^2cv = 0.86$; RPDcv = 2.70) using the NIR FOSS DS 2500 spectrometer (850–2500 nm) (Table 3). The PoliSPEC-NIR spectrometer (900–1680 nm) had the best predicting performances for fat content ($R^2cv = 0.82$; RPDcv = 2.34) in ground samples; whereas, for the other traits, the portable device showed lower performances compared to the benchtop (Table 3).

To deeper investigate if the divergences between the devices might depend on the different spectral range used, a further prediction equation was performed for the benchtop using the same spectral range (900–1680 nm, every 2 nm) of the portable tool. In the comparison with the performance obtained considering the whole spectrum, a greater performance of predictions was observed for ash ($R^2cv = 0.90$; RPDcv = 3.20), protein ($R^2cv = 0.93$; RPDcv = 3.84), DM ($R^2cv = 0.94$; RPDcv = 4.16), and lipids ($R^2cv = 0.83$; RPDcv = 2.43). However, although the TPC remained unpredictable, an increment was observed in the new prediction equation ($R^2cv = 0.46$; RPDcv = 1.37). Although a good predictive capability was maintained, lower performance prediction was observed for TA ($R^2cv = 0.85$; RPDcv = 2.60) and pH ($R^2cv = 0.82$; RPDcv = 2.34).

Thus, to comprehend the origin of the performance divergences between devices, the component loadings were developed for each tool to assess and compare the interactions between wavelengths and functional groups (Figure 2). The loading plots permit to better understand which wavelengths are more informative for a specific trait variability, showing the range which is mostly considered to develop the model. A strong similarity between portable and benchtop devices were overall observed for chemical parameters directly quantified.

In particular, although the same ranges and performance of prediction ($R^2cv = 0.83$) were obtained in both devices for lipid loading plot, the highest loadings were observed in the spectral region between the 1212 and 1232 nm and 1368 and 1398 nm for portable and benchtop, respectively.

Such association between those range and lipid variability has been confirmed by [64] in cereal food products. Similar patterns for the protein loading plot were observed between the two devices; however, the high loadings observed between 1200 to 1400 nm were related to C–H second overtone and N–H stretching first overtone of protein, respectively [65,66]. Moreover, a high loading was observed around 1100 nm exclusively for the benchtop device; this is probably due to the higher sensitivity of the device that is reflected in the best performance of prediction ($R^2cv = 0.93$) the range 1100–1400 nm being considered as an essential spectral region for the protein quantification analysis [67].

A comparable loading plot was also observed for pH in which the highest trait variability was explained by the 910 [68] and 1398 nm for both devices. Divergences in titratable acidity loading patterns were found; however, the most informative wavelengths (930–950; 1106; 1390–1400 nm) are related to the second combination region of the carboxylic acids [69]. The loading plot of DM showed notable peaks between 1200–1224 and 1373–1394 nm, mainly related to the water [69].

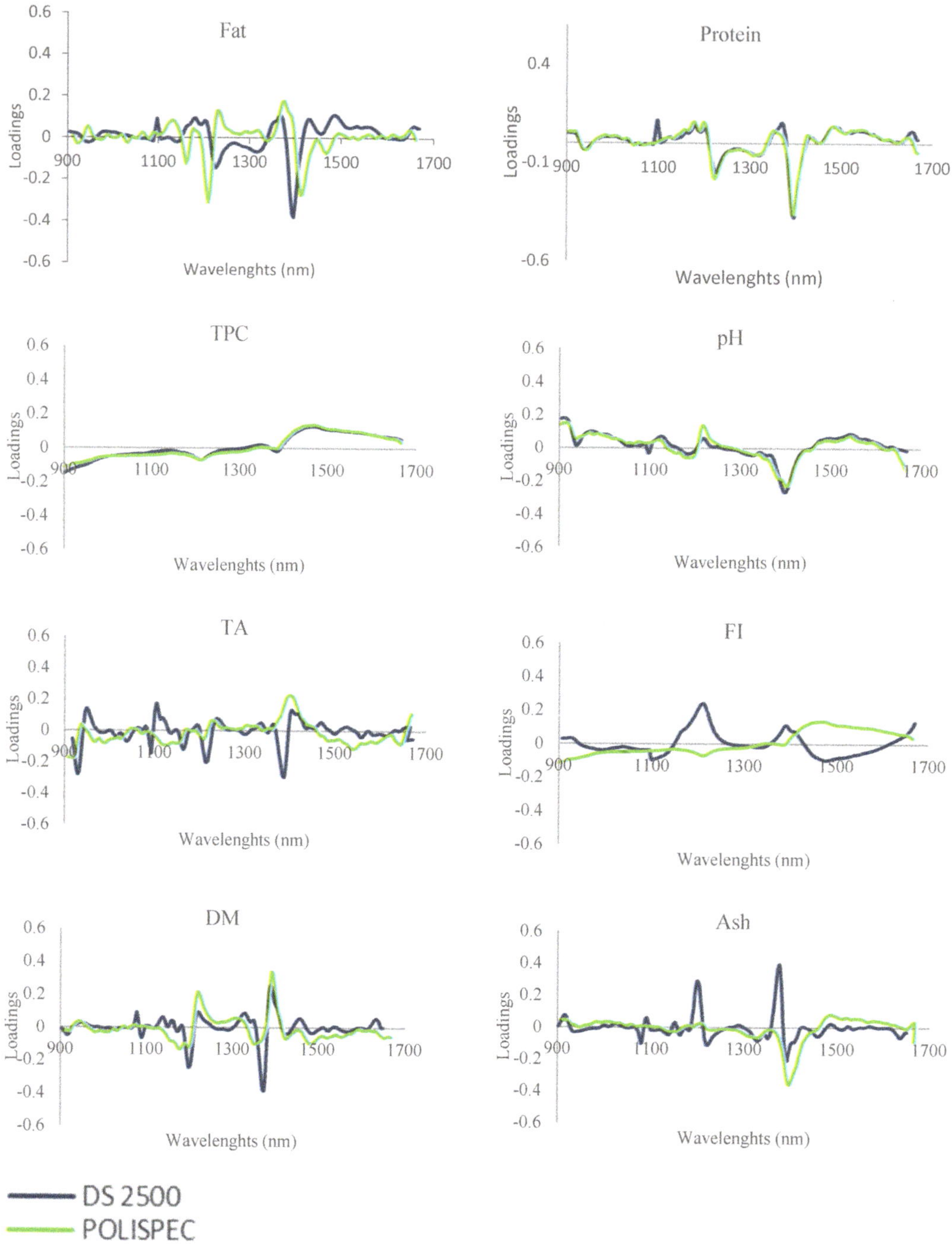

Figure 2. Loadings for the first principal component of fat, protein, pH, titratable acidity (TA), total phenolic compound (TPC), fermentation index (FI), dry matter (DM), and ash for NIR spectra of the ground cocoa samples for DS 2500 (blue line) and PoliSPEC NIR (green line).

Ash being an inorganic matter cannot be directly detected by NIRS; its amount is indirectly measured by the association with organic bonds, thus the loadings plot and the highest variability observed for ash is 1200 nm and 1376 nm for the benchtop, and 1396 nm for the portable device account for other organic components. Otherwise, loadings plots observed for TPC and FI were not strictly related to a specific spectrum range, probably due to the lower variability collected with the samples considered. In general, the performance divergences between the two NIRS devices could be explained by the difference in the detector equipment; in detail, the semiconductors included in portable (PoliSPEC-NIR) and benchtop (NIR FOSS DS 2500) devices are Indium gallium arsenide (InGaAs) and silicon lead sulfide, respectively, which affect the spectral response and the prediction capability Lin, et al. [70].

In our study, the accuracy of prediction for both FI and TPC was not satisfactory for any of the instruments and for both whole and ground cocoa bean samples. The influence of variable fermentation degrees of cocoa samples can be crucial in the prediction of FI and TPC, which are strictly related to the fermentation level of cocoa beans. Sunoj, Igathinathane and Visvanathan [32] showed how factors such as pod storage duration (before the fermentation process), and fermentation time, had a significant effect on the fermentation index, which was seen to increase together with the increment of these two parameters. The authors argued that these parameters are indirectly affected by the samples' chemical composition, thus the accuracy of prediction models are generally lower than those reported for major components. The reason might fall on the fact that our samples included only commercial cocoa beans which were supposed to be well-fermented, although with some natural variations, thus reducing the variability for the TPC and FI. Moreover, there could have been a negative influence of lipid absorbances in the models for TPC: fat has been indicated as a disturbance factor as beans with higher relative fat content have lower non-fat solids, where polyphenols are concentrated [61].

Although the FI was not correctly predicted by the constructed models, the estimations in ground samples of parameters related to correct fermentation such as pH and TA were approximative and good with the portable and benchtop devices, respectively, in line with previous results [25,49]. This method could provide a rapid and low-cost multiparametric analysis for cocoa evaluation. Portable instruments are usually less expensive than benchtop solutions (about a fifth) [71], and the cost of analyses are mainly related to the development and upgrade of calibration curves. Moreover, compared to wet analyses, through the application of spectrometric methods the cost of the analytical determination is drastically reduced as the number of examined samples increases.

The presented prediction models might be the basis for an overall cocoa bean quality evaluation based on NIR spectra. However, despite the presented parameters being good indicators of cocoa bean quality, a grading classification of cocoa beans' was beyond the scope of the present work, as it would require the investigation of other indicators, also related to the sensorial profile of the beans, as reported in previous studies on cocoa quality indexes (CQI) [72,73].

4. Conclusions

The results of this paper demonstrated that NIRS portable and benchtop devices coupled with chemometrics methods could be adopted for the chemical evaluation of commercial cocoa beans. The performances of predictions are affected by the presence of shell and the sample particle sizes of cocoa beans. The current study has successfully demonstrated that NIR, as a nondestructive analytical method, can be considered as rapid and reliable option to traditional methods to quantify lipids, protein, pH, titratable acidity, dry matter and ash in cocoa ground beans.

The NIRS benchtop instrument provided better performance of quantification considering the whole (800–2500 nm) and the reduced spectrum (900–1680 nm) than the portable device. Variable selection through iPLS or PCR did not improve prediction models compared to full spectra analyses. Benchtop instrument showed excellent prediction capability in DM (R^2cv = 0.94), protein (R^2cv = 0.93) and ash (R^2cv = 0.90), whereas lipids (R^2cv = 0.83), TA (R^2cv = 0.86) and pH (R^2cv = 0.88) were well predicted on ground beans considering wavelengths between 900–1680 nm. Those results indicate that models developed for benchtop devices are applicable for cocoa quality control as an excellent option to substitute conventional methods.

On the other hand, the NIRS portable device showed lower but valuable performance of prediction than benchtop spectrometer. The prediction obtained for handheld device represents an appealing strategy for food business operators to apply in the field to control and check the product in every phase of trade and transportation, and also to segregate whole cocoa beans targeted to a specific transformation in different supply chains.

Based on these results, further studies including a wider variability of fermentation phases, cocoa bean varieties and origins as well as additional production steps of the cocoa supply chain could be investigated to support the fair-trade cocoa sector.

Supplementary Materials: The following supporting information can be downloaded at: https://www.mdpi.com/xxx/s1, Table S1: Cocoa origin of different commercial lot analysed in this study Table S2: Fitting statistics of prediction models for ground cocoa traits developed using full spectra and principal component regression (PCR) and cross-validation results for benchtop (NIR FOSS DS 2500); Table S3. Fitting statistics of prediction models for ground cocoa traits developed using selected wavelengths through the interval PLS (iPLS) and cross-validation results for benchtop (NIR FOSS DS 2500).

Author Contributions: Conceptualization, P.C., M.F., K.D. and L.F.; methodology, D.V.d.W., K.D., M.M.; formal analysis, M.F., S.C. and M.M.; resources, P.C. and L.F.; data curation, M.F., S.C. and M.M.; writing—original draft preparation, M.F. and S.C.; writing—review and editing, S.C., D.V.d.W., K.D., L.F. and P.C.; supervision, P.C. and L.F. All authors have read and agreed to the published version of the manuscript.

Funding: This research received no external funding.

Data Availability Statement: All related data and methods are presented in this paper. Additional inquiries should be addressed to the corresponding author.

Acknowledgments: The authors gratefully acknowledge Roberta Marcuzzi for the valuable help in performing the analyses.

Conflicts of Interest: The authors declare no conflict of interest.

References

1. McNeil, C.L. *Chocolate in Mesoamerica: A Cultural History of Cacao*; University Press of Florida: Gainesville, FL, USA, 2009.
2. Afoakwa, E.O. *Chocolate Science and Technology*; John Wiley & Sons: Hoboken, NJ, USA, 2016.
3. Benea-Popușoi, E.; Casapu, M. The forms of international specialization and ensuing economic flows. Case study on the cocoa products. In *Center for Studies in European Integration Working Papers Series*; Center for Studies in European Integration: Chisnau, Moldova, 2021.
4. Sayago-Ayerdi, S.; García-Martínez, D.L.; Ramírez-Castillo, A.C.; Ramírez-Concepción, H.R.; Viuda-Martos, M. Tropical Fruits and Their Co-Products as Bioactive Compounds and Their Health Effects: A Review. *Foods* **2021**, *10*, 1952. [CrossRef]
5. Badrie, N.; Bekele, F.; Sikora, E.; Sikora, M. Cocoa agronomy, quality, nutritional, and health aspects. *Crit. Rev. Food Sci. Nutr.* **2015**, *55*, 620–659. [CrossRef] [PubMed]
6. Gayi, S.K.; Tsowou, K. *Cocoa Industry: Integrating Small Farmers into the Global Value Chain*; United Nations: New York, NY, USA, 2017.
7. Beg, M.S.; Ahmad, S.; Jan, K.; Bashir, K. Status, supply chain and processing of cocoa–A review. *Trends Food Sci. Technol.* **2017**, *66*, 108–116. [CrossRef]
8. Dzelagha, B.F.; Ngwa, N.M.; Nde Bup, D. A Review of Cocoa Drying Technologies and the Effect on Bean Quality Parameters. *Int. J. Food Sci.* **2020**, *2020*, 8830127. [CrossRef]
9. Voora, V.; Bermúdez, S.; Larrea, C. *Global Market Report: Cocoa*; International Institute for Sustainable Development: Winnipeg, MB, Canada, 2019.
10. Villacis, A.H.; Alwang, J.R.; Barrera, V.; Dominguez, J. Prices, specialty varieties, and postharvest practices: Insights from cacao value chains in Ecuador. *Agribusiness* **2021**, *38*, 426–458. [CrossRef]

11. Lima, L.J.; Almeida, M.H.; Nout, M.R.; Zwietering, M.H. *Theobroma cacao* L., "The food of the Gods": Quality determinants of commercial cocoa beans, with particular reference to the impact of fermentation. *Crit. Rev. Food Sci. Nutr.* **2011**, *51*, 731–761. [CrossRef] [PubMed]
12. Hernandez, C.E.; Granados, L. Quality differentiation of cocoa beans: Implications for geographical indications. *J. Sci. Food Agric.* **2021**, *101*, 3993–4002. [CrossRef]
13. Santander Muñoz, M.; Rodríguez Cortina, J.; Vaillant, F.E.; Escobar Parra, S. An overview of the physical and biochemical transformation of cocoa seeds to beans and to chocolate: Flavor formation. *Crit. Rev. Food Sci. Nutr.* **2020**, *60*, 1593–1613. [CrossRef]
14. Opoku-Ameyaw, K.; Baah, F.; Gyedu-Akoto, E.; Anchirinah, V.; Dzahini-Obiatey, H.; Cudjoe, A.; Opoku, S. *Cocoa Manual—A Source Book for Sustainable Cocoa Production*; Cocoa Research Institute of Ghana: Tafo, Ghana, 2010.
15. Saltini, R.; Akkerman, R.; Frosch, S. Optimizing chocolate production through traceability: A review of the influence of farming practices on cocoa bean quality. *Food Control.* **2013**, *29*, 167–187. [CrossRef]
16. Beckett, S.T. *Industrial Chocolate Manufacture and Use*; John Wiley & Sons: Hoboken, NJ, USA, 2011.
17. Jalil, A.M.; Ismail, A. Polyphenols in cocoa and cocoa products: Is there a link between antioxidant properties and health? *Molecules* **2008**, *13*, 2190–2219. [CrossRef]
18. Kongor, J.E.; Hinneh, M.; Van de Walle, D.; Afoakwa, E.O.; Boeckx, P.; Dewettinck, K. Factors influencing quality variation in cocoa (*Theobroma cacao*) bean flavour profile—A review. *Food Res. Int.* **2016**, *82*, 44–52. [CrossRef]
19. CAOBISCO; ECA; FCC. *Cocoa Beans: Chocolate and Cocoa Industry Quality Requirements*; End, M.J., Dand, R., Eds.; ECA-Caobisco-FCC Cocoa Research Fund: Brussels, Belgium, 2015.
20. Scheu, J.; Dand, R. *Cocoa. A Guide to Trade Practices*; International Trade Centre UNCTAD/WTO: Geneva, Switzerland, 2001.
21. Hashimoto, J.C.; Lima, J.C.; Celeghini, R.M.; Nogueira, A.B.; Efraim, P.; Poppi, R.J.; Pallone, J.A. Quality control of commercial cocoa beans (*Theobroma cacao* L.) by near-infrared spectroscopy. *Food Anal. Methods* **2018**, *11*, 1510–1517. [CrossRef]
22. Fountain, A.C.; Huetz-Adams, F. *Cocoa Barometer 2020*; VOICE Network: Ede, The Netherlands, 2020.
23. Teye, E.; Anyidoho, E.; Agbemafle, R.; Sam-Amoah, L.K.; Elliott, C. Cocoa bean and cocoa bean products quality evaluation by NIR spectroscopy and chemometrics: A review. *Infrared Phys. Technol.* **2020**, *104*, 103127. [CrossRef]
24. Quelal-Vasconez, M.A.; Lerma-Garcia, M.J.; Perez-Esteve, E.; Talens, P.; Barat, J.M. Roadmap of cocoa quality and authenticity control in the industry: A review of conventional and alternative methods. *Compr. Rev. Food Sci. Food Saf.* **2020**, *19*, 448–478. [CrossRef]
25. Anyidoho, E.K.; Teye, E.; Agbemafle, R. Differentiation of Organic Cocoa Beans and Conventional Ones by Using Handheld NIR Spectroscopy and Multivariate Classification Techniques. *Int. J. Food Sci.* **2021**, *2021*, 1844675. [CrossRef]
26. Anyidoho, E.K.; Teye, E.; Agbemafle, R.; Amuah, C.L.Y.; Boadu, V.G. Application of portable near infrared spectroscopy for classifying and quantifying cocoa bean quality parameters. *J. Food Process. Preserv.* **2021**, *45*, e15445. [CrossRef]
27. Hayati, R.; Zulfahrizal, Z.; Munawar, A.A. Robust prediction performance of inner quality attributes in intact cocoa beans using near infrared spectroscopy and multivariate analysis. *Heliyon* **2021**, *7*, e06286. [CrossRef]
28. Singleton, V.L.; Rossi, J.A. Colorimetry of total phenolics with phosphomolybdic-phosphotungstic acid reagents. *Am. J. Enol. Vitic.* **1965**, *16*, 144–158.
29. Romero-Cortes, T.; Salgado-Cervantes, M.A.; Garcia-Alamilla, P.; Garcia-Alvarado, M.A.; Rodriguez-Jimenes Gdel, C.; Hidalgo-Morales, M.; Robles-Olvera, V. Relationship between fermentation index and other biochemical changes evaluated during the fermentation of Mexican cocoa (*Theobroma cacao*) beans. *J. Sci. Food Agric.* **2013**, *93*, 2596–2604. [CrossRef]
30. Racine, K.C.; Lee, A.H.; Wiersema, B.D.; Huang, H.; Lambert, J.D.; Stewart, A.C.; Neilson, A.P. Development and Characterization of a Pilot-Scale Model Cocoa Fermentation System Suitable for Studying the Impact of Fermentation on Putative Bioactive Compounds and Bioactivity of Cocoa. *Foods* **2019**, *8*, 102. [CrossRef]
31. Afoakwa, E.O.; Jennifer, Q.; Agnes, S.B.; Jemmy, S.T.; Saalia, F. Influence of pulp-preconditioning and fermentation on fermentative quality and appearance of ghanaian cocoa (*theobroma cacao*) beans. *Int. Food Res. J.* **2012**, *19*, 127–133.
32. Sunoj, S.; Igathinathane, C.; Visvanathan, R. Nondestructive determination of cocoa bean quality using FT-NIR spectroscopy. *Comput. Electron. Agric.* **2016**, *124*, 234–242. [CrossRef]
33. Hue, C.; Gunata, Z.; Bergounhou, A.; Assemat, S.; Boulanger, R.; Sauvage, F.X.; Davrieux, F. Near infrared spectroscopy as a new tool to determine cocoa fermentation levels through ammonia nitrogen quantification. *Food Chem.* **2014**, *148*, 240–245. [CrossRef] [PubMed]
34. Nazaruddin, R.; Seng, L.K.; Hassan, O.; Said, M. Effect of pulp preconditioning on the content of polyphenols in cocoa beans (*Theobroma Cacao*) during fermentation. *Ind. Crops Prod.* **2006**, *24*, 87–94. [CrossRef]
35. Teye, E.; Huang, X.; Sam-Amoah, L.K.; Takrama, J.; Boison, D.; Botchway, F.; Kumi, F. Estimating cocoa bean parameters by FT-NIRS and chemometrics analysis. *Food Chem.* **2015**, *176*, 403–410. [CrossRef]
36. Andersen, C.M.; Bro, R. Variable selection in regression—A Tutorial. *J. Chemom.* **2010**, *24*, 728–737. [CrossRef]
37. Xie, L.H.; Tang, S.Q.; Chen, N.; Luo, J.; Jiao, G.A.; Shao, G.N.; Wei, X.J.; Hu, P.S. Optimisation of near-infrared reflectance model in measuring protein and amylose content of rice flour. *Food Chem.* **2014**, *142*, 92–100. [CrossRef]
38. Williams, P. Interpretation of Calibration Evaluation II. The Spectral Basics of Interpretation. *NIR News* **2013**, *24*, 21–25. [CrossRef]

39. Chadalavada, K.; Anbazhagan, K.; Ndour, A.; Choudhary, S.; Palmer, W.; Flynn, J.R.; Mallayee, S.; Pothu, S.; Prasad, K.; Varijakshapanikar, P.; et al. NIR Instruments and Prediction Methods for Rapid Access to Grain Protein Content in Multiple Cereals. *Sensors* **2022**, *22*, 3710. [CrossRef]
40. Manuelian, C.L.; Currò, S.; Penasa, M.; Cassandro, M.; De Marchi, M. Prediction of minerals, fatty acid composition and cholesterol content of commercial cheeses by near infrared transmittance spectroscopy. *Int. Dairy J.* **2017**, *71*, 107–113. [CrossRef]
41. Albanell, E.; Miñarro, B.; Carrasco, N. Detection of low-level gluten content in flour and batter by near infrared reflectance spectroscopy (NIRS). *J. Cereal Sci.* **2012**, *56*, 490–495. [CrossRef]
42. Moghimi, A.; Aghkhani, M.H.; Sazgarnia, A.; Sarmad, M. Vis/NIR spectroscopy and chemometrics for the prediction of soluble solids content and acidity (pH) of kiwifruit. *Biosyst. Eng.* **2010**, *106*, 295–302. [CrossRef]
43. Williams, P.C.; Sobering, D.C. Comparison of Commercial near Infrared Transmittance and Reflectance Instruments for Analysis of Whole Grains and Seeds. *J. Near Infrared Spectrosc.* **1993**, *1*, 25–32. [CrossRef]
44. Karoui, R.; Mouazen, A.M.; Dufour, É.; Schoonheydt, R.; De Baerdemaeker, J. A comparison and joint use of VIS-NIR and MIR spectroscopic methods for the determination of some chemical parameters in soft cheeses at external and central zones: A preliminary study. *Eur. Food Res. Technol.* **2006**, *223*, 363–371. [CrossRef]
45. Sinnaeve, G.; Dardenne, P.; Agneessens, R.; Biston, R. The Use of near Infrared Spectroscopy for the Analysis of Fresh Grass Silage. *J. Near Infrared Spectrosc.* **1994**, *2*, 79–84. [CrossRef]
46. Quelal-Vásconez, M.A.; Lerma-García, M.J.; Pérez-Esteve, É.; Arnau-Bonachera, A.; Barat, J.M.; Talens, P. Fast detection of cocoa shell in cocoa powders by near infrared spectroscopy and multivariate analysis. *Food Control.* **2019**, *99*, 68–72. [CrossRef]
47. Okiyama, D.C.; Navarro, S.L.; Rodrigues, C.E. Cocoa shell and its compounds: Applications in the food industry. *Trends Food Sci. Technol.* **2017**, *63*, 103–112. [CrossRef]
48. Afoakwa, E.O.; Quao, J.; Takrama, J.; Budu, A.S.; Saalia, F.K. Chemical composition and physical quality characteristics of Ghanaian cocoa beans as affected by pulp pre-conditioning and fermentation. *J. Food Sci. Technol.* **2013**, *50*, 1097–1105. [CrossRef]
49. Krahmer, A.; Engel, A.; Kadow, D.; Ali, N.; Umaharan, P.; Kroh, L.W.; Schulz, H. Fast and neat–determination of biochemical quality parameters in cocoa using near infrared spectroscopy. *Food Chem.* **2015**, *181*, 152–159. [CrossRef]
50. Barbin, D.F.; Maciel, L.F.; Bazoni, C.H.V.; da Silva Ribeiro, M.; Carvalho, R.D.S.; da Silva Bispo, E.; Miranda, M.d.P.S.; Hirooka, E.Y. Classification and compositional characterization of different varieties of cocoa beans by near infrared spectroscopy and multivariate statistical analyses. *J. Food Sci. Technol.* **2018**, *55*, 2457–2466. [CrossRef]
51. Agus, B.A.P.; Mohamad, N.N.; Hussain, N. Composition of unfermented, unroasted, roasted cocoa beans and cocoa shells from Peninsular Malaysia. *J. Food Meas. Charact.* **2018**, *12*, 2581–2589. [CrossRef]
52. Samaniego, I.; Espín, S.; Quiroz, J.; Rosales, C.; Carrillo, W.; Mena, P.; García-Viguera, C. Effect of the growing area on the fat content and the fatty acid composition of Ecuadorian cocoa beans. *Int. J. Food Sci. Nutr.* **2021**, *72*, 901–911. [CrossRef] [PubMed]
53. Wollgast, J.; Anklam, E. Review on polyphenols in *Theobroma cacao*: Changes in composition during the manufacture of chocolate and methodology for identification and quantification. *Food Res. Int.* **2000**, *33*, 423–447. [CrossRef]
54. Anyidoho, E.K.; Teye, E.; Agbemafle, R. Nondestructive authentication of the regional and geographical origin of cocoa beans by using a handheld NIR spectrometer and multivariate algorithm. *Anal. Methods* **2020**, *12*, 4150–4158. [CrossRef]
55. Djikeng, F.T.; Teyomnou, W.T.; Tenyang, N.; Tiencheu, B.; Morfor, A.T.; Touko, B.A.H.; Houketchang, S.N.; Boungo, G.T.; Karuna, M.S.L.; Ngoufack, F.Z.; et al. Effect of traditional and oven roasting on the physicochemical properties of fermented cocoa beans. *Heliyon* **2018**, *4*, e00533. [CrossRef]
56. Ndukwu, M.; Udofia, M. Kinetics of change in colour and some bio-chemical composition during fermentation of cocoa bean. *Cogent Food Agric.* **2016**, *2*, 1268743. [CrossRef]
57. Hii, C.L.; Abdul Rahman, R.; Jinap, S.; Che Man, Y.B. Quality of cocoa beans dried using a direct solar dryer at different loadings. *J. Sci. Food Agric.* **2006**, *86*, 1237–1243. [CrossRef]
58. Huang, X.; Teye, E.; Owusu-Sekyere, J.D.; Takrama, J.; Sam-Amoah, L.K.; Yao, L.; Firempong, C.K. Simultaneous measurement of titratable acidity and fermentation index in cocoa beans by electronic tongue together with linear and non-linear multivariate technique. *Food Anal. Methods* **2014**, *7*, 2137–2144. [CrossRef]
59. Kutsanedzie, F.Y.H.; Chen, Q.; Hassan, M.M.; Yang, M.; Sun, H.; Rahman, M.H. Near infrared system coupled chemometric algorithms for enumeration of total fungi count in cocoa beans neat solution. *Food Chem.* **2018**, *240*, 231–238. [CrossRef]
60. Teye, E.; Huang, X. Novel prediction of total fat content in cocoa beans by FT-NIR spectroscopy based on effective spectral selection multivariate regression. *Food Anal. Methods* **2015**, *8*, 945–953. [CrossRef]
61. Caporaso, N.; Whitworth, M.B.; Fowler, M.S.; Fisk, I.D. Hyperspectral imaging for non-destructive prediction of fermentation index, polyphenol content and antioxidant activity in single cocoa beans. *Food Chem.* **2018**, *258*, 343–351. [CrossRef]
62. Satriyo, P.; Munawar, A.A. Data analysis on near infrared spectroscopy as a part of technology adoption for cocoa farmer in Aceh Province, Indonesia. *Data Brief* **2020**, *29*, 105251.
63. Hernández-Hernández, C.; Fernández-Cabanás, V.M.; Rodríguez-Gutiérrez, G.; Fernández-Prior, Á.; Morales-Sillero, A. Rapid screening of unground cocoa beans based on their content of bioactive compounds by NIR spectroscopy. *Food Control.* **2022**, *131*, 108347. [CrossRef]
64. Vines, L.L.; Kays, S.E.; Koehler, P.E. Near-Infrared Reflectance Model for the Rapid Prediction of Total Fat in Cereal Foods. *J. Agric. Food Chem.* **2005**, *53*, 1550–1555. [CrossRef] [PubMed]

55. Saha, U.; Endale, D.; Tillman, P.G.; Johnson, W.C.; Gaskin, J.; Sonon, L.; Schomberg, H.; Yang, Y. Analysis of Various Quality Attributes of Sunflower and Soybean Plants by Near Infrared Reflectance Spectroscopy: Development and Validation Calibration Models. *Am. J. Anal. Chem.* **2017**, *8*, 462–492. [CrossRef]
56. Williams, P.; Manley, M.; Antoniszyn, J. *Near Infrared Technology: Getting the Best out of Light*; African Sun Media: Stellenbosch, South Africa, 2019.
57. Delwiche, S.R. Protein Content of Single Kernels of Wheat by Near-Infrared Reflectance Spectroscopy. *J. Cereal Sci.* **1998**, *27*, 241–254. [CrossRef]
58. Shao, Y.; He, Y. Nondestructive measurement of the internal quality of bayberry juice using Vis/NIR spectroscopy. *J. Food Eng.* **2007**, *79*, 1015–1019. [CrossRef]
59. Workman, J.J.; Weyer, L. *Practical Guide and Spectral Atlas for Interpretive Near-Infrared Spectroscopy*, 2nd ed.; CRC Press: Boca Raton, FL, USA, 2012. [CrossRef]
70. Lin, M.; A Rasco, B.; G Cavinato, A.; Al-Holy, M. Chapter 6—Infrared (IR) Spectroscopy—Near-Infrared Spectroscopy and Mid-Infrared Spectroscopy. In *Infrared Spectroscopy for Food Quality Analysis and Control*; Sun, D.-W., Ed.; Academic Press: San Diego, CA, USA, 2009; pp. 119–143. [CrossRef]
71. Sandak, J.; Sandak, A.; Zitek, A.; Hintestoisser, B.; Picchi, G. Development of Low-Cost Portable Spectrometers for Detection of Wood Defects. *Sensors* **2020**, *20*, 545. [CrossRef]
72. Gil, M.; Jaramillo, Y.; Bedoya, C.; Llano, S.M.; Gallego, V.; Quijano, J.; Londono-Londono, J. Chemometric approaches for postharvest quality tracing of cocoa: An efficient method to distinguish plant material origin. *Heliyon* **2019**, *5*, e01650. [CrossRef]
73. Araujo, Q.R.d.; Loureiro, G.A.H.d.A.; Baligar, V.C.; Ahnert, D.; Faria, J.C.; Valle, R.R. Cacao quality index for cacao agroecosystems in Bahia, Brazil. *Int. J. Food Prop.* **2019**, *22*, 1799–1814. [CrossRef]

foods

MDPI

Article

Effect of Moisture Content Difference on the Analysis of Quality Attributes of Red Pepper (*Capsicum annuum* L.) Powder Using a Hyperspectral System

Ji-Young Choi [1], Jeong-Seok Cho [1], Kee Jai Park [2], Jeong Hee Choi [1] and Jeong Ho Lim [1,*]

1 Food Safety and Distribution Research Group, Korea Food Research Institute, Wanju 55365, Republic of Korea
2 Smart Food Manufacturing Project Group, Korea Food Research Institute, Wanju 55365, Republic of Korea
* Correspondence: jhlim@kfri.re.kr; Tel.: +82-63-219-9331

Abstract: The variety of characteristics of red pepper makes it difficult to analyze at the production field through hyperspectral imaging. The importance of pretreatment to adjust the moisture content (MC) in the process of predicting the quality attributes of red pepper powder through hyperspectral imaging was investigated. Hyperspectral images of four types of red pepper powder with different pungency levels and MC were acquired in the visible near-infrared (VIS-NIR) and short-wave infrared (SWIR) regions. Principal component analysis revealed that the powders were grouped according to their pungency level, color value, and MC (VIS-NIR, Principal Component 1 = 95%; SWIR, Principal Component 1 = 91%). The loading plot indicated that 580–610, 675–760, 870–975, 1020–1130, and 1430–1520 nm are the key wavelengths affected by the presence of O-H and C-H bonds present in red pigments, capsaicinoids, and water molecules. The R^2 of the partial least squares model for predicting capsaicinoid and free sugar in samples with a data MC difference of 0–2% was 0.9 or higher, and a difference of more than 2% in MC had a negative effect on prediction accuracy. The color value prediction accuracy was barely affected by the difference in MC. It was demonstrated that adjusting the MC is essential for capsaicinoid and free sugar analysis of red pepper.

Keywords: red pepper powder; hyperspectral imaging; multivariate analysis; moisture adjustment

Citation: Choi, J.-Y.; Cho, J.-S.; Park, K.J.; Choi, J.H.; Lim, J.H. Effect of Moisture Content Difference on the Analysis of Quality Attributes of Red Pepper (*Capsicum annuum* L.) Powder Using a Hyperspectral System. *Foods* **2022**, *11*, 4086. https://doi.org/10.3390/foods11244086

Academic Editors: Mourad Kharbach and Samuli Urpelainen

Received: 30 November 2022
Accepted: 10 December 2022
Published: 17 December 2022

1. Introduction

Red pepper (*Capsicum annuum* L.) is a single crop belonging to the Solanaceae family. It has a spicy taste and red color [1] and it is often dried and processed into a powder and used as a spice for food additives [2]. Preference for the quality of red pepper is ultimately determined by the taste components (mixed with spicy, sweet, and other flavor components) contained in red pepper powder. Homologs of capsaicinoids, which are components of hot pepper, include capsaicin, dihydrocapsaicin, nordihydrocapsaicin, and glucose and fructose, which are reducing sugars, and are particularly closely related to the overall preference of red pepper powder. In particular, sweet flavor is negatively correlated with capsaicin content and stinging pain [3].

It is cultivated in different varieties and even in the same variety, and the capsaicinoid and sugar contents differ depending on the cultivation conditions, such as sunlight, precipitation, soil characteristics, or the difference in harvest time [4]. The survey report of the Consumers Federation of Korea (2013) noted that 80% of consumers responded that a label on the taste of red pepper powder is necessary, which affects product purchases. Therefore, real-time quality monitoring is required to label objective information on products [5]. High-performance liquid chromatography (HPLC) and gas chromatography/mass spectrometry (GC/MS) have been used to measure the content of capsaicinoid in red pepper [6–9]. However, these methods have some disadvantages as they are time-consuming, destructive, and lack capable real time detection systems. Among alternative methods, hyperspectral imaging (HSI) technology, which combines spectroscopy and cameras, can simultaneously provide spectral and spatial information regarding the external and internal qualities of agricultural products and are advantageous as they are fast, non-destructive, and cost-effective [10,11]. To enhance the applicability of HSI, chemometric methods such as principal component analysis (PCA) and partial least squares (PLS) regression are widely used for spectral analysis of foods with complex characteristics, because they offer better flexibility in conditions such as multicollinearity and when the number of variables exceeds the number of samples [12].

Previously, various spectroscopic trials and chemometrics were performed to analyze quality characteristics including capsaicinoids, free sugars, and moisture content of red pepper and red pepper powder [13–15]. Because the water content and particle distribution of the powder affect the light penetration depth and reflective ability, which influence spectroscopic signals such as any physical interference and chemical signals [16–18], it has been conjectured that ensuring uniformity can improve the measurement accuracy of components such as capsaicinoid in red pepper powder [19]. Compared with sieving to make the particle size of red pepper powder uniform, it is practically difficult to apply the manufacturing process to ensure that the water content is the same in the field. Therefore, by confirming the prediction accuracy according to the range of the difference in moisture content between samples, no previous study has shown the need for moisture control in the spectroscopic analysis of red pepper powder or established the moisture distribution conditions for sample preparation.

In this study, the moisture content of red pepper powder with different levels of spiciness produced in Gochang-gun, Shintaein-eup, Gwanchon-myeon, and Jeongeup-si was adjusted to 7, 8, 9, 10, 11, and 12%, respectively. By extracting Vis-NIR (400–1000 nm) and SWIR (900–1700 nm) image spectrum information and performing multivariate analysis, the capsaicinoid content, free sugar content, and color prediction accuracy of red pepper powder were determined according to the range of moisture content difference (7–8%, 7–9%, 7–10%, 7–11%, and 7–12%). It was hypothesized that this process would be able to prove the extent of which the moisture content difference has a high reliability for each quality prediction model. This study provides a basis for application in the field of red pepper powder production by overcoming the limitations of hyperspectral image analysis, which is strongly influenced by the bonding of water molecules. It can be a useful reference for determining the range of moisture content in samples in hyperspectral analysis studies of various agricultural foods, as well as red pepper powder.

2. Materials and Methods

2.1. Sample Preparation

Red peppers produced in Gochang-gun (GC), Sintaein-eup (ST), Kwanchon-myeon (KC), and Jeongeup-si (JU) regions, Jeollabuk-do, Korea, in 2021 were purchased as samples. Red peppers were ground after hot air drying (50–60 °C), and the particle size of the red pepper powder was uniformly prepared with a particle size of 425–850 μm using a standard sieve. Samples were prepared based on the particle size of red pepper powder for seasoning, which is most commonly used in Korea, according to Korean Industrial standards (KS). To ensure that the moisture content of each sample was 7%, 8%, 9%, 10%, 11%, and 12%, the following process was performed. KS presents less than 13% as the appropriate moisture content of red pepper powder, and the average moisture content of red pepper powder sold on the market is 7–12%.

First, the moisture contents of the GC, ST, KC, and JU powder samples were measured using the atmospheric pressure drying method in a drying oven. The samples were dried at 100 °C for 4 h, and the moisture content was calculated using the weight differences before and after drying. The moisture contents of the GC, ST, KC, and JU were 11.12%, 11.36%, 10.12%, and 11.09%, respectively. To adjust the initial moisture content to 12%, it was necessary to seal the samples in plastic bags and humidify by spraying additional 10.75 mL, 7.81 mL, 22.57 mL, and 11.08 mL of water on 950 g of GC, ST, KC, and JU samples.

75 g of the red pepper powder whose moisture content was adjusted to 12% was dried in a dry oven set at 55 °C, and the weight of the red pepper powder was measured every 15 min. A graph was prepared as shown in Figure 1. The moisture content was calculated as the change between the initial weight and the weight after drying, and red pepper powder samples with moisture contents of 7%, 8%, 9%, 10%, 11%, and 12% were prepared. According to the production area and moisture content of the samples, Gochang samples were GC7, GC8, GC9, GC10, GC11, and GC12, Shintaein samples were ST7, ST8, ST9, ST10, ST11, ST12. Kwanchon samples were KC7, KC8, KC9, KC10, KC11, and KC12 and Jeongeup samples were named JU7, JU8, JU9, JU10, JU11, and JU12.

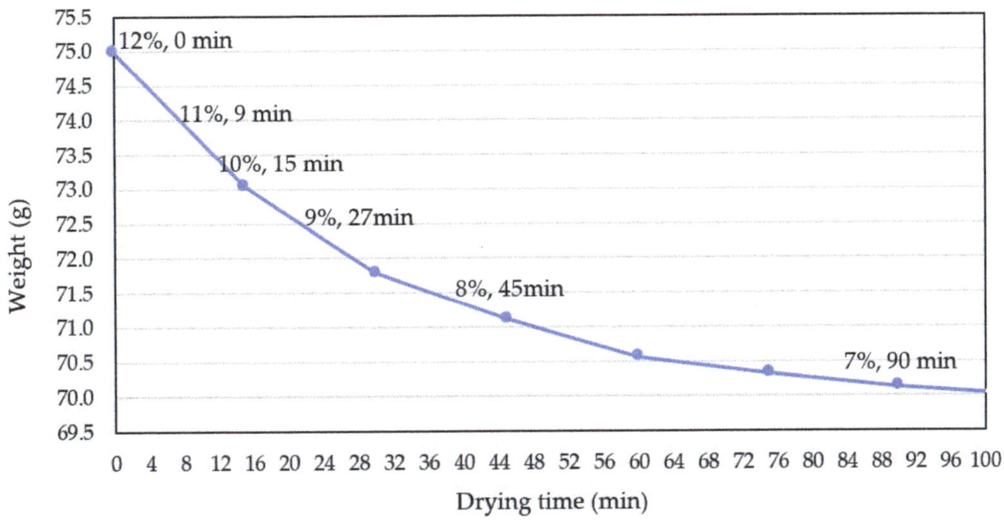

Figure 1. Weight change of red pepper powder according to drying time and calculated moisture content.

2.2. *Determination of Quality Indicators*

To analyze the capsaicinoid and free sugar content of the samples, pretreatment was required to make the particle size uniform. The samples were finely ground using a food mixer (SNSG-1002SS, Hanil Electric, Seoul, Korea), filtered through a 30 mesh sieve (pore size, 0.6 mm), and then used for analysis.

2.2.1. Moisture Content Measurement and American Spice Trade Association (ASTA) Color

The moisture content of the red pepper powder was measured by drying for 6 h in a vacuum oven dryer (OV-11, Jeio Tech, Daejeon, Republic of Korea) set at 70 °C, according to ASTA analytical method 2.1. The ASTA color value measurement method was based on AOAC official method 971.26, and acetone was filled in 0.1 g of the sample, shaken for 1 min, and left in the dark for 16 h to prepare a test solution. The absorbance of the test solution was measured at 460 nm using a UV spectrophotometer (Thermo Fisher Scientific, Vantaa, Finland), and the results were substituted into the equation below to calculate the ASTA color value.

$$\text{ASTA value} = \frac{\text{A} \times 16.4}{\text{W}} \tag{1}$$

A: absorbance at 460 nm; W: sample weight (g).

2.2.2. Capsaicinoid

Capsaicin and dihydrocapsaicin contents were analyzed by referring to the methods of Ku et al. [20] and Namgung et al. [21]. The extraction method for capsaicinoid analysis was as follows: Methanol (10 mL) and a boiling chip were added to 2 g of the sample and heated on a dry heating block (MaXtable H10, Daehan, Incheon, Korea) set at 90 °C for 1 h, and then cooled to room temperature. The extract was filtered with Whatman No. 1 and then filtered again with a 0.2 μm syringe filter. Capsaicinoid content was analyzed using an HPLC system (Agilent 1260 infinity II, Agilent Technology, Santa Clara, CA, USA). An XTerraTMRP18 (5 μm, 4.6 × 150 mm id., Waters, Milford, MA, USA) column was used, and the mobile phase (A: acetic acid, B: acetonitrile) was applied in a gradient method (A: B = 60:40, 38:62, and 20:80) at a rate of 1 mL/min. The column temperature was set at 35 °C and the injection volume was 10 μL. A variable-wavelength detector was used, and the absorbance was measured at 280 nm. Capsaicin and dihydrocapsaicin were used as standards to prepare calibration curves.

2.2.3. Free Sugar

The free sugar content of the red pepper powder was analyzed by high-performance liquid chromatography (HPLC, Agilent 1260 infinity II, Agilent Technology, CA, USA). 40 mL of 80% ethanol was added to 2 g of the sample, extracted for 1 min with a vortex mixer, filtered through a 0.2 μm membrane filter, and 20 μL was injected into the 1260 II Infinity HPLC-Refractive Index (RI) detector for analysis. Fructose, glucose, and sucrose (Sigma-Aldrich, St. Louis, MO, USA) dissolved in 80% ethanol were used as the standards. For the mobile phase, a solvent mixture of acetonitrile and water at a ratio of 75:25 (v/v) was separated in the isocratic mode at a flow rate of 1 mL/min. The column temperature was set to 30 °C, and the temperature of the RI detector was set to 35 °C. All analysis processes were performed by referring to the methods of Ku et al. [2].

2.2.4. Statistics Analysis

All experimental measurements of 24 samples were performed three times, and the results are presented as means and standard deviations ($n = 72$, mean ± SD). The results were analyzed by ANOVA and Duncan's multiple range test ($p < 0.05$) using the SPSS software package (version 20, IBM SPSS Statistics, Inc., Chicago, IL, USA).

2.3. *Hyperspectral Image Analysis*

2.3.1. Hyperspectral Image Acquisition and Data Extraction

Hyperspectral images in the VIS-NIR region (400–1000 nm) were acquired using the line scan method (pushbroom) using a SPECIM FX10 spectrometer (Spectral Imaging Ltd., Oulu, Finland) equipped with three halogen light sources. It was operated by obtaining the reflection intensity from the sample, and image data with a spectral resolution of 1.3 nm were acquired for a total of 448 bands. A white plate made of polytetrafluoroethylene and the sample were scanned together, and the acquired image was normalized using the IDL Virtual Machine Application program (8.8.0, L3Harris Geospatial, Boulder, CO, USA).

HSI data of the red pepper powders were acquired using an ImSpector N17E (Specim, Spectral Imaging Ltd., Oulu, Finland) in the short-wave infrared (SWIR) region, 900–1700 nm. The light source consisted of two halogen lamps (1400 nm long-pass filter). The system consisted of an NIR camera with an indium gallium arsenide (InGaAs) sensor operated in reflectance mode with line-by-line scanning (pushbroom) to obtain intensity images at 5 nm intervals through a 30 μm slit (256 images per scene). A white plate was used as the reference material and was scanned before each sample was scanned. The samples were scanned line-by-line along the *Y*-axis and moved along the *X*-axis to obtain a three-dimensional (3D) hypercube containing both spatial and spectral information.

The powder (3.5 g) was placed in a transparent Petri dish (5 cm diameter) and spread flat to cover the bottom of the Petri dish. To reduce the diffuse reflection that may have been caused by the particle surface, the surface of the sample was compressed with a presser to make it as level as possible. Fifty hyperspectral images were acquired per sample for a total of 1200 images. All the hyperspectral imaging systems were operated using Microsoft Windows. To obtain the necessary information from the acquired images, image spectrum data for the inner area of the Petri dish were obtained using the region of interest function of the ENVI (version 5.4, Exelis Visual Information Solutions, Boulder, CO, USA) program.

2.3.2. Chemometrics

Chemometrics is a method of high-level interpretation of one-dimensional data obtained through chemical analysis using computers, mathematics, and statistics and was used in this study to link quality-related factors and measurement technology. Multivariate statistical analysis consists of unsupervised learning, which finds data patterns or relationships between data when the characteristics of the data are unknown, and supervised learning, which predicts results by finding the optimal model by learning through an algorithm set with input and output values [22].

In this study, principal component analysis (PCA), a representative unsupervised learning method, was performed to visualize the overall clustering tendency according to the sourness and moisture content of the red pepper powder samples. Two-dimensional and three-dimensional PCA score plots were derived from the spectral data in the 400–1000 nm and 900–1700 nm regions. As the number of principal components increases, overfitting occurs, and the reliability of the predictive model decreases [23], so the maximum principal component was set to 7. Principal component analysis was performed using Unscrambler statistics program (version 10.5, CAMO, Trondheim, Norway).

To predict capsaicinoid content, partial least squares regression (PLSR) analysis, a supervised learning method, was attempted. The PLS statistical method combines the functions of principal component analysis and multiple regression analysis and aims to predict the independent variable by expressing the relationship between the predictor variable X (spectral data) and the independent variable Y (measured capsaicinoid content) in a linear model [24]. The predicted value of Y was calculated using the following equation:

$$Y = \beta X + b \tag{2}$$

β: vector of regression coefficient; b: model offset.

The PLS model showed more stable characteristics than the principal component model, considering only the independent variables. Of the total spectral data, 70% were used to develop the calibration model, and the remaining 30% were used for testing to verify the developed model. To evaluate the performance of all developed PLS models, the coefficient of determination (R_c^2) in the calibration model, coefficient of determination (R_v^2) in the cross-validation model, root mean square error of calibration (RMSEC), cross-validation model, and root mean square error of validation (RMSEV) value were considered. Table 1 shows the PLS model names developed in this study and the data samples (spectral and physicochemical data) inserted into each model. The entire model developed using samples with uniform moisture content was named Model A, and the entire model developed with samples having different moisture contents was named Model B.

Table 1. Developed partial least square model.

Model Name		N	Inserted Data
Model A	A7	200	GC7, ST7, KC7, JU7
	A8	200	GC8, ST8, KC8, JU8
	A9	200	GC9, ST9, KC9, JU9
	A10	200	GC10, ST10, KC10, JU10
	A11	200	GC11, ST11, KC11, JU11
	A12	200	GC12, ST12, KC12, JU12

Table 1. *Cont.*

Model Name		N	Inserted Data
Model B	B7-8	400	GC7, ST7, KC7, JU7, GC8, ST8, KC8, JU8
	B7-9	600	GC7, ST7, KC7, JU7, GC8, ST8, KC8, JU8, GC9, ST9, KC9, JU9
	B7-10	800	GC7, ST7, KC7, JU7, GC8, ST8, KC8, JU8, GC9, ST9, KC9, JU9, GC10, ST10, KC10, JU10
	B7-11	1000	GC7, ST7, KC7, JU7, GC8, ST8, KC8, JU8, GC9, ST9, KC9, JU9, GC10, ST10, KC10, JU10, GC11, ST11, KC11, JU11
	B7-12	1200	GC7, ST7, KC7, JU7, GC8, ST8, KC8, JU8, GC9, ST9, KC9, JU9, GC10, ST10, KC10, JU10, GC11, ST11, KC11, JU11, GC12, ST12, KC12, JU12

3. Results and Discussion

3.1. Quality Indicators Analysis and Correlation between Physicochemical Properties

Table 2 shows the analysis results of physicochemical characteristics of red pepper powder. The moisture content showed an error of 0.42–8.00% compared to the intended moisture content, but it was confirmed that the sample was prepared with an increase in moisture content with an R^2 of 0.99 or more. The capsaicinoid content of the red pepper powders is listed in Table 1, indicating that the capsaicin content of all samples was higher than the dihydrocapsaicin content. The pungent substances in red pepper are capsaicin homologs, and the main components of capsaicinoids are capsaicin, dihydrocapsaicin, and nodihydrocapsaicin, each at approximately 70%, 21–40%, and 2–12% composition, respectively [25]. For total capsaicinoid content, GC ranged from 156.80–165.57 mg/kg, ST ranged from 252.14–269.10 mg/kg, KC ranged from 510.44–544.65 mg/kg, and JU ranged from 676.04–731.92 mg/kg. According to the Korean Industrial Standard, GC and ST are classified as 'Slight Hot' and KC and JU as 'Medium Hot'. There was a slight difference in the capsaicin, dihydrocapsaicin, and total capsaicinoid content depending on the water content, but no significant differences were observed.

Park et al. [26] and Choi et al. [3] stated that fructose and glucose account for 70% of the total sugars in red pepper, and the sweetness of red pepper is in the order of fructose, glucose, and sucrose. All red pepper powders were composed of free sugars in the order of fructose > glucose > sucrose content, and the free sugar content was not affected by the water or capsaicinoid content of red pepper powder.

The American Spice Trade Association (ASTA) color values were calculated as 83.90–86.92 for JU, 75.95–79.14 for GC, 62.93–65.75, ST, and 57.72–59.65 for KC. JU, GC, ST, and KC were dark red. The ASTA color, which is a criterion for the color of red pepper powder [2] and the pigment content of red pepper powder are known to fluctuate depending on the variety, cultivation area, and drying method, such as sun drying and hot air drying [27–29]. Therefore, it is difficult to determine the degree of spiciness and sweetness by observing the appearance of red pepper powder with the naked eye without analysis.

The pungency components, including capsaicin, dihydrocapsaicin, and capsaicinoid, showed a low correlation with moisture content, ASTA value, and free sugars (fructose, glucose, sucrose, and total free sugar) indicating that there was no significant effect on pungency level. Therefore, when predicting the pungency level of red pepper powder using spectral information, it is proven that pungency components can show independent spectral characteristics without mutual influence between physicochemical characteristics.

Table 2. Physicochemical properties of red pepper powder according to pungency level and moisture contents.

Sample [(1)]	Moisture Content (%)	ASTA Value	Capsaicinoid (mg/kg)			Free Sugar (%)			
			Capsaicin	Dihydrocapsaicin	Total	Fructose	Glucose	Sucrose	Total
GC7	7.24 ± 0.06 [b(2)]	77.33 ± 0.16 [f]	98.04 ± 2.66 [a]	64.84 ± 1.37 [a]	162.89 ± 4.00 [a]	6.86 ± 0.02 [ab]	3.19 ± 0.08 [ab]	0.95 ± 0.02 [i]	11.00 ± 0.12 [ab]
GC8	8.29 ± 0.05 [e]	76.37 ± 0.88 [f]	96.84 ± 2.65 [a]	63.62 ± 0.97 [a]	160.46 ± 3.59 [a]	6.88 ± 0.04 [ab]	3.17 ± 0.09 [ab]	0.98 ± 0.01 [ij]	11.03 ± 0.09 [ab]
GC9	9.35 ± 0.05 [h]	77.27 ± 0.88 [f]	99.30 ± 0.65 [a]	66.26 ± 0.38 [a]	165.57 ± 1.03 [a]	7.09 ± 0.05 [bc]	3.23 ± 0.05 [b]	1.00 ± 0.02 [jk]	11.32 ± 0.06 [b]
GC10	10.39 ± 0.04 [k]	77.25 ± 0.31 [f]	94.41 ± 1.02 [a]	63.58 ± 0.28 [a]	158.99 ± 1.17 [a]	6.92 ± 0.06 [ab]	3.10 ± 0.04 [ab]	1.01 ± 0.01 [jk]	11.03 ± 0.09 [ab]
GC11	11.54 ± 0.05 [m]	79.14 ± 0.44 [g]	94.35 ± 2.07 [a]	63.85 ± 1.76 [a]	158.20 ± 3.80 [a]	6.95 ± 0.25 [ab]	3.03 ± 0.16 [ab]	0.99 ± 0.03 [jk]	10.97 ± 0.44 [ab]
GC12	12.35 ± 0.08 [p]	75.95 ± 2.99 [f]	93.51 ± 0.63 [a]	63.28 ± 0.76 [a]	156.80 ± 1.26 [a]	6.72 ± 0.03 [a]	2.98 ± 0.04 [ab]	1.00 ± 0.02 [jk]	10.69 ± 0.09 [a]
ST7	7.13 ± 0.02 [a]	64.52 ± 0.19 [de]	159.22 ± 2.17 [d]	105.91 ± 1.79 [bc]	265.13 ± 3.96 [cd]	9.09 ± 0.14 [ij]	5.11 ± 0.08 [j]	0.62 ± 0.03 [a]	14.82 ± 0.24 [hi]
ST8	8.16 ± 0.02 [d]	64.39 ± 0.63 [de]	160.98 ± 4.65 [d]	108.13 ± 3.55 [c]	269.10 ± 8.16 [d]	9.14 ± 0.14 [ij]	5.13 ± 0.06 [j]	0.66 ± 0.02 [bc]	14.94 ± 0.21 [hi]
ST9	9.41 ± 0.04 [h]	62.93 ± 0.48 [c]	157.00 ± 1.57 [cd]	105.67 ± 0.61 [bc]	262.68 ± 2.14 [bcd]	8.82 ± 0.15 [efghi]	4.84 ± 2.00 [ghi]	0.64 ± 0.01 [ab]	14.30 ± 0.25 [defgh]
ST10	10.34 ± 0.05 [k]	63.49 ± 0.42 [cd]	159.82 ± 0.25 [d]	108.12 ± 0.36 [c]	267.94 ± 0.56 [d]	9.15 ± 0.21 [ij]	5.00 ± 1.57 [ij]	0.68 ± 0.01 [c]	14.83 ± 0.28 [hi]
ST11	11.30 ± 0.06 [l]	65.75 ± 0.96 [e]	149.91 ± 3.58 [b]	102.24 ± 2.39 [b]	252.14 ± 5.97 [b]	8.96 ± 0.12 [hij]	4.82 ± 1.59 [fghi]	0.67 ± 0.03 [bc]	14.46 ± 0.22 [fghi]
ST12	12.05 ± 0.02 [o]	65.30 ± 0.41 [e]	150.40 ± 1.74 [bc]	102.73 ± 1.38 [bc]	253.13 ± 3.12 [bc]	9.17 ± 0.13 [ij]	4.93 ± 1.92 [hij]	0.68 ± 0.01 [c]	14.79 ± 0.22 [hi]
KC7	7.56 ± 0.13 [c]	59.91 ± 0.71 [b]	280.00 ± 5.44 [g]	251.80 ± 5.38 [fg]	531.80 ± 10.81 [fg]	9.79 ± 0.10 [k]	5.39 ± 0.08 [k]	0.74 ± 0.01 [e]	15.92 ± 0.17 [j]
KC8	8.64 ± 0.06 [f]	59.65 ± 0.38 [b]	285.14 ± 3.05 [gh]	257.28 ± 2.30 [hi]	542.42 ± 5.34 [gh]	9.27 ± 0.41 [j]	5.06 ± 0.24 [j]	0.74 ± 0.04 [de]	15.07 ± 0.69 [i]
KC9	9.18 ± 0.02 [g]	59.45 ± 0.38 [b]	286.09 ± 3.84 [h]	258.56 ± 3.00 [hi]	544.65 ± 6.84 [h]	8.55 ± 0.13 [def]	4.59 ± 0.08 [de]	0.70 ± 0.01 [cd]	13.84 ± 0.21 [de]
KC10	10.19 ± 0.04 [j]	59.50 ± 0.36 [b]	278.72 ± 10.04 [fg]	251.81 ± 9.08 [fg]	530.53 ± 19.12 [fg]	8.58 ± 0.06 [defg]	4.57 ± 0.02 [de]	0.71 ± 0.01 [cde]	13.87 ± 0.06 [def]
KC11	11.37 ± 0.03 [l]	57.72 ± 0.35 [a]	267.65 ± 4.85 [e]	242.79 ± 3.87 [d]	510.44 ± 8.72 [e]	8.53 ± 0.08 [de]	4.44 ± 0.05 [d]	0.74 ± 0.01 [e]	13.72 ± 0.13 [d]
KC12	12.08 ± 0.04 [o]	59.27 ± 0.37 [b]	272.20 ± 2.01 [ef]	247.76 ± 1.69 [def]	519.96 ± 3.70 [ef]	8.91 ± 0.05 [fghij]	4.65 ± 0.05 [defg]	0.79 ± 0.02 [f]	14.35 ± 0.12 [efgh]
JU7	7.54 ± 0.03 [c]	85.24 ± 0.49 [hi]	459.26 ± 5.14 [k]	262.04 ± 2.81 [ij]	721.30 ± 7.95 [k]	8.83 ± 0.09 [efghi]	4.74 ± 0.06 [efgh]	1.01 ± 0.03 [jk]	14.58 ± 0.16 [ghi]
JU8	8.36 ± 0.02 [e]	85.66 ± 0.44 [ij]	465.90 ± 5.59 [k]	266.02 ± 3.13 [j]	731.92 ± 8.71 [k]	8.93 ± 0.03 [ghij]	4.77 ± 0.06 [efgh]	1.03 ± 0.01 [k]	14.73 ± 0.08 [hi]
JU9	9.56 ± 0.04 [i]	85.47 ± 0.98 [i]	438.91 ± 2.58 [j]	250.88 ± 1.95 [efg]	689.79 ± 4.53 [j]	8.69 ± 0.29 [defgh]	4.58 ± 0.22 [de]	1.01 ± 0.04 [jk]	14.29 ± 0.53 [defgh]
JU10	10.24 ± 0.03 [j]	86.92 ± 0.64 [j]	430.31 ± 3.54 [i]	245.73 ± 1.82 [de]	676.04 ± 5.37 [i]	8.43 ± 0.56 [d]	4.61 ± 0.33 [def]	0.92 ± 0.01 [h]	13.96 ± 0.90 [defg]
JU11	11.66 ± 0.03 [n]	85.95 ± 0.67 [ij]	461.74 ± 6.55 [k]	264.00 ± 3.25 [j]	725.75 ± 9.79 [k]	8.55 ± 0.21 [def]	4.46 ± 0.11 [d]	1.03 ± 0.01 [k]	14.04 ± 0.28 [defg]
JU12	12.55 ± 0.08 [q]	83.90 ± 0.38 [h]	443.85 ± 5.30 [j]	253.51 ± 2.77 [gh]	697.36 ± 8.07 [j]	7.40 ± 0.19 [c]	3.82 ± 0.09 [c]	0.88 ± 0.03 [g]	12.10 ± 0.26 [c]

[(1)] GC, red pepper powder produced in Gochang-gun; ST, red pepper powder produced in Sintaein-eup; KC, red pepper powder produced in Kwanchon-myeon; JU, red pepper powder produced in Jeongeup-si. [(2)] Mean ± standard deviation (n = 3) with different superscript letters is significantly different at 5% level.

3.2. Spectral Characteristics

Figure 2 shows the hyperspectral mean spectra obtained from the GC, KC, ST, and JU red pepper powders with different pungency levels and moisture contents. Red pepper powder is composed of 50–60% carbohydrates, 10–15% crude protein, 10% crude fat, and 5% ash [30]. Therefore, as a result of observing the spectra, the shapes of all spectra were similar, except for the difference in the overall reflection intensity depending on the sample. In the observation of the characteristics of the average reflectance spectrum in the VIS-NIR region without any chemometrics analysis (Figure 2A), the reflectance intensity was relatively low in the sample with high moisture content, whereas differences in reflectance by pungency level, ASTA color, and free sugar were not observed.

Red pepper powder absorbs light at approximately 1130, 1200, 1425–1440, and 1515 nm in the SWIR band (Figure 2B), which is similar to the results reported by Mo et al. [4]. Each peak represents the 2nd overtone region of the CH bond (1200 nm), 1st overtone combination of CH and OH bonds (1425 nm) and 1st overtone of the NH bond (1520 nm), respectively [31–34].

In the band of approximately 1410–1540 nm, which is common in GC, ST, KC, and JU, the reflectance intensity was low in samples with high moisture content, and it seems that the absorption phenomenon was strengthened by a large number of OH bonds. However, since it is difficult to quantify the sweetness and spiciness of red pepper only by observing the average spectrum, additional chemometrics analysis is required. Therefore, by attempting multivariate analysis of hyperspectral data, there is a possibility of evaluating the quality of red pepper powder and expressing it numerically.

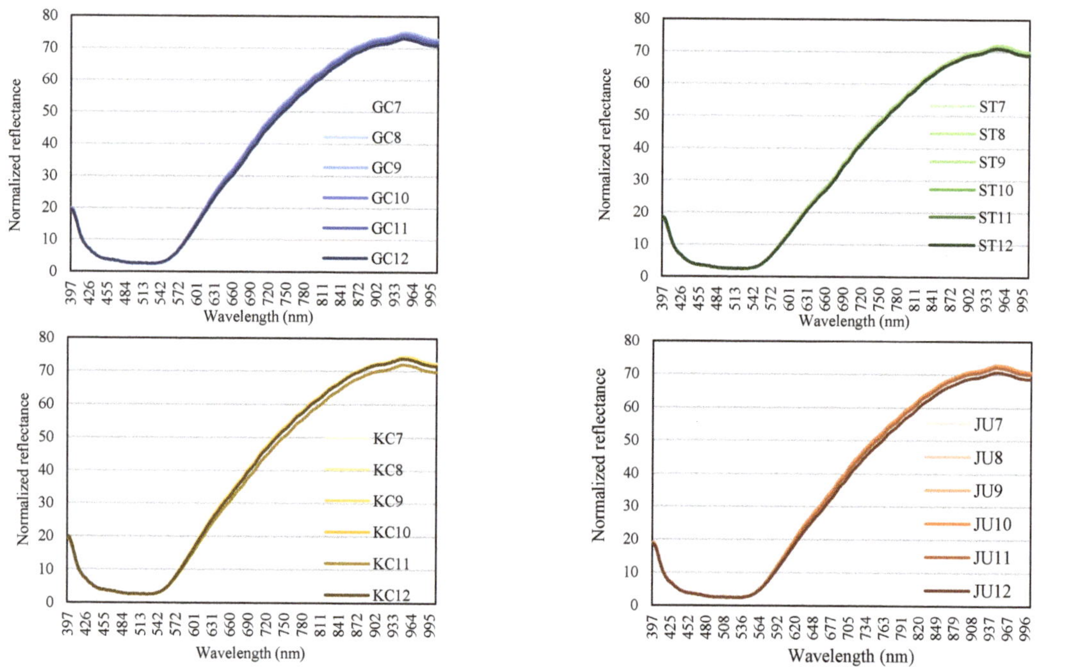

(**A**) Mean spectra of red pepper powders in VIS-NIR band (400–900 nm)

Figure 2. *Cont.*

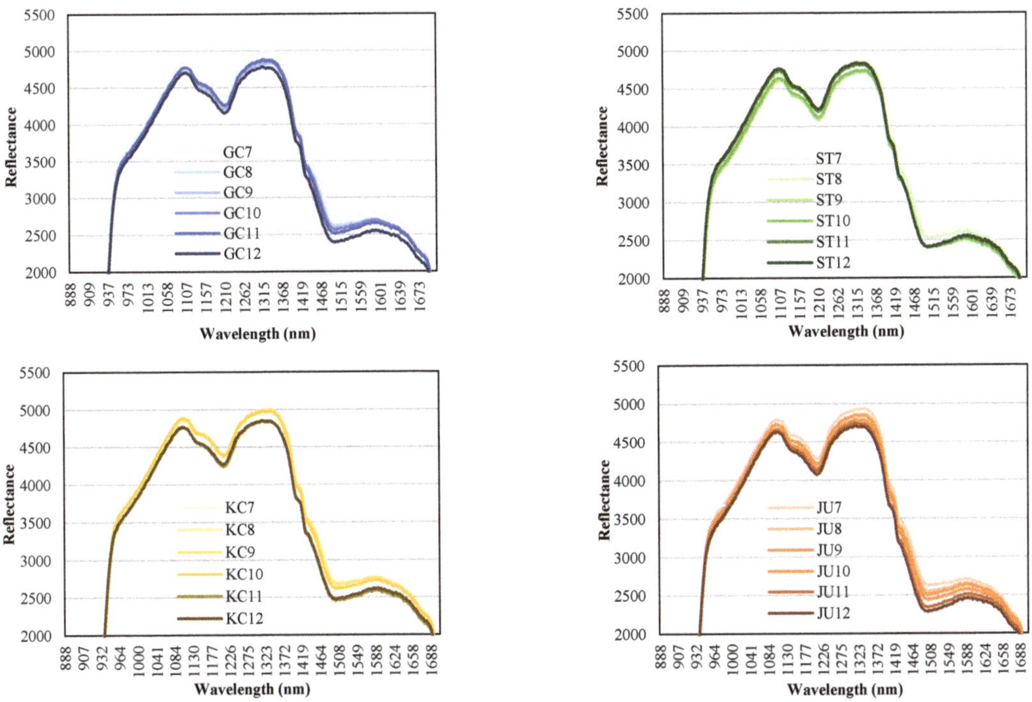

(B)Mean spectra of red pepper powders in SWIR band (900–1700 nm)

Figure 2. Mean spectra of red pepper powders in the Vis-NIR (**A**) and SWIR (**B**) wavelength ranges according to pungency levels and moisture contents.

3.3. Chemometrics

3.3.1. Principal Component Analysis

The original reflectance spectral data matrix was reduced to a system of coordinate axes, where samples were located according to principal component analysis (PCA) scores instead of intensities in the wavelength space [35]. Therefore, samples with similar spectral properties tend to project to the same location in principal component space. A clear differentiation according to capsaicinoid content and moisture content is indicated in the PCA score plots shown in Figure 3, which are expressed in two dimensions and three dimensions by the principal component factors based on the hyperspectral spectra. In the score plot, GC is shown in blue, ST in green, KC in yellow, and JU in orange; the higher the moisture content, the darker the color. PC−1, PC−2, and PC−3 contributed 95%, 3%, and 1% of the hyperspectral image data of red pepper powder obtained in the VIS-NIR region, respectively (Figure 3A,B). As indicated by the dotted circle, it is clearly classified according to the production area of red pepper powder, which may mean that it is classified according to the degree of spiciness or ASTA color; therefore, additional interpretation is needed through the loading plot result. In addition, the distribution of darker markers closer to the upper left corner of the score plot indicates that PCA analysis using hyperspectral data in the VIS−NIR region can visually show the difference in the moisture content of red pepper powder.

PCA results of the SWIR region showed that the first principal component (PC1) and the second principal component (PC2) accounted for 91% and 6% of the spectral variance, respectively. Because the first two principal components can explain 97% of the data, this data reveals the high feasibility of discrimination among red pepper powders. In the two-dimensional plot, it was sequentially distributed according to the moisture content, which can be the basis for the hyperspectral spectrum to represent the relative moisture content distribution of red pepper powder. In the three-dimensional plot, separate grouping was performed according to the sample and moisture content. Therefore, PCA analysis using hyperspectral data in the SWIR region can be a method that can effectively show the difference in the distribution of moisture content and other quality characteristics of red pepper powder. This plot only demonstrates the qualitative differences between the examined samples without referring to their quantitative attributes [35].

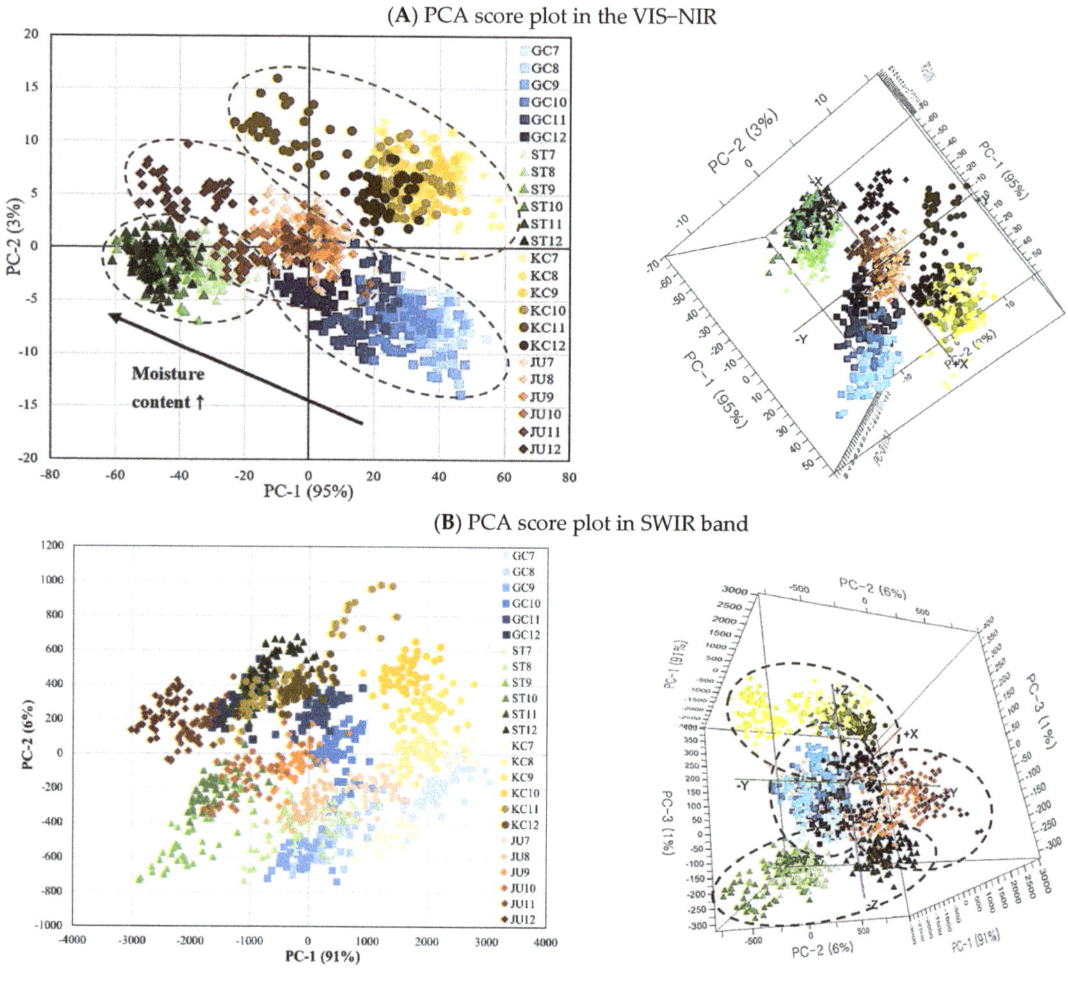

Figure 3. PCA score plot of hyperspectral spectra in the VIS−NIR (**A**) and SWIR band (**B**).

3.3.2. Loading Plot

The first two PCs accounted for 97% or more of the spectral variation in the tested samples; therefore, these five PCs can be used as alternatives to the variables for the classification of red pepper powder (Figure 4). In this study, to identify the key wavelengths that are highly correlated with each PC for VIS−NIR and SWIR systems, the PC loadings were plotted against their spectral ranges, and all characteristic wavelengths were marked. PC loading can be used to identify wavelengths highly correlated with each PC [36]. In addition, the PCA results of the spectral data of all tested red pepper powder spectra loadings are the regression coefficients for each wavelength in each principal component, indicating which wavelength has a dominant effect on identification.

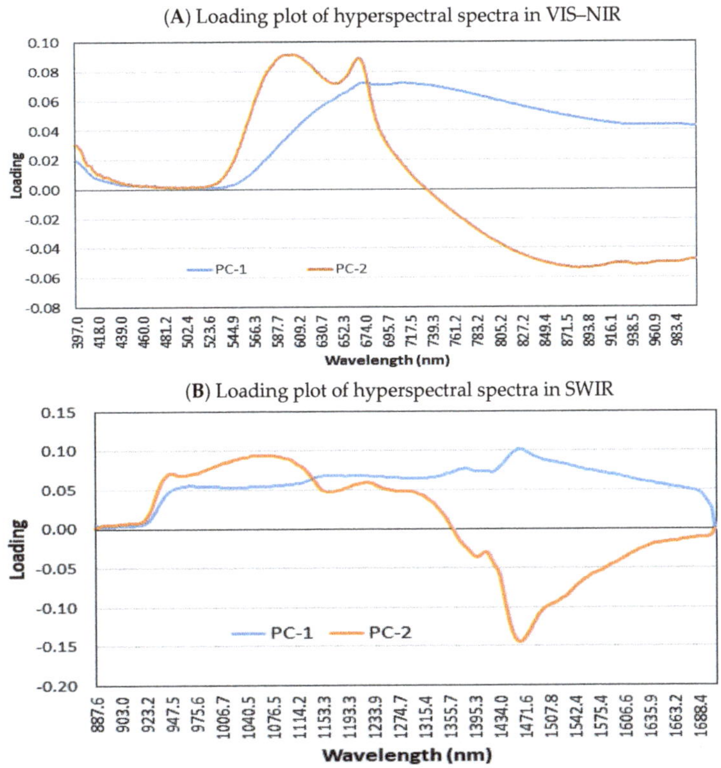

Figure 4. Loading plot of PC1 and PC2 derived from PCA of hyperspectral spectra in VIS–NIR (**A**) and SWIR band (**B**).

As a result of observing the PCA loading plot of VIS-NIR data (Figure 4A), PC1 explained 95% of the total variance in the samples. Key wavelengths (675–760 nm) were shown from this component, and key peaks were observed in the 580–610 nm, 675 nm, and 870–970 nm bands from PC2. Among the key wavelengths (580–610, 675–760, 870–975 nm) shown by PCA loadings, a peak observed in the red region (675 nm) might also be related to the presence of carotenoids [37]. The high absorbance observed at 625–740 nm is associated with red absorbing pigments, mainly chlorophyll absorption [38,39]. Absorption at 750 and 974 nm is due to water absorption bands related to O–H stretching second overtones [40,41]. Owing to the obvious difference in ASTA color value and moisture content between the samples in Table 1, VIS–NIR spectroscopic images can be used to compare the moisture content and color of red pepper powder.

As a result of observing the PCA loading plot of the SWIR data, PC–1 showed a prominent peak only at 1460 nm, and PC-2 showed peaks at 1020–1130 nm and 1430–1520 nm (Figure 4B). Capsaicin and dihydrocapsaicin are alkaloids with molecular formulas of $C_{18}H_{27}NO_3$ and $C_{18}H_{29}NO_3$, respectively, and the capsaicin molecule can be divided into three regions: aromatic rings, amide bonds, and hydrophobic side chains [42]. The chemical bonds that are read include O–H str. 1st overtone was detected in the wavelength range of 1395–1452 nm and this chemical bond in the form of the C–H stretch 1st overtone is due to the presence of aromatic and alkene functional groups, which are also known to be constituents of capsaicin [34]. The 2nd overtone occurred because of the presence of a hydroxyl group (-OH) derived from several sources of antioxidants in red chili, such as capsanthin and capsaicin.

Therefore, it is foreseen that wavelengths at water absorption bands and capsaicinoid absorption bands are important for discrimination of pungency level and moisture content within each red pepper powder.

3.3.3. Prediction of Quality Attribute in Red Pepper Powder

The prediction results of capsaicinoid, free sugar, and ASTA color by PLS modeling in VIS-NIR and SWIR are shown in Figures 5 and 6. The average R_p^2 of Model A in VIS-NIR for capsaicinoid was 0.98, and the average R^2 value decreased to approximately 0.92 in B7-10, B7-11 and B7-12 models, respectively: A decrease in R_p^2 of approximately 5.9% occurred. The SWIR R_p^2 values of the B7-10, B7-11, and B7-12 prediction models for the capsaicinoid were 0.85–0.87, a decrease of approximately 8.7% from the average R_p^2 value of A7–A12. Referring to Figure 4A, the loading peaks at 590 nm and 670 nm, which can explain the red color, were about 0.04 higher than those at 750 nm and 970 nm related to moisture. On the other hands, there is a peak that stands out more than other bands at 1450 nm where the vibration of OH bond in water molecules is revealed in Figure 4B. Therefore, the SWIR spectra were more sensitive to the moisture content of the sample compared to VIS-NIR spectra, which hindered the prediction of capsaicinoid content by difference of water contents.

The modeling results for free sugars are as follows. In Figure 5, the prediction Model A with uniform moisture content had an R_p^2 value of 0.96 or more. However, R_p^2 decreased in the order of B7-8 (0.94), B7-9 (0.90), B7-10 (0.90), B7-11 (0.85), and B7-12 (0.80) models. In Figure 6, it can be observed that the average R_p^2 of Model A is 0.951, whereas that of Model B is 0.839, a decrease of about 12%. As shown in Figure 5, the fact that the R_p^2 value did not decrease sequentially can be interpreted as a slight error according to the resolution of the SWIR system itself and the number of measurement bands. As a result, it means that the adjustment of the water content of the sample has a significant effect on the accuracy of the PLS model in predicting the free sugar content in both the VIS-NIR and SWIR regions.

The training, and prediction model of the ASTA color value in VIS-NIR maintained an R_c^2, R_{cv}^2 and R_p^2 of 0.97 or more regardless of the moisture content distribution. In the SWIR region, it was observed that the R^2 values of the B7-11 and B7-12 models slightly decreased below 0.95 in the ASTA prediction model, but the prediction accuracy was still high. Although capsanthin, zeaxanthin, cryptoxanthin, and betacarotene are responsible for the red color in red pepper powders [43], the use of VIS-NIR region, which was based on the external color values of red peppers was better for developing the prediction model of ASTA color value than the use of SWIR region, which was based on the chemical structure of red peppers by water molecules (OH bond). Therefore, the hyperspectral imaging system is more useful and convenient for estimating ASTA values because there is less need to adjust the moisture content of the sample.

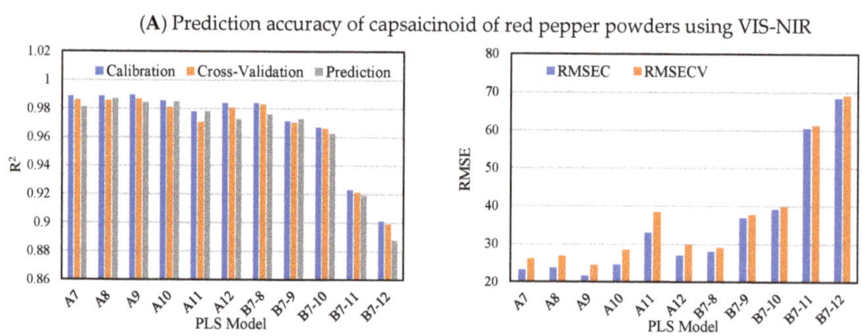

Figure 5. *Cont.*

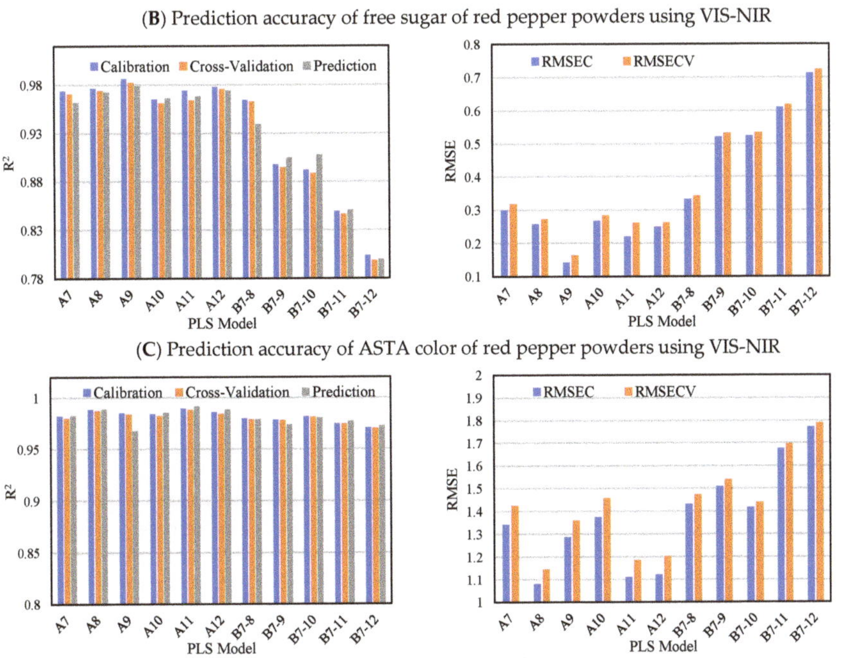

Figure 5. Prediction accuracy of capsaicinoid (**A**), free sugar (**B**) and ASTA color (**C**) of red pepper powders using VIS-NIR wavelength range in accordance with moisture content. RMSEC, root mean square error of calibration; RMSECV, root mean square error of cross-validation.

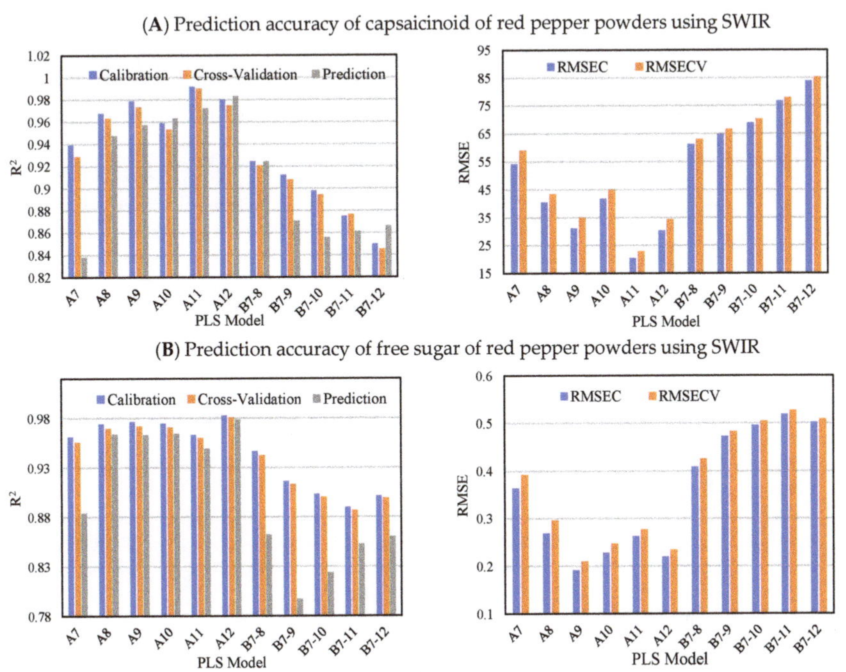

Figure 6. *Cont.*

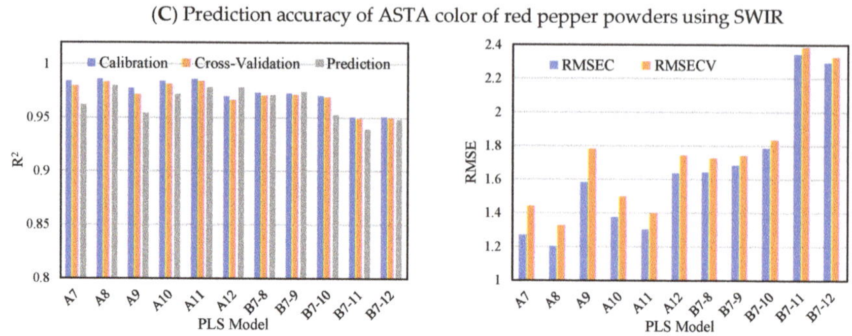

Figure 6. Prediction accuracy of capsaicinoid (**A**), free sugar (**B**) and ASTA color (**C**) of red pepper powders using SWIR wavelength range in accordance with moisture content. RMSEC, root mean square error of calibration; RMSECV, root mean square error of cross-validation.

4. Conclusions

The present study predicted the capsaicinoid and free sugar content through hyperspectral imaging and PLS analysis of red pepper powder with different moisture contents and different pungency levels. There is an explicit tendency for the RMSE value to increase as the difference in moisture content of the modeling sample increases for all predicted quality attributes. Finally, a difference of more than 2% in MC had a negative effect on prediction accuracy for capsaicinoid and free sugar. Therefore, this study demonstrated that it is essential to adjust the moisture content difference of red pepper powder samples to be used for modeling within 2% using a hyperspectral imaging system. It is expected that this will be used as a basis for the development of automated systems for the rapid grading of pungency levels and sweetness.

Author Contributions: Conceptualization, J.-Y.C. and J.H.L.; methodology, J.-S.C.; validation, K.J.P. and J.H.C.; Formal analysis, J.-Y.C.; data curation, J.-S.C.; writing—original draft preparation, J.-Y.C. and J.-S.C.; writing—review and editing, K.J.P., J.H.C. and J.H.L.; visualization, J.-Y.C.; supervision, J.H.L.; project administration, J.H.L. All authors have read and agreed to the published version of the manuscript.

Funding: This work was supported by Korea Institute of Planning and Evaluation for Technology in Food, Agriculture, Forestry (IPET) through (High Value-added Food Technology Development Program), funded by Ministry of Agriculture, Food and Rural Affairs (MAFRA) (321049-05).

Data Availability Statement: Not applicable.

Conflicts of Interest: The authors declare no potential conflict of interest.

References

1. Hwang, S.Y.; An, Y.H.; Shin, G.M. A study on the quality of commercial red pepper powder. *Korean J. Food Nutr.* **2001**, *14*, 424–428.
2. Ku, K.H.; Kim, N.Y.; Park, J.B.; Park, W.S. Characteristics of color and pungency in the red pepper for Kimchi. *Korean J. Food Sci. Technol.* **2001**, *33*, 231–237.
3. Choi, S.M.; Jeon, Y.S.; Park, K.Y. Comparison of quality of red pepper powders produced in Korea. *Korean J. Food Sci. Technol.* **2000**, *32*, 1251–1257.
4. Mo, C.Y.; Kang, S.W.; Lee, K.J.; Lim, J.G.; Cho, B.K.; Lee, H.D. Development of prediction model for capsaicinoids content in red pepper powder using near-infrared spectroscopy-particle size effect. *Food Eng. Prog.* **2011**, *15*, 48–55.
5. Kang, J.H.; Son, H.J.; Hong, E.J.; Noh, B.S. Discrimination of grading pungency for red peppers spice using electronic nose based on mass spectrometer. *Food Eng. Prog.* **2010**, *14*, 35–40.
6. Ku, K.H.; Kim, N.Y.; Park, W.S. Characteristics analysis for the standardization of commercial kimchi. *Korean J. Food Sci. Technol.* **2003**, *35*, 316–319.
7. Yu, J.O.; Choi, W.S.; Lee, U.S. Determination of capsaicin and dihydrocapsaicin in various species of red peppers and their powdered products in market by GC-MS analysis. *Food Eng. Prog.* **2009**, *13*, 38–43.
8. Mo, C.Y.; Lee, K.J.; Lim, J.G.; Kang, S.K.; Lee, H.D.; Cho, B.K. Development of non-destructive pungency measurement technique for red-pepper powder produced in different domestic origins. *Korean J. Agric. Sci.* **2012**, *39*, 603–612.

9. Othman, Z.A.A.; Ahmed, Y.B.H.A.; Habila, M.A.; Ghafar, A.A. Determination of capsaicin and dihydrocapsaicin in Capsicum fruit samples using high performance liquid chromatography. *Molecules* **2011**, *16*, 8919–8929. [CrossRef]
10. Diago, M.P.; Fernandes, A.M.; Melo-Pinto, P.; Tardaguila, J. Use of visible and short-wave near-infrared hyperspectral imaging to fingerprint anthocyanins in intact grape berries. *J. Agric. Food Chem.* **2016**, *64*, 7658–7666. [CrossRef]
11. Li, D.; Li, L.; Xiao, G.; Limwachiranon, J.; Xu, Y.; Lu, H.; Yang, D.; Luo, Z. Effects of elevated CO2 on energy metabolism and γ-aminobutyric acid shunt pathway in postharvest strawberry fruit. *Food Chem.* **2018**, *265*, 281–289. [CrossRef] [PubMed]
12. Cheng, J.H.; Sun, D.W.; Pu, H.; Zhu, Z. Development of hyperspectral imaging coupled with chemometric analysis to monitor K value for evaluation of chemical spoilage in fish fillets. *Food Chem.* **2015**, *185*, 245–253. [CrossRef] [PubMed]
13. Johnson, J.B.; Mani, J.S.; Naiker, M. Infrared spectroscopy for the quality assessment of habanero chilli: A proof-of-concept study. *Eng. Proc.* **2021**, *8*, 19.
14. Rahman, A.; Lee, H.; Kim, M.S.; Cho, B.K. Mapping the Pungency of Green Pepper Using Hyperspectral Imaging. *Food Anal. Methods* **2018**, *11*, 3042–3052. [CrossRef]
15. Jiang, J.; Cen, H.; Zhang, C.; Lyu, X.; Weng, H.; Xu, H.; He, Y. Nondestructive quality assessment of chili peppers using near-infrared hyperspectral imaging combined with multivariate analysis. *Postharvest Biol. Technol.* **2018**, *146*, 147–154. [CrossRef]
16. Fu, X.; Chen, J.; Zhang, J.; Fu, F.; Wu, C. Effect of penetration depth and particle size on detection of wheat flour adulterant using hyperspectral imaging. *Biosyst. Eng.* **2021**, *204*, 64–78. [CrossRef]
17. Wang, H.; Mann, C.K.; Vickers, T.J. Effect of powder properties on the intensity of Raman scattering by crystalline solids. *Appl. Spectrosc.* **2002**, *56*, 1538–1544. [CrossRef]
18. Myers, T.L.; Brauer, C.S.; Su, Y.-F.; Blake, T.A.; Tonkyn, R.G.; Ertel, A.B.; Johnson, T.J.; Richardson, R.L. Quantitative reflectance spectra of solid powders as a function of particle size. *Appl. Opt.* **2015**, *54*, 4863–4875. [CrossRef]
19. Choi, J.Y.; Cho, J.S.; Park, K.J.; Kim, S.S.; Lim, J.H. Grading the pungency of red pepper powder using hyperspectral imaging coupled with multivariate analysis. *Korean J. Food Preserv.* **2022**, *29*, 918–931. [CrossRef]
20. Ku, K.H.; Lee, K.A.; Park, J.B. Physicochemical properties and sensory evaluation for the heat level (hot taste) of Korean red pepper powder. *Prev. Nutr. Food. Sci.* **2012**, *17*, 29–35. [CrossRef]
21. Namgung, B.; Lee, Y.; Ha, J. Determination of capsaicinoids in red pepper powder using ultra high performance liquid chromatography. *Anal. Sci. Technol.* **2013**, *26*, 256–261. [CrossRef]
22. Lee, K.B.; Park, S.; Sung, S.; Park, D. A study on the prediction of CNC tool wear using machine learning technique. *J. Korean Converg. Soc.* **2019**, *10*, 15–21.
23. Jang, H.J.; Choi, C.H.; Choi, T.H.; Kim, J.H.; Kwon, G.H.; Oh, S.I.; Kim, H.; Kim, Y.J. The analysis of oat chemical properties using visible-near infrared spectroscopy. *Korean J. Agric. Sci.* **2016**, *43*, 715–722.
24. Abdi, H. Partial least squares (PLS) regression. In *Encyclopedia of Social Sciences Research Methods*; Sage Publications: Thousand Oaks, CA, USA, 2003; pp. 792–795.
25. Suzuki, T.; Iwai, K. *The Alkaloids: Chemistry and Pharmacology*; Academic Press: Cambridge, MA, USA, 1984; p. 228.
26. Park, S.H.; Koo, H.J.; Lim, H.S.; Yoo, J.H.; Hwang, S.Y.; Shin, E.H.; Park, Y.H.; Lee, J.H.; Cho, J.S. The physicochemical changes during storage of red pepper powder dried in hot-air various processing methods. *J. Korean Soc. Food Sci. Nutr.* **2003**, *32*, 876–881.
27. Kim, S.; Park, J.H. Composition of main carotenoids in Korean red pepper (*Capsicum annuum* L.) and change of pigment stability during the drying and storage process. *J. Food Sci.* **2004**, *69*, 39–44. [CrossRef]
28. Kim, J.B.; Ko, H.C.; Lee, J.Y.; Ha, S.H.; Kim, J.B.; Kim, H.H.; Gwag, J.G.; Kim, T.S. Change in the carotenoid content of the Korean pepper (*Capsicum annuum* var. subicho.) during ripening stages. *J. Korean Soc. Int. Agric.* **2009**, *21*, 276–281.
29. Vega-Gálvez, A.; Scala, K.D.; Rodríguez, K.; Lemus-Mondaca, R.; Miranda, M.; López, J.; Perez-Won, M. Effect of air-drying temperature on physico-chemical properties, antioxidant capacity, colour and total phenolic content of red pepper (*Capsicum annuum*, L. var. Hungarian). *Food Chem.* **2009**, *117*, 647–653. [CrossRef]
30. Jeong, E.J.; Bang, B.H.; Kim, K.P. The characteristics of kimchi by the degree of hotness of powdered red pepper. *Korean J. Food Nutr.* **2005**, *18*, 88–93.
31. Park, T.S.; Bae, Y.M.; Sim, M.J.; Kim, D.E.; Cho, S.I. Analysis of capsaicinoids from hot red pepper powder by near-infrared spectroscopy. *ASABE* **2008**, 083703.
32. Alessandrini, L.; Romani, S.; Pinnavaia, G.; Rosa, M.D. Near infrared spectroscopy: An analytical tool to predict coffee roasting degree. *Anal. Chimca Acta* **2008**, *625*, 95–102. [CrossRef]
33. Yin, W.; Zhang, C.; Zhu, H.; Zhao, Y.; He, Y. Application of near-infrared hyperspectral imaging to discriminate different geographical origins of Chinese wolfberries. *PLoS ONE* **2017**, *12*, e0180534. [CrossRef] [PubMed]
34. Aprilia, Y.I.; Khuriyati, N.; Sukartiko, A.C. Classification of chili powder (*Capsicum annuum* L.) antioxidant activity based on near infrared spectra. *Food Res.* **2021**, *5*, 51–56. [CrossRef]
35. Kamruzzaman, M.; ElMasry, G.; Sun, D.W.; Allen, P. Application of NIR hyperspectral imaging for discrimination of lamb muscles. *J. Food Eng.* **2011**, *104*, 332–340. [CrossRef]
36. Lawrence, K.C.; Windham, W.R.; Park, B.; Smith, D.P.; Poole, G.H. Comparison between visible/NIR spectroscopy and hyperspectral imaging for detecting surface contaminants on poultry carcasses. *Monit. Food Saf. Agric. Plant Health* **2004**, *5271*, 35–42.
37. Gómez, A.H.; He, Y.; Pereira, A.G. Non-destructive measurement of acidity, soluble solids and firmness of Satsuma mandarin using Vis/NIR-spectroscopy techniques. *J. Food Eng.* **2006**, *77*, 313–319. [CrossRef]
38. Sahoo, R.N.; Ray, S.S.; Manjunath, K.R. Hyperspectral remote sensing of agriculture. *Curr. Sci.* **2015**, *108*, 848–859.

39. Giovenzana, V.; Beghi, R.; Romaniello, R.; Tamborrino, A.; Guidetti, R.; Leone, A. Use of visible and near infrared spectroscopy with a view to on-line evaluation of oil content during olive processing. *Biosyst. Eng.* **2018**, *172*, 102–109. [CrossRef]
40. Li, M.Z. *Spectroscopy Analysis Technology and Application*; Science Press: Beijing, China, 2006.
41. Cozzolino, D.; Smyth, H.E.; Gishen, M. Feasibility study on the use of visible and near-infrared spectroscopy together with chemometrics to discriminate between commercial white wines of different varietal origins. *J. Agric. Food Chem.* **2003**, *51*, 7703–7708. [CrossRef]
42. Reyes-Escogido, M.D.L.; Gonzalez-Mondragon, E.G.; Vazquez-Tzompantzi, E.V. Chemical and pharmacological aspects of capsaicin. *Molecules* **2011**, *16*, 1253–1270. [CrossRef]
43. Gómez-García, M.D.R.; Ochoa-Alejo, N. Biochemistry and Molecular Biology of Carotenoid Biosynthesis in Chili Peppers (*Capsicum* spp.). *Int. J. Mol. Sci.* **2013**, *14*, 19025–19053. [CrossRef]

foods

MDPI

Article

Time-Resolved Laser-Induced Breakdown Spectroscopy for Accurate Qualitative and Quantitative Analysis of Brown Rice Flour Adulteration

Honghua Ma [1,2,†], Shengqun Shi [1,†], Deng Zhang [1], Nan Deng [1], Zhenlin Hu [1], Jianguo Liu [1] and Lianbo Guo [1,*]

[1] Wuhan National Laboratory for Optoelectronics (WNLO), Huazhong University of Science and Technology, Wuhan 430074, China
[2] School of Physics and Electronic Information Engineering, Hubei Engineering University, Xiaogan 432000, China
* Correspondence: lbguo@hust.edu.cn
† These authors contributed equally to this work.

Abstract: To solve the adulteration problem of brown rice flour in the commodity market, a novel, accurate, and stable detection method based on time-resolved laser-induced breakdown spectroscopy (TR-LIBS) is proposed. Qualitative and quantitative analysis was used to detect five adulterants and seven different adulterant ratios in brown rice flour. Being able to excavate more information from plasma by obtaining time-resolved spectra, TR-LIBS has a stronger performance, which has been further verified by experiments. For the qualitative analysis of adulterants, the traditional machine learning models based on TR-LIBS, linear discriminant analysis (LDA), naïve Bayes (NB) and support vector machine (SVM) have significantly better classification accuracy than those based on traditional LIBS, increasing by 3–11%. The deep learning classification model based on TR-LIBS also achieved the same results, with an accuracy increase of more than 8%. For the quantitative analysis of the adulteration ratio, compared with traditional LIBS, the quantitative model based on TR-LIBS reduces the limit of detection (LOD) of five adulterants from about 8–51% to 4–19%, which effectively improves the quantitative detection performance. Moreover, t-SNE visualization proved that there were more obvious boundaries between different types of samples based on TR-LIBS. These results demonstrate the great prospect of TR-LIBS in the identification of brown rice flour adulteration.

Keywords: laser-induced breakdown spectroscopy; brown rice flour adulteration; time-resolved spectra; machine learning; deep learning

Citation: Ma, H.; Shi, S.; Zhang, D.; Deng, N.; Hu, Z.; Liu, J.; Guo, L. Time-Resolved Laser-Induced Breakdown Spectroscopy for Accurate Qualitative and Quantitative Analysis of Brown Rice Flour Adulteration. *Foods* **2022**, *11*, 3398. https://doi.org/10.3390/foods11213398

Academic Editors: Mourad Kharbach and Samuli Urpelainen

Received: 30 September 2022
Accepted: 25 October 2022
Published: 27 October 2022

1. Introduction

Food adulteration refers to artificially and purposefully adding diverting ingredients to food, such as for preservation, color enhancement, improvement of appearance, texture and masking, to achieve the purpose of improving economic benefits [1]. Rice, the staple food for more than 3.5 billion people, can be processed to produce brown and white rice [2,3]. Compared with white rice, brown rice is rich in starch, protein, fat, vitamins, minerals, and functional health care ingredients, and also contains dietary fiber, oryzanol, glutathione, γ-aminobutyric acid, rice bran polysaccharide. It can provide more comprehensive nutrition, which is very suitable for people with obesity, gastrointestinal dysfunction, anemia, constipation and diabetes [4,5]. Therefore, the sales of brown rice are on a continuous rise worldwide [6]. However, many unscrupulous suppliers adulterate brown rice in pursuit of more profits, which may lead to decreased nutritional value, human health problems, and even death in serious cases [7]. Therefore, rice flour adulteration detection is an important way to ensure food safety.

Plenty of studies have provided various approaches for adulteration identification of rice products, including a DNA-based method combined with High Resolution Melting (HRM) [8], inductively coupled plasma mass spectrometry (ICP-MS) and isotope ratio mass spectrometry (IRMS) [9–12]. Similarly, several studies have shown that gas chromatography-mass spectrometry (GC-MS) and liquid chromatography-mass spectrometry (LC-MS) are useful as certification methods for organic rice to detect adulterated components in high-quality rice [13–17]. However, they require sample digestion, which means a long detection period, high requirements for the operation environment, and environmental pollution caused by the use of chemical reagents [18]. In addition, recent studies have shown that spectrum-based detection methods are also applicable to food safety monitoring. For instance, Attaviroj et al. proposed the application of Fourier-transform near-infrared spectroscopy (FT-NIR) to identify moist brown rice varieties [19]. Li et al. proposed terahertz (THz) spectroscopy combined with support vector machine (SVM) to identify adulterated rice [20]. Hyperspectral imaging technology (HSI) can also identify both spatial and chemical information of wheat flour products [21]. Nevertheless, the above spectral detection methods also have some disadvantages. The liquid composition in the sample affects the analytical performance of the NIR and THz spectra. In addition, NIR requires a large number of representative known chemical values for modeling. HSI data is large and redundant, and processing is very complex. Therefore, a simple, rapid, and in situ method for distinguishing brown rice from adulterated samples is urgently needed to ensure food safety in the grain market.

Laser-induced breakdown spectroscopy (LIBS) has become a well-established and powerful optical emission spectroscopy analysis technology after more than 60 years of development. It has been praised as the "future super star" in analytical chemistry with the advantages of multi-element analysis, fast response, remote detection and no or simple preparation [22]. In recent years, LIBS has been gradually applied to the field of food detection, including monitoring calcium content in comminuted poultry meat [23], adulteration detection of milk powder [24,25], and classification of red wine based on its protected designation of origin (PDO) [26]. However, few LIBS works were reported on the quality detection of brown rice products. Ribeiro et al. analyzed the composition of rice varieties by LIBS combined with Fourier-transform infrared spectroscopy (FTIR) [27]. Yang et al. proved that LIBS combined with chemometrics can distinguish the geographical origin of different rice [28]. Pérez-Rodríguez et al. carried out significant research on brown rice detection. They used spark discharge-LIBS (SD-LIBS) combined with the k-nearest neighbor (KNN) to distinguish the PDO certification of brown rice [7]. These representative works on LIBS detection of rice demonstrate the potential for rapid identification of adulterated products without chemical digestion. However, the LIBS spectra are unstable due to the influence of sample morphology, laser energy and other factors, and this fluctuation leads to a decrease in the accuracy of qualitative analysis. In addition, detection methods similar to SD-LIBS will lead to increased operational complexity. The above factors seriously restrict the further development of LIBS in the field of rice flour adulteration identification. Therefore, the detection of different adulterants in rice products by LIBS needs to be further improved.

In this work, a new simplified time-resolved LIBS (TR-LIBS) was developed for accurate qualitative and quantitative analysis of brown rice flour adulteration based on [25]. TR-LIBS technology does not require sample digestion. Compared to NIR and HSI technologies, it has a lower data dimension and does not require a large number of representative chemical values for modeling. Moreover, TR-LIBS can effectively mine the key information such as intensity and its evolution over delay time by obtaining time-resolved spectra, so as to enhance the accuracy and robustness of the analytical models. To verify the effectiveness of TR-LIBS, we compared it with traditional LIBS from the perspective of qualitative and quantitative analysis of adulteration. The results proved that TR-LIBS is an accurate, reliable, and stable analytical method.

2. Materials and Method

2.1. Materials and Sample Preparation

In the commodity market, products with low nutritional value are often mixed with products with high nutritional value. As a kind of food with high nutritional value, brown rice flour is always mixed with some low-priced food to obtain additional economic benefits, such as sorghum flour, talc powder and corn flour. Samples used in this work include sorghum flour (SF), talc powder (TP), corn flour (CF), buckwheat flour (BF), gypsum powder (GP), and brown rice flour (BRF), all of which were purchased from the market in China. ICP-MS was used as a reference method to determine the elemental composition of these six kinds of samples, and the results in Table 1 show that TP and GP are significantly different from SF, CF, BF, and BRF, while SF, CF, BF, and BRF are similar. Appropriate amounts of SF, TP, CF, BF and GP were mixed into BRF to achieve the target concentration: 1%, 3%, 5%, 8%, 10%, 15%, 20% and 25% (*w*/*w*). Five grams of mixed powder was pelleted by electric tablets press applying a pressure of 30 tons for 1 min. The thickness and diameter of the pellets were about 3 mm and 40 mm, respectively. To eliminate individual differences in samples, two repeated samples were made for each concentration gradient. A total of 82 pressed pellets were prepared for LIBS measurement without further treatment.

Table 1. The elemental concentration of six kinds of samples were measured by ICP-MS.

Samples	Elemental Concentration (mg/kg)							
	Ca	Cu	Fe	K	Mn	Na	P	Zn
SF	93.88	2.14	28.04	2332.01	12.68	14.61	2600.04	17.00
TP	16,387.98	1.08	3655.71	207.02	102.65	118.67	171.52	4.21
CF	34.15	0.59	8.95	1127.84	0.63	9.48	773.68	6.70
BF	131.17	3.14	21.14	2965.93	8.78	11.61	3138.66	16.78
GP	249,581.83	0.70	47.09	62.84	0.31	49.86	4.01	7.50
BRF	97.07	2.19	10.66	1376.19	35.31	14.51	1702.35	16.23

2.2. LIBS Setup and Measurement

The traditional LIBS device was used in this work. A laser pulse from a Q-switched Nd: YAG laser (Nimma-400, wavelength: 532 nm; pulse duration: 8 ns; flattened Gaussian beam, Beamtech Optronics Co., Ltd., Beijing, China) is focused on the sample, which is placed on a three-dimensional electric displacement platform (DZY110TA-3Z, Beijing Jiangyun Juli Technology Co., Ltd., Beijing, China), after passing through a quartz lens to generate plasma. The movement of the x-y-z platform is controlled by the laser rangefinder (CDX-85A, Aotaisi Industrial Automation Control Equipment Co., Ltd., Guangdong, China). The plasma emission is detected by the spectrometer (Mechelle 5000, resolution: $\lambda/\Delta\lambda$ = 5000; spectral range: 200–950 nm; Andor Technology Ltd., Belfast, United Kingdom) via a light collector and UV-enhanced fiber optic with a 50 μm core. The whole system is controlled by using a self-designed LIBS digital delay generator (LDG 3.0, Wuhan N&D Laser Engineering Co., Ltd., Wuhan, China). LIBS detection system device schematic is shown in Figure 1.

In this work, the optimal experimental parameters were as follows, the laser energy was set as 40 mJ and the repetition frequency was 1 Hz. The gate width and exposure time of the spectrometer were fixed to 2 μs and 0.101 s, respectively. The gate delay was set from 1 μs to 4.5 μs, with 0.5 μs as a step to obtain the time-resolved spectra. All samples were measured in the atmosphere. The LIBS spectra (gate delay: 1 μs; gate width: 2 μs) of 25% adulteration were shown in Figure 2a. In addition, time-resolved spectra of BRF and its adulterated samples are also shown in Figure 2. For each pellet, 15 groups of time-resolved spectra (120 LIBS spectra without accumulation and average) were obtained. Different adulterations (1%, 3%, 5%, 8%, 10%, 15% and 25%) of the same mixture were regarded as the same class, and all samples were divided into six classes including BRF + SF (class 1), BRF + TP (class 2), BRF + CF (class 3), BRF + BF (class 4), BRF + GP (class 5) and BRF (class 6). A total of 1230 groups of time-resolved spectra were used for subsequent qualitative and quantitative analysis.

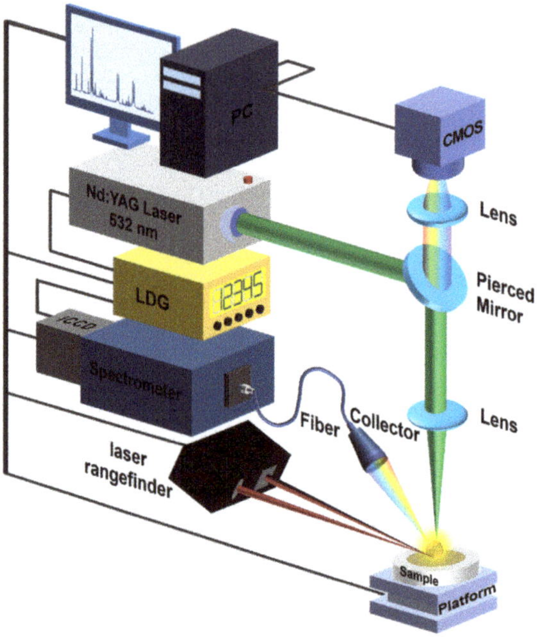

Figure 1. LIBS rice flour detection system device schematic.

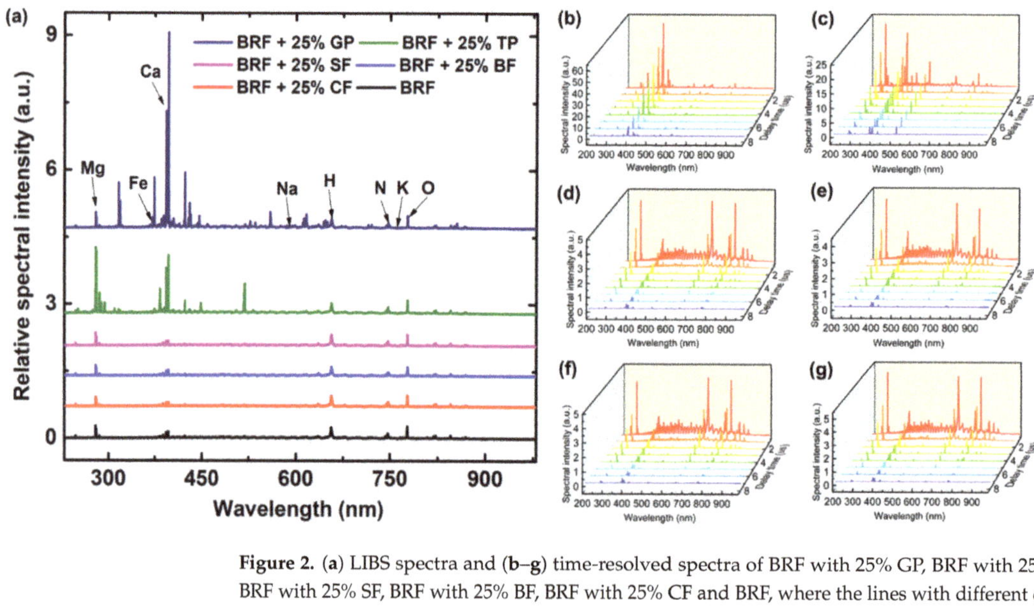

Figure 2. (**a**) LIBS spectra and (**b**–**g**) time-resolved spectra of BRF with 25% GP, BRF with 25% TP, BRF with 25% SF, BRF with 25% BF, BRF with 25% CF and BRF, where the lines with different colors in (**b**–**g**) represent the spectra with different time delay.

2.3. Data Analysis

Traditional LIBS analysis only focuses on the spectra radiated by plasma at a certain gate delay and gate width, which results in only the element intensity information under this state being obtained, while a large amount of useful information of plasma is lost. To solve this problem, a novel time-resolved laser-induced breakdown spectroscopy (TR-LIBS) is proposed in this work. After obtaining LIBS spectra under multiple delay times, this method extracted features and spliced them, and finally inputted them into the analysis models. Next, the process of TR-LIBS is described in detail. As can be seen from Figure 2, the spectra of these samples are sparse, and most wavelengths are redundant. Therefore, spectral feature selection or dimensionality reduction is needed. In this work, 64 lines were selected, mainly including lines of nitrogen (N), oxygen (O), hydrogen (H), carbon (C), calcium (Ca), sodium (Na), potassium (K), magnesium (Mg) and other elements according to ICP-MS results and National Institute of Standards and Technology (NIST) atomic spectral database. After feature selection, eight time-resolved spectra were spliced to form a 512-length one-dimensional sequence, which was used as the input of the analysis models.

To verify the effectiveness of TR-LIBS in improving qualitative model analysis performance, LIBS and TR-LIBS were compared based on the traditional machine learning model and deep learning model. For traditional machine learning models, three commonly used models, linear discriminant analysis (LDA), Naive Bayes (NB), and support vector machines (SVM) were selected. LDA is a supervised machine learning algorithm. It can find the optimal representation of data in low dimensions by maximizing the inter-class divergence matrix and minimizing the intra-class divergence matrix, and can effectively extract classification features. The NB algorithm is a method based on Bayes theorem and independence hypothesis of characteristic condition, which needs fewer parameters and is not easy to disturb outliers. The SVM algorithm is based on statistical learning theory and can maximize the interval between data while minimizing the empirical error. It is also a classical supervised classification algorithm. It uses a kernel function to transform the linear inseparable problem of low dimensional space into the linear separable problem of high dimensional space and then realizes the accurate classification of data. These methods are also the most common and effective analytical methods in spectral analysis [29,30]. For the deep learning model, a one-dimensional convolutional neural network (1D-CNN) was constructed. The core idea of CNN is the sparse connection, weight sharing and pooling sampling. Through convolution operation, the original signal features can be enhanced and the noise can be reduced. The pooling operation uses the principle of local correlation of the image to downsample the image, which can reduce the amount of data processing while retaining useful information. CNN also achieves excellent performance in LIBS analysis [31].

In addition, to verify the effectiveness of TR-LIBS in improving quantitative analysis performance, a partial least squares regression (PLSR) model [32] was established on seven different adulteration ratios of five adulterants. PLSR is a linear multivariate data analysis method based on factor analysis, which extracts components from the independent variables that have both higher generalization of information on the independent variable system and better interpretation of the dependent variable and determines the number of principal factors, and then builds a regression model of the principal factors and the dependent variable.

2.4. Evaluation Indexes

The qualitative and quantitative performance of the traditional LIBS and our proposed TR-LIBS method were compared in detail through several evaluation indexes. Qualitative analysis is mainly evaluated by accuracy and confusion matrix. Different from qualitative analysis, the performance evaluation indexes for quantitative analysis mainly include the determination coefficient (R^2), the root-mean-square error (RMSE), and the average relative error (ARE) [33]. This work also calculated the limit of detection (LOD) of five adulterants under PLSR [34]. Their expressions are as follows:

$$Accuracy = \frac{100}{n}\sum_{i=1}^{n}\delta_i \tag{1}$$

$$R^2 = 1 - \frac{\sum_{i=1}^{n}(\hat{y}_i - y_i)^2}{\sum_{i=1}^{n}(\hat{y}_i - \overline{y})^2} \tag{2}$$

$$\text{RMSE} = \sqrt{\frac{\sum_{i=1}^{n}(\hat{y}_i - y_i)^2}{n}} \tag{3}$$

$$\text{ARE} = \frac{100}{n}\sum_{i=1}^{n}\frac{|\hat{y}_i - y_i|}{y_i} \tag{4}$$

$$\text{LOD} = 3.3s_{pu}^{-1}\left[\left(1 + h_{0min} + \frac{1}{n}\right)var_{pu}\right]^{\frac{1}{2}} \tag{5}$$

where n is the number of samples, δ_i is either 1 when a spectrum is classified correctly or 0 otherwise, y_i is the certified concentration of the i_{th} sample, $\overline{y}$ is the average value of y_i over n sample, $\hat{y}_i$ is the predicted concentration of the i_{th} sample, s_{pu} is the slope of the Pseudounivariate line, h_{0min} is the minimum projected leverage for a blank sample, var_{pu} is the variance of the regression residuals.

3. Results of Qualitative and Quantitative Analysis

In this work, the validity of TR-LIBS was verified by spectral analysis of BRF and its adulterated samples. To solve data problems such as imbalance, the Synthetic Minority Over-sampling Technique (SMOTE) [35] approach was utilized. Before modeling, the spectra were randomly divided into the training set and test set according to the ratio of 7:3. The training set was used to train the models and optimize the hyperparameters of the models. The test set was used to validate the optimized models. To eliminate the influence of the randomness of sample division, 20 random segmentation operations were performed, and the average accuracy of models was obtained by averaging the 20 test results. The qualitative and quantitative analysis results of LIBS and TR-LIBS in brown rice flour and adulterants were introduced below.

3.1. *Qualitative Analysis of Adulterants*

3.1.1. The Results of Traditional Machine Learning Models

To make the results universal, three traditional machine learning classification models, LDA, NB, and SVM, are used for analysis. First, these six types of samples are classified based on traditional LIBS. Traditional LIBS only uses a single spectrum of each sample for classification. In this work, LIBS spectra are obtained at eight delay times, such as 1 µs, 1.5 µs, 2 µs, 2.5 µs, 3 µs, 3.5 µs, 4 µs, and 4.5 µs. Therefore, after feature selection, eight classification models are established based on the spectra obtained under different delay times, and the results are shown in Figure 3a,b. It can be seen that the performance of classification models established under different delay times is obviously diverse. The training set accuracy of LDA classification models is established under different delay times from 62.12% to 80.66%, and the test set accuracy from 62.82% to 80.95%. The training set accuracy of NB classification models is 72.92–80.42%, and the test set accuracy is 73.01–80.20%. The training set accuracy of SVM classification models is 83.40–91.29%, and the test set accuracy is 85.98–92.48%. It can be seen from the results that the accuracy of the classification models increases first and then decreases with the increase in delay time. This is due to the continuum emission of the initial plasma being very strong, where the atomic and ion spectral lines are submerged. With the evolution of the plasma, the continuum emission intensity decreases, the intensity of atomic and ionic spectral lines increases, and the effective information of plasma can be fully mined. However, with further evolution, the plasma gradually annihilates, and the intensity of atomic and ionic spectral lines decreases or even disappears. The classification performance of SVM is significantly higher than LDA and NB. Among them, the SVM model established under the delay time of 1.5 μs has the optimal performance. The training set accuracy of this model is 91.29%, and the accuracy of the test set is 92.48%. The training set confusion matrix is shown in Figure 2c. The misclassification rate between class 1, 3, and 4 is very high. It can be seen from the ICP-MS results that the very similar type and concentration of elements among SF, CF, and BF are the reasons for this phenomenon.

Furthermore, TR-LIBS is utilized to classify these six kinds of samples. After feature selection, spectral features under different delay times are spliced and finally input into LDA, NB, and SVM classification models. The results of classification models based on TR-LIBS are shown in Figure 3a,b. The training set accuracy of the LDA classification model is 89.63%, and the test set accuracy is 90.59%. The training set accuracy of the NB classification model is 91.11%, and the test set accuracy is 91.27%. The training set accuracy of the SVM classification model is 94.34%, and the test set accuracy is 95.40%. The optimal classification model based on TR-LIBS is still SVM. The accuracy of classification models based on TR-LIBS is obviously better than those based on traditional LIBS, which is increased by about 3–11%. As can be seen from the confusion matrix of the test set in Figure 3d, the misclassifications among the six classes were significantly reduced, especially between classes 1, 3, and 4. These results preliminarily demonstrate the effectiveness of TR-LIBS in improving qualitative analysis performance.

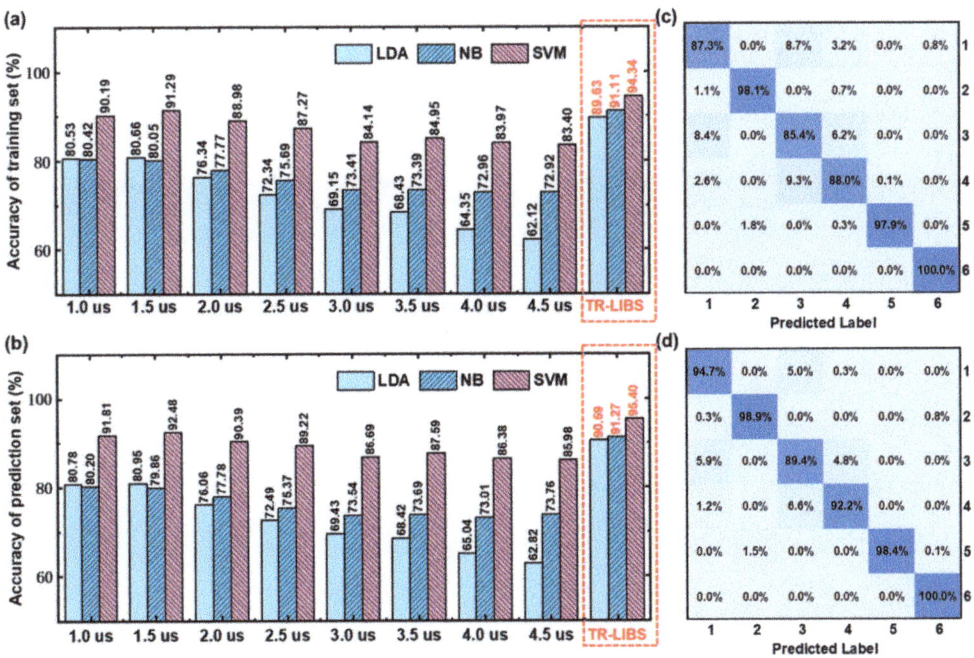

Figure 3. (**a**) Training set and (**b**) test set results of LDA, NB and SVM models based on LIBS spectra and time-resolved spectra, and the test set confusion matrix of SVM models based on (**c**) LIBS spectra under 1.5 μs and (**d**) time-resolved spectra.

3.1.2. The Results of Deep Learning Models

Furthermore, the convolutional neural network in the deep learning model is used to compare the performance of traditional LIBS and TR-LIBS. As shown in Figure 4a, an

18-layer one-dimensional convolutional neural network (1D CNN) is built after structure and parameter optimization in this work. Three convolution layers, three average pooling layers, and three Relu activation layers are used. The size of the convolution layer is 1 × 3, and the size of the average pooling layer is 1 × 5. The number of convolution kernel is set as 8, 8, and 8, respectively. To prevent over-fitting, the batch normalization layer is placed before the activation layer. A flatten layer and two full connection layers are used to map the feature space computed by the front layer to the sample marker space. The number of neurons in the full connection layer is set to 20 and 6, respectively. The activation function of the output layer is Softmax. The hyperparameters of the 1D CNN model are Epoch of 60, Batchsize of 128, and Learning Rate of 0.0005.

This constructed 1D CNN is used to identify these six types of samples based on traditional LIBS. Similarly, 1D CNN classification models are established based on the spectra obtained under different delay times, and the results are shown in Figure 4b,c. The accuracy of the training set ranges from 69.82% to 81.37%, and that of the test set ranges from 65.30% to 77.37%. The evolution of 1D CNN classification model accuracy with delay time is consistent with that of the traditional machine learning model. However, the delay time corresponding to the optimal model is different. The 1D CNN model at 2.5 μs has the best performance. The accuracy of the training set and the test set is 81.37%, and 77.37%, respectively. Compared with traditional machine learning models, 1D CNN model has poor performance, which may be caused by the small amount of data. Convolutional Neural Network is a deep model with numerous parameters, requiring more data for parameter optimization and learning. The test set confusion matrix is shown in Figure 4d, and the misclassification between classes 1, 3, and 4 is very obvious.

Furthermore, a 1D CNN classification model is trained based on TR-LIBS, and the results are shown in Figure 4b–e. The accuracy of the training set and test set of this 1D CNN classification model is 90.71% and 85.66%, respectively. It can be seen that compared with traditional LIBS, the performance of 1D CNN model is significantly improved, and the test set accuracy is increased by more than 8%. Finally, we compare the classification accuracy of the traditional LIBS and TR-LIBS combined with machine learning models and deep learning models on the training and prediction set, respectively, in Table 2. The above experimental results fully verify the effectiveness of TR-LIBS in improving qualitative analysis performance from two aspects of traditional machine learning models and deep learning models.

Table 2. Classification accuracy of traditional LIBS and TR-LIBS combined with machine learning models and deep learning models on the training and prediction set.

Method	Delay	Accuracy of the Training Set (%)				Accuracy of the Prediction Set (%)			
		LDA	NB	SVM	1D-CNN	LDA	NB	SVM	1D-CNN
Traditional LIBS	1 μs	80.53	80.42	90.19	69.82	80.78	80.20	91.81	65.30
	1.5 μs	80.66	80.05	91.29	77.88	80.95	79.86	92.48	74.57
	2 μs	76.34	77.77	88.98	80.69	76.06	77.78	90.39	76.79
	2.5 μs	72.34	75.69	87.27	81.37	72.49	75.37	89.22	77.37
	3 μs	69.15	73.41	84.14	79.82	69.43	73.54	86.69	75.06
	3.5 μs	68.43	73.39	84.95	79.40	68.42	73.69	87.59	73.66
	4 μs	64.35	72.96	83.97	79.42	65.04	73.01	86.38	73.27
	4.5 μs	62.12	72.92	83.40	79.32	62.82	73.76	85.98	73.87
TR-LIBS	—	**89.63**	**91.11**	**94.34**	**90.71**	**90.59**	**91.27**	**95.40**	**85.66**

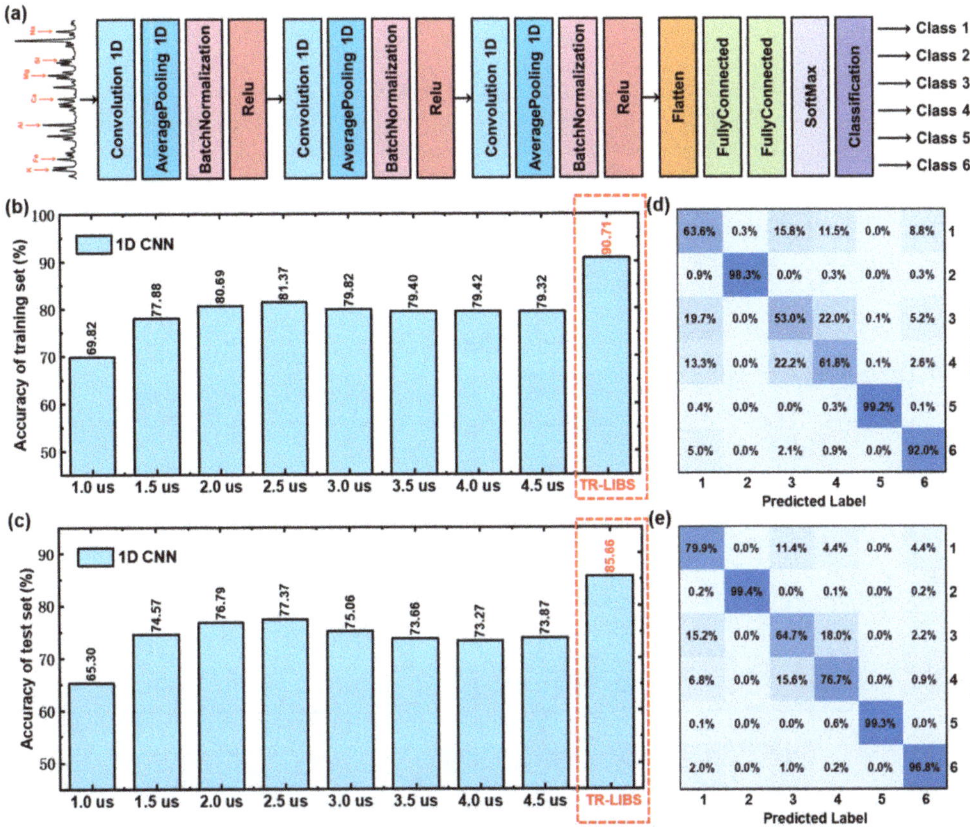

Figure 4. (**a**) Structural diagram of 1D CNN model, (**b**) training set and (**c**) test set results of 1D CNN models based on LIBS spectra and time-resolved spectra, and the test set confusion matrix of 1D CNN models based on (**d**) LIBS spectra under 2.5 μs and (**e**) time-resolved spectra.

3.2. Quantitative Analysis of the Adulteration Ratio

Based on the above qualitative analysis of adulterants, seven adulterant ratios (1, 3, 5, 8, 10, 15, and 25%) of five adulterants are further quantitatively analyzed in this section. The adulteration is composed of different elements. Therefore, the quantitative analysis of the ratio is a multivariate regression process. In this work, the PLSR model is used to fit the ratio of adulteration.

Data preprocessing for quantitative analysis for LIBS and TR-LIBS is the same process as the qualitative analysis, but the supervision label changes from the adulteration category to the adulteration ratio. Based on the above evaluation metrics, the quantitative performance of traditional LIBS and TR-LIBS on the training set and prediction set of five adulterations is compared in detail in Table 3. For the LIBS method, this section selects the best results under eight different delays. The data show that the quantitative results of TP are the best. The LOD of the PLSR model based on LIBS is 7.930%, and the determination coefficient of prediction (R^2_P), root-mean-square error of prediction ($RMSE_P$) and average relative error of prediction (ARE_P) are 0.963, 1.563% and 27.578%, respectively. The corresponding results of the PLSR model based on TR-LIBS are 4.175%, 0.977, 1.312% and 24.509%, respectively. However, the quantitative results of CF are relatively poor, the LOD of the PLSR model based on LIBS is 51.148%, and the R^2_P, $RMSE_P$ along with ARE_P are 0.308, 5.945% and 127.231%, respectively. The corresponding results of the PLSR model based on TR-LIBS are 18.680%, 0.701, 4.011% and 61.823%, respectively. The quantitative results of the PLSR model based on TR-LIBS and LIBS are consistent with the ICP-MS element detection results. The results of ICP-MS show that the content of Ca in TP and GP is higher, which is significantly different from that of BRF, while the element levels of SF, CF and BF are less different from those of BRF. In addition, the quantitative results also show that, compared with LIBS, the model based on TR-LIBS has improved all evaluation indexes on the training set and the prediction set, and the model performance is better.

To further analyze the quantitative performance of TR-LIBS and LIBS, the quantitative performance of PLSR on the two methods is compared by linear fitting with the adulteration TP as the representative in this section. The results are shown in Figure 5. The above experimental results prove the effectiveness of TR-LIBS in quantitative analysis.

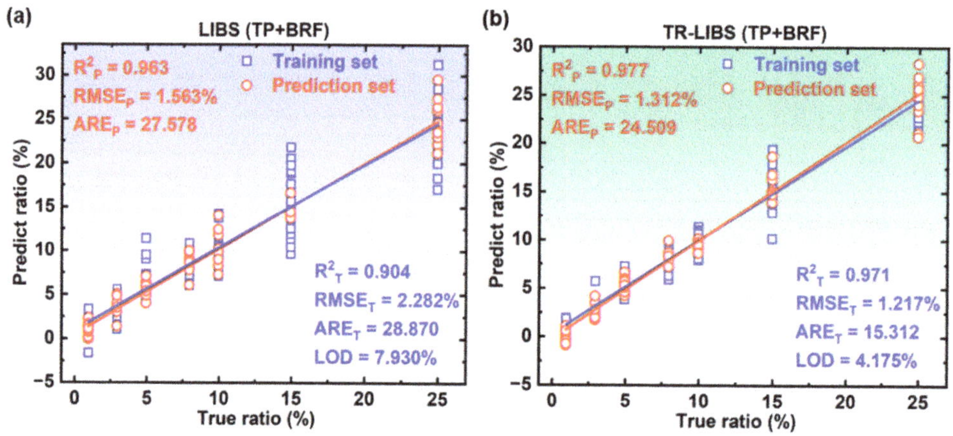

Figure 5. Linear fitting results of the PLSR model based on (**a**) LIBS and (**b**) TR-LIBS in the training set and prediction set of adulterant TP.

Table 3. Comparison of quantitative analysis results of five adulterants in BRF by LIBS and TR-LIBS. The subscript value is the corresponding optimal delay.

Adulteration Samples	Method	LOD (%)	Training Set			Prediction Set		
			R^2_T	$RMSE_T$ (%)	ARE_T (%)	R^2_P	$RMSE_P$ (%)	ARE_P (%)
sorghum flour (SF)	$LIBS_{(4.5)}$	20.357	0.662	4.340	58.528	0.606	5.017	69.303
	TR-LIBS	**12.665**	0.832	3.137	56.326	0.782	3.507	70.675
talc powder (TP)	$LIBS_{(2.0)}$	7.930	0.904	2.282	28.870	0.963	1.563	27.578
	TR-LIBS	**4.175**	0.971	1.217	15.312	0.977	1.312	24.509
corn flour (CF)	$LIBS_{(1.5)}$	51.148	0.414	5.992	124.777	0.308	5.945	127.231
	TR-LIBS	**18.680**	0.721	4.095	88.590	0.701	4.011	61.823
buckwheat flour (BF)	$LIBS_{(1.5)}$	19.138	0.722	4.103	74.164	0.536	4.945	87.205
	TR-LIBS	**11.184**	0.861	2.886	46.862	0.822	3.027	44.806
gypsum powder (GP)	$LIBS_{(1.5)}$	8.419	0.885	2.556	37.408	0.895	2.547	32.365
	TR-LIBS	**4.492**	0.968	1.311	19.374	0.974	1.321	21.409

4. Discussion

This section will discuss in detail the main reasons why TR-LIBS is better than traditional LIBS. Firstly, in addition to the spectral intensity information of elements, the evolution of spectral intensity is one of the important information in plasma. TR-LIBS can effectively mine this information by acquiring time-resolved spectra. Secondly, TR-LIBS can effectively reduce the influence of plasma fluctuations caused by external disturbances by acquiring multiple spectra, thus having stronger robustness. In summary, TR-LIBS improves qualitative and quantitative analysis performance by excavating more information from plasmas. To make the explanation more convincing, the t-distributed stochastic neighbor embedding (t-SNE) algorithm is utilized for the visual analysis of traditional LIBS and TR-LIBS data. The visualization results are shown in Figure 6. It can be seen that for the traditional LIBS, there is a serious overlap between the spectra of different samples and no obvious class boundaries. For TR-LIBS, the spectra of the same sample are more aggregated, and the spectra of different samples are more dispersed with obvious boundaries.

The LIBS spectral levels of the same element in different matrices are different. Therefore, the matrix effect has always been considered as one of the problems in the qualitative and quantitative analysis of LIBS [36]. However, in TR-LIBS, the matrix can be regarded as effective information as a potential feature for qualitative and quantitative analysis. Since the time-resolved spectra of the same element in different matrices have different trends. In addition, compared with the LIBS spectrum collected under the traditional single time series, the time-resolved LIBS collected under one time series has stronger resistance to fluctuation. These factors are of great significance to the qualitative and quantitative analysis of substances. The above principle analysis once again proves the effectiveness of TR-LIBS in improving the performance of the qualitative and quantitative analysis, which is of positive significance for promoting the development of LIBS in other fields such as environmental pollution detection.

According to the results of qualitative and quantitative analysis, this study also shows some limitations. The element level of the adulterated material is too close to that of the original blank sample, which will have a negative impact on the qualitative and quantitative performance of the model. This issue is also one of the directions that LIBS needs to further research in the field of food safety control in the future.

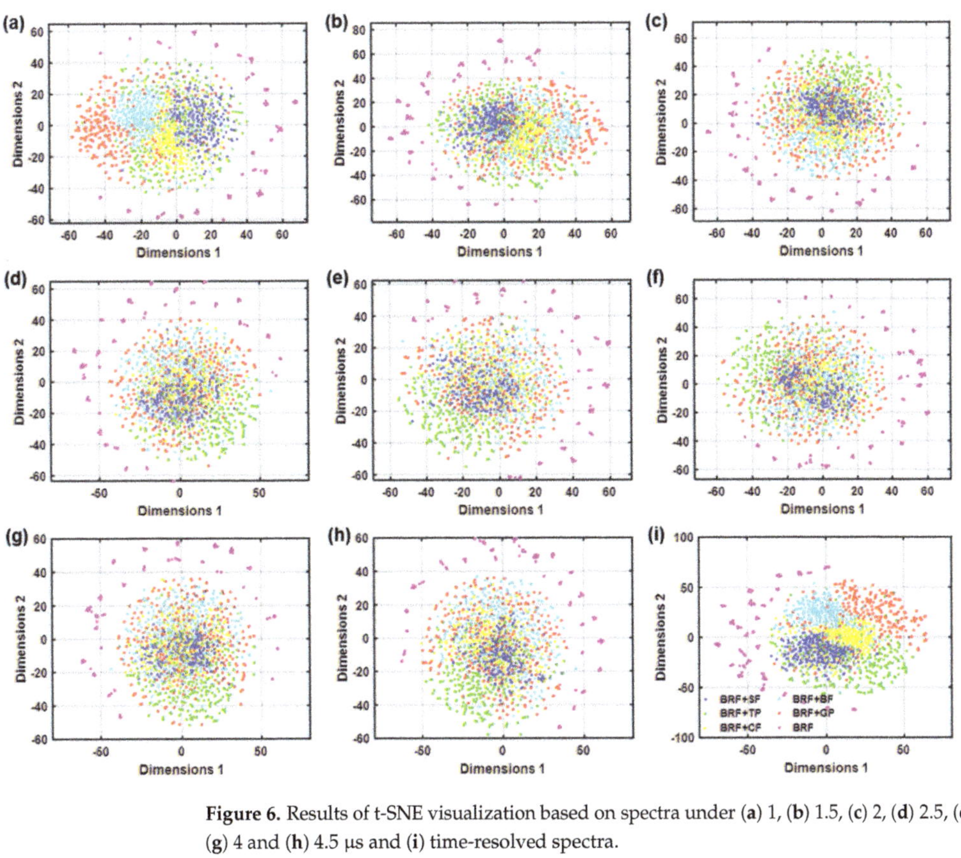

Figure 6. Results of t-SNE visualization based on spectra under (**a**) 1, (**b**) 1.5, (**c**) 2, (**d**) 2.5, (**e**) 3, (**f**) 3.5, (**g**) 4 and (**h**) 4.5 μs and (**i**) time-resolved spectra.

5. Conclusions

To realize the high precision identification of brown rice flour adulteration, this work proposes a novel method named TR-LIBS. TR-LIBS can excavate more effective information from plasma by obtaining time-resolved spectra, to improve the performance of the qualitative and quantitative analysis. This study fully verifies the effectiveness of the method from two aspects of adulterant classification and adulterant proportion quantification. For the qualitative classification of adulterants, the results of three traditional machine learning models (LDA, NB and SVM) are compared. The results show that the accuracy of the machine learning models based on TR-LIBS is significantly better than that of the machine learning models based on traditional LIBS, which is improved by about 3–11%. Moreover, for the qualitative classification of deep learning models, the test set accuracy of 1D CNN based on TR-LIBS is improved from 77.37% to 85.66% compared with traditional LIBS, an increase of more than 8%. For the quantitative analysis of the proportion of adulteration, the results of PLSR models based on traditional LIBS and TR-LIBS are compared. The results show that all the performance evaluation indexes of the model based on TR-LIBS are significantly better than those based on traditional LIBS. The LOD based on LIBS was about 8% to 51%, while the LOD of the five adulterants based on TR-LIBS was reduced from about 4% to 19%. This indicates that TR-LIBS significantly improves detection performance. Finally, visual analysis of spectra based on the t-SNE algorithm shows that there are obvious boundaries between different types of samples in TR-LIBS, while traditional LIBS does not. These results demonstrate that TR-LIBS is a reliable and stable high-precision qualitative and quantitative analysis method, which is of great significance for promoting the further application of LIBS in various fields.

Author Contributions: Conceptualization, L.G. and J.L.; methodology, H.M. and S.S.; software, Z.H.; validation, H.M., S.S. and N.D.; formal analysis, D.Z. and Z.H.; investigation, H.M. and N.D.; resources, L.G. and J.L.; data curation, S.S., D.Z. and Z.H.; writing—original draft preparation, H.M., S.S. and D.Z.; writing—review and editing, H.M., S.S. and D.Z.; visualization, S.S. and D.Z.; supervision, L.G. and J.L.; project administration, L.G. and J.L.; funding acquisition, L.G. All authors have read and agreed to the published version of the manuscript.

Funding: This research was financially supported by the National Natural Science Foundation of China [grant number 62075069].

Institutional Review Board Statement: Not applicable.

Informed Consent Statement: Not applicable.

Data Availability Statement: The datasets generated for this study are available on request to the corresponding author.

Conflicts of Interest: The authors declare that they have no known competing financial interests or personal relationships that could have appeared to influence the work reported in this paper.

References

1. Zhang, D.; Ouyang, S.; Cai, M.; Zhang, H.; Ding, S.; Liu, D.; Cai, P.; Le, Y.; Hu, Q.-N. FADB-China: A Molecular-Level Food Adulteration Database in China Based on Molecular Fingerprints and Similarity Algorithms Prediction Expansion. *Food Chem.* **2020**, *327*, 127010. [CrossRef]
2. Bhar, S.; Bose, T.; Dutta, A.; Mande, S.S. A Perspective on the Benefits of Consumption of Parboiled Rice over Brown Rice for Glycaemic Control. *Eur. J. Nutr.* **2022**, *61*, 615–624. [CrossRef] [PubMed]
3. Vemireddy, L.R.; Satyavathi, V.V.; Siddiq, E.A.; Nagaraju, J. Review of Methods for the Detection and Quantification of Adulteration of Rice: Basmati as a Case Study. *J. Food Sci. Technol.* **2015**, *52*, 3187–3202. [CrossRef]
4. Lee, Y.-M.; Kim, S.-A.; Lee, I.-K.; Kim, J.-G.; Park, K.-G.; Jeong, J.-Y.; Jeon, J.-H.; Shin, J.-Y.; Lee, D.-H. Effect of a Brown Rice Based Vegan Diet and Conventional Diabetic Diet on Glycemic Control of Patients with Type 2 Diabetes: A 12-Week Randomized Clinical Trial. *PLoS ONE* **2016**, *11*, e0155918. [CrossRef] [PubMed]
5. Li, S.; Xu, H.; Sui, Y.; Mei, X.; Shi, J.; Cai, S.; Xiong, T.; Carrillo, C.; Castagnini, J.M.; Zhu, Z.; et al. Comparing the LC-MS Phenolic Acids Profiles of Seven Different Varieties of Brown Rice (*Oryza Sativa* L.). *Foods* **2022**, *11*, 1552. [CrossRef]
6. Sim, J.H.; Tian, F.; Jung, S.Y.; Auh, J.-H.; Chun, H.S. Multiplex Polymerase Chain Reaction Assays for the Detection of the Zearalenone Chemotype of Fusarium Species in White and Brown Rice. *Int. J. Food Microbiol.* **2018**, *269*, 120–127. [CrossRef]

7. Pérez-Rodríguez, M.; Dirchwolf, P.M.; Silva, T.V.; Villafañe, R.N.; Neto, J.A.G.; Pellerano, R.G.; Ferreira, E.C. Brown Rice Authenticity Evaluation by Spark Discharge-Laser-Induced Breakdown Spectroscopy. *Food Chem.* **2019**, *297*, 124960. [CrossRef] [PubMed]
8. Ganopoulos, I.; Argiriou, A.; Tsaftaris, A. Adulterations in Basmati Rice Detected Quantitatively by Combined Use of Microsatellite and Fragrance Typing with High Resolution Melting (HRM) Analysis. *Food Chem.* **2011**, *129*, 652–659. [CrossRef]
9. Ariyama, K.; Shinozaki, M.; Kawasaki, A. Determination of the Geographic Origin of Rice by Chemometrics with Strontium and Lead Isotope Ratios and Multielement Concentrations. *J. Agric. Food Chem.* **2012**, *60*, 1628–1634. [CrossRef]
10. Barbosa, R.M.; de Paula, E.S.; Paulelli, A.C.; Moore, A.F.; Souza, J.M.O.; Batista, B.L.; Campiglia, A.D.; Barbosa, F. Recognition of Organic Rice Samples Based on Trace Elements and Support Vector Machines. *J. Food Compos. Anal.* **2016**, *45*, 95–100. [CrossRef]
11. Liu, Z.; Yuan, Y.; Xie, T.; Zhang, Y.; Shao, S.; Nie, J.; Xia, W.; Rogers, K.M.; Zhang, W. Long-Term Agricultural Effects on the Authentication Accuracy of Organic, Green, and Conventional Rice Using Isotopic and Elemental Chemometric Analyses. *J. Agric. Food Chem.* **2020**, *68*, 1213–1225. [CrossRef] [PubMed]
12. Yuan, Y.; Zhang, W.; Zhang, Y.; Liu, Z.; Shao, S.; Zhou, L.; Rogers, K.M. Differentiating Organically Farmed Rice from Conventional and Green Rice Harvested from an Experimental Field Trial Using Stable Isotopes and Multi-Element Chemometrics. *J. Agric. Food Chem.* **2018**, *66*, 2607–2615. [CrossRef] [PubMed]
13. Arslan, M.; Zareef, M.; Tahir, H.E.; Zhang, J.; Ahmad, W.; Rakha, A.; Shi, J.; Xiaobo, Z.; Khan, M.R. Discrimination of Basmati Rice Adulteration Using Colorimetric Sensor Array System. *Food Control* **2022**, *132*, 108513. [CrossRef]
14. Chung, I.-M.; Kim, J.-K.; An, Y.-J.; Kwon, C.; Kim, S.-Y.; Yang, Y.-J.; Yarnes, C.T.; Chi, H.-Y.; Kim, S.-H. Compound-Specific Δ13C and Δ15N Analyses of Fatty Acids and Amino Acids for Discrimination of Organic, Pesticide-Free, and Conventional Rice (*Oryza Sativa* L.). *Food Chem.* **2019**, *283*, 305–314. [CrossRef]
15. Griglione, A.; Liberto, E.; Cordero, C.; Bressanello, D.; Cagliero, C.; Rubiolo, P.; Bicchi, C.; Sgorbini, B. High-Quality Italian Rice Cultivars: Chemical Indices of Ageing and Aroma Quality. *Food Chem.* **2015**, *172*, 305–313. [CrossRef]
16. Prabakaran, M.; Chung, I.-M.; Son, N.-Y.; Chi, H.-Y.; Kim, S.-Y.; Yang, Y.-J.; Kwon, C.; An, Y.-J.; Ahmad, A.; Kim, S.-H. Analysis of Selected Phenolic Compounds in Organic, Pesticide-Free, Conventional Rice (*Oryza Sativa* L.) Using LC-ESI-MS/MS. *Molecules* **2019**, *24*, 67. [CrossRef]
17. Sabir, A.; Rafi, M.; Darusman, L.K. Discrimination of Red and White Rice Bran from Indonesia Using HPLC Fingerprint Analysis Combined with Chemometrics. *Food Chem.* **2017**, *221*, 1717–1722. [CrossRef]
18. Guo, L.; Zhang, D.; Sun, L.-X.; Yao, S.-C.; Zhang, L.; Wang, Z.-Z.; Wang, Q.-Q.; Ding, H.-B.; Lu, Y.; Hou, Z.-Y.; et al. Development in the Application of Laser-Induced Breakdown Spectroscopy in Recent Years: A Review. *Front. Phys.* **2021**, *16*, 22500. [CrossRef]
19. Attaviroj, N.; Kasemsumran, S.; Noomhorm, A. Rapid Variety Identification of Pure Rough Rice by Fourier-Transform Near-Infrared Spectroscopy. *Cereal Chem.* **2011**, *88*, 490–496. [CrossRef]
20. Li, C.; Li, B.; Ye, D. Analysis and Identification of Rice Adulteration Using Terahertz Spectroscopy and Pattern Recognition Algorithms. *IEEE Access* **2020**, *8*, 26839–26850. [CrossRef]
21. Zheng, L.; Bao, Q.; Weng, S.; Tao, J.; Zhang, D.; Huang, L.; Zhao, J. Determination of Adulteration in Wheat Flour Using Multi-Grained Cascade Forest-Related Models Coupled with the Fusion Information of Hyperspectral Imaging. *Spectrochim. Acta Part A Mol. Biomol. Spectrosc.* **2022**, *270*, 120813. [CrossRef]
22. Markiewicz-Keszycka, M.; Cama-Moncunill, X.; Casado-Gavalda, M.P.; Dixit, Y.; Cama-Moncunill, R.; Cullen, P.J.; Sullivan, C. Laser-Induced Breakdown Spectroscopy (LIBS) for Food Analysis: A Review. *Trends Food Sci. Technol.* **2017**, *65*, 80–93. [CrossRef]
23. Andersen, M.-B.S.; Frydenvang, J.; Henckel, P.; Rinnan, Å. The Potential of Laser-Induced Breakdown Spectroscopy for Industrial at-Line Monitoring of Calcium Content in Comminuted Poultry Meat. *Food Control* **2016**, *64*, 226–233. [CrossRef]
24. Huang, W.; Guo, L.; Kou, W.; Zhang, D.; Hu, Z.; Chen, F.; Chu, Y.; Cheng, W. Identification of Adulterated Milk Powder Based on Convolutional Neural Network and Laser-Induced Breakdown Spectroscopy. *Microchem. J.* **2022**, *176*, 107190. [CrossRef]
25. Zhang, D.; Nie, J.; Niu, X.; Chen, F.; Hu, Z.; Wen, X.; Li, Y.; Guo, L. Time-Resolved Spectral-Image Laser-Induced Breakdown Spectroscopy for Precise Qualitative and Quantitative Analysis of Milk Powder Quality by Fully Excavating the Matrix Information. *Food Chem.* **2022**, *386*, 132763. [CrossRef] [PubMed]
26. Moncayo, S.; Rosales, J.D.; Izquierdo-Hornillos, R.; Anzano, J.; Caceres, J.O. Classification of Red Wine Based on Its Protected Designation of Origin (PDO) Using Laser-Induced Breakdown Spectroscopy (LIBS). *Talanta* **2016**, *158*, 185–191. [CrossRef]
27. Ribeiro, M.C.S.; Senesi, G.S.; Cabral, J.S.; Cena, C.; Marangoni, B.S.; Kiefer, C.; Nicolodelli, G. Evaluation of Rice Varieties Using LIBS and FTIR Techniques Associated with PCA and Machine Learning Algorithms. *Appl. Opt.* **2020**, *59*, 10043–10048. [CrossRef]
28. Yang, P.; Zhou, R.; Zhang, W.; Tang, S.; Hao, Z.; Li, X.; Lu, Y.; Zeng, X. Laser-Induced Breakdown Spectroscopy Assisted Chemometric Methods for Rice Geographic Origin Classification. *Appl. Opt.* **2018**, *57*, 8297–8302. [CrossRef]
29. Esteki, M.; Shahsavari, Z.; Simal-Gandara, J. Use of Spectroscopic Methods in Combination with Linear Discriminant Analysis for Authentication of Food Products. *Food Control* **2018**, *91*, 100–112. [CrossRef]
30. Zdunek, R.; Nowak, M.; Plinski, E. Statistical Classification of Soft Solder Alloys by Laser-Induced Breakdown Spectroscopy: Review of Methods. *J. Eur. Opt. Soc. Rapid Publ.* **2016**, *11*, 16006i. [CrossRef]
31. Li, L.-N.; Liu, X.-F.; Yang, F.; Xu, W.-M.; Wang, J.-Y.; Shu, R. A Review of Artificial Neural Network Based Chemometrics Applied in Laser-Induced Breakdown Spectroscopy Analysis. *Spectroc. Acta Part B Atom. Spectr.* **2021**, *180*, 106183. [CrossRef]
32. Temiz, H.T.; Sezer, B.; Berkkan, A.; Tamer, U.; Boyaci, I.H. Assessment of Laser Induced Breakdown Spectroscopy as a Tool for Analysis of Butter Adulteration. *J. Food Compos. Anal.* **2018**, *67*, 48–54. [CrossRef]

33. Wang, X.Z.; Hao, Z.Q.; Guo, L.B.; Li, X.Y.; Zeng, X.Y.; Lu, Y.F. Quantitative Analysis of Manganese in Low Alloy Steel with Micro-Laser-Induced Breakdown Spectroscopy. *Spectrosc. Spectr. Anal.* **2017**, *37*, 1254–1258. [CrossRef]
34. Allegrini, F.; Olivieri, A.C. IUPAC-Consistent Approach to the Limit of Detection in Partial Least-Squares Calibration. *Anal. Chem.* **2014**, *86*, 7858–7866. [CrossRef] [PubMed]
35. Chawla, N.V.; Bowyer, K.W.; Hall, L.O.; Kegelmeyer, W.P. SMOTE: Synthetic Minority Over-Sampling Technique. *J. Artif. Intell. Res.* **2002**, *16*, 321–357. [CrossRef]
36. Zhang, D.; Chu, Y.; Ma, S.; Zhang, S.; Cui, H.; Hu, Z.; Chen, F.; Sheng, Z.; Guo, L.; Lu, Y. A Plasma-Image-Assisted Method for Matrix Effect Correction in Laser-Induced Breakdown Spectroscopy. *Anal. Chim. Acta* **2020**, *1107*, 14–22. [CrossRef]

foods

MDPI

Article

Classification of *Prunus* Genus by Botanical Origin and Harvest Year Based on Carbohydrates Profile

Marius Gheorghe Miricioiu [1], Roxana Elena Ionete [1], Diana Costinel [1] and Oana Romina Botoran [1,2,*]

1 ICSI Analytics Group, National Research and Development Institute for Cryogenics and Isotopic Technologies—ICSI, 240050 Râmnicu Vâlcea, Romania
2 Academy of Romanian Scientists, Splaiul Independentei 54, 050094 Bucharest, Romania
* Correspondence: oana.dinca@icsi.ro; Tel.: +4-0250-732744

Abstract: The ^{1}H-NMR carbohydrates profiling was used to discriminate fruits from *Rosaceae* family in terms of botanical origin and harvest year. The classification was possible by application of multivariate data analysis, such as principal component analysis (PCA), linear discriminant analysis (LDA) and Pearson analysis. Prior, a heat map was created based on ^{1}H-NMR signals which offered an overview of the content of individual carbohydrates in plum, apricot, cherry and sour cherry, highlighting the similarities. Although, the PCA results were almost satisfactory, based only on carbohydrates signals, the LDA reached 94.39% and 100% classification of fruits according to their botanical origin and growing season, respectively. Additionally, a potential association with the relevant climatic data was explored by applying the Pearson analysis. These findings are intended to create an efficient NMR-based solution capable of differentiating fruit juices based on their basic sugar profile.

Keywords: ^{1}H-NMR; carbohydrates; fruits; PCA; LDA

Citation: Miricioiu, M.G.; Ionete, R.E.; Costinel, D.; Botoran, O.R. Classification of *Prunus* Genus by Botanical Origin and Harvest Year Based on Carbohydrates Profile. *Foods* **2022**, *11*, 2838. https://doi.org/10.3390/foods11182838

Academic Editors: Mourad Kharbach and Samuli Urpelainen

Received: 10 August 2022
Accepted: 9 September 2022
Published: 14 September 2022

Publisher's Note: MDPI stays neutral with regard to jurisdictional claims in published maps and institutional affiliations.

1. Introduction

Nowadays, people pay particular attention to a balanced and controlled diet. Thus, there is a high demand for fresh fruits and derived juices which are considered, along with vegetables, the healthiest foods. People have also begun to recognize their beneficial contribution to health by protecting the human body against different type of illnesses through their vitamins which usually result in an immunity increase [1]. Unfortunately, this high demand for quality fruit juices has also attracted some frauds as shown in several studies on this topic [2–5]. Traceability, establishing chemical profile, appropriate physical attributes, adequate textural properties, controlled toxins and microbial contamination, as well as processing and storing method all represent characteristics of high quality, particularly for agricultural products. For example, a common fraudulent act is represented by the adulteration of juices by addition of other types of cheaper juices obtained from less expensive or more common fruits in the respective area [6]. Among others, dilution with water, sugar syrup and colorants, production method (conventional, organic, traditional techniques) as well as non-declared processing technologies (freezing, irradiation) can be mentioned [7]. Fruit juices have a high added value, and they are also more vulnerable to being subjected to different adulteration techniques because of their acknowledged attributes. Accordingly, authorities must be able to determine the compliance of a suspect product based on the product description, identify fraudulent processing practices, prevent adulteration, and control any other practices that may deceive the consumer. As is well known, it is not only consumers who benefit from food authenticity assessment but also the food industry parties which rely on the ability to ensure their commodities' label conformity and brand protection. In order to verify the authenticity of food items and guide and assist law enforcement, official bodies are periodically requesting an updated list of analytical procedures. In this scope, many analytical techniques such as molecular techniques

(DNA-based approaches) [8], isotopic approaches [9], ultraviolet-visible spectrophotometry (UV-VIS), high performance liquid chromatography (HPLC), gas chromatography (GC), inductively coupled plasma mass spectrometry (ICP-MS), atomic absorption spectroscopy (AAS), and infrared spectroscopy (IR) are used to control and to detect the food products' adulteration [2,6]. A single analyte or approach may rarely be associated with overall quality compliance because of the fruit juice matrices' complexity. As a result, their quality is derived from a unique combination of characteristics. To acquire the defined quality markers and control the critical production parameters, it is often essential to use multivariate data analysis. It is easier to discriminate between fruit samples and establish their authenticity when multivariate analysis is used in conjunction with different spectroscopic or chromatographic based methods. Data fusion may offer more precise information about a sample and better interpretation than a single approach, but usually the use of several techniques is not economically profitable, doesn't respect green chemistry principles, and is time consuming. In this respect, nuclear magnetic resonance (NMR) is a non-destructive technique providing high analytical precision, enabling simultaneously compounds identification, exposing complex frauds, and, in conjunction with chemometric analysis, revealing possible markers of fruits authenticity based on their composition profile [10]. Moreover, through NMR, spectra fruits can be classified according to their varietal and geographical origin [11–14], this being possible by application of some dedicated instruments, such as unsupervised machine learning techniques (principal component analysis—PCA) and pattern-recognition tools (discriminant analysis—DA).

The ^{1}H-NMR spectra in combination with PCA and DA have been applied to evaluate changes in the composition and metabolic profile of juices during thermal concentration process [15] for the correlation of different varieties of fruits [16,17] to reveal the juices' adulteration [18], and to evaluate the plant growth regulator in strawberries [19]. Also, the metabolomic analysis of ^{1}H NMR results obtained from fruit juices investigation give an overview about the relationships between the major metabolites and the sensory characteristics of the fruits. Among these metabolites, the carbohydrate content is the most relevant for the maturity level of fruits and for the consumer perception, and its domain is represented by three predominant components: glucose, fructose and sucrose [20,21]. Generally, the glucose and fructose are presented in lower quantities than sucrose and their relation affects the taste of the fruits [22]. Thus, the sweetness of the fruits is directly influenced by the fructose quantity which is 2.3 and 1.7 times sweeter than glucose and sucrose, respectively [20]. Beside this, fructose is highly appreciated for its therapeutic effect on the gastrointestinal tract [22,23]. Also, sucrose is appreciated as sweetener, energy source, and antioxidant [22], but its excessive consumption is closely related to the risk of caries, obesity, and diabetes [24].

In this study, the NMR method coupled with multivariate statistical analysis was used to obtain information about the variation of carbohydrate content in different fruits (plum, cherry, sour cherry, and apricot) from the *Rosaceae* family, *Prunus* genus, harvested in three different years. Furthermore, discrimination models based on DA were developed for samples classification. Another approach was the evaluation of different climatic conditions' (temperature and precipitation) influence over the three investigated years on the fructose, sucrose, glucose, and total carbohydrates content. These data will represent a contribution to regional horticultural varieties' characterization as well as provide useful information for industries which use fruit-derived nutrients in food production.

2. Materials and Methods

2.1. Chemicals

All analytical standards and reactants used for samples preparation and data interpretation were purchased from Sigma-Aldrich (St. Louis, MO, USA) and are hereafter listed: HCl (5N), NaOH (5N), D2O, and TMSP.

2.2. Sample Collection and Pre-Treatment

The fruits analyzed in this study were chosen from the *Rosaceae* family, *Prunus* genus, namely plum, cherry, sour cherry, and apricot. In particular, 76 samples were analyzed from Romania. The fruits were provided by Vâlcea Fruit Growing Research and Development Station, Romania, harvested in a state of consumption maturity, and the collected fruit samples were cooled and transported to the laboratory, assuring the maintenance of the cold chain. Then, the fruits (approximately 5 kg per variety) were washed with water, kept frozen, and stored at −20 °C in a freezer until sample preparation.

The juice was obtained by squeezing the whole fruit using a juicer (Moulinex, Jinan, China), and then approximately 100 g of sample was centrifuged with 10,000 rotations/min (Hettich ROTINA 420, Tuttlingen, Germany) for 10 min and filtered through a filter with 45 μm porosity, leading to about 20 mL being obtained. Further, the samples were pH adjusted to 2.65 by using 5 N HCl and 5 N NaOH. For NMR analysis, 700 μL of each sample was combined with 70 μL of deuterium oxide (D_2O), 99.9% D containing, and 0.05 wt. % of 3-(trimethylsilyl) propionic-2,2,3, 3-d 4 acid sodium salt as an internal standard (TMSP). The mixtures were transferred to 5 mm NMR tubes.

2.3. NMR Analysis and Data Processing

All ^{1}H-NMR spectra were recorded at 300 K temperature on a 400 MHz Bruker Avance spectrometer (Bruker France SAS, Wissembourg, France), operating at 9.4 T, equipped with a 5 mm BBO probe and ATM (Automatic Tuning Matching). In addition, the instrument was fitted with an autosampler from Bruker controlled by Icon NMR software which allows a loading of 60 samples. To complete the temperature equilibration, a time delay of 5 min between sample injection and preacquisition calibrations was set. The suppression of H_2O signals was assured through Bruker standard pulses sequence, noesygppr-1d, by applying continuous waves during the relaxation delay (10 s) with a mixing time of 10 ms. Each spectrum is the result of 8 scans and 32768 (33 k) data points. The spectral width was adjusted to 6402 Hz with an acquisition time of 2.559 s per scan. Spectra were Fourier transformed, manually phased, baseline corrected, and referenced to TMSP signal at 0 ppm using TopSpin 3.2 software (Bruker Biospin, Rheinstetten, Germany). Principal component analysis (PCA) and discriminant analysis (DA) were performed with LSTAT Addinsoft 2014.5.03 software version (Addinsoft, New York, NY, USA) in order to evaluate some potential variables and their influence on fruit juices' discrimination. Also, to reveal the possible relationships between sugars and climatic conditions, the Pearson correlations coefficient at $p \leq 0.05$ was used.

3. Results and Discussion

3.1. Fruits ^{1}H-NMR Spectra and Assignment of the Interest Peaks

In order to investigate the potential classification of fruit juices according to their botanical origin (apricot, cherry, sour cherry, and plum) and harvest year (2015, 2016, and 2017) their metabolic profile was obtained by using the ^{1}H-NMR spectroscopy. A typical 400 MHz ^{1}H 1D-NOESY NMR spectrum for a juice sample is shown in Figure 1.

Generally, the fruit juices' spectra are dominated by α-glucose, β-glucose, sucrose, and fructose signals, having the highest concentration among all the other metabolites. These belong to the carbohydrates region of the spectrum, which is placed in the middle-frequency region of the spectrum between 3.0 and 6.0 ppm, and it is followed by, in terms of signal intensities, amino acids, aliphatic region (from 0.5 and 3.0 ppm), and phenolic region (from 6.0 to 8.5 ppm) [16,17,25–29]. When the 5.40 to 4.10 ppm region was investigated, a pair of duplets were observed at 5.22 and 4.63 ppm that correspond to α and β glucose hydrogen in position 1. In the same range was identified one more pair of duplets at 5.39 and 4.20 ppm, which were assigned to hydrogen in position 1 from glucose and position 2 from fructose found in the composition of a sucrose molecule. Among the investigated spectra, sucrose signals were also detected at 5.40, from 4.19 to 4.21 ppm, 3.75 to 3.85 ppm, 3.67 and 3.54 ppm, while glucose signals were observed at 5.23, 5.22, 3.76, 3.71, 3.51, 3.505,

3.43, and 3.41 ppm, and from 4.64 to 3.21 ppm. Fructose signals were in the range of 3.57 to 3.60, 4.10 to 3.99 ppm, and 3.68 to 3.77 ppm, while the peaks were found at 3.60 and 4.10 ppm. The identification of carbohydrates signals was possible by consulting the literature [15,18,25–29]. Thus, the relevant ^{1}H-NMR peaks obtained for the carbohydrates region are reported in Table 1, and, for fruit juices' classification, each signal was taken into account, this region summing a total of 38 signals. These signals were different for all the studied fruit juices. For example, sucrose showed higher intensities in plum and apricot juices, while α-glucose, β-glucose, and fructose signals were more intense in the case of cherry and sour cherry juices.

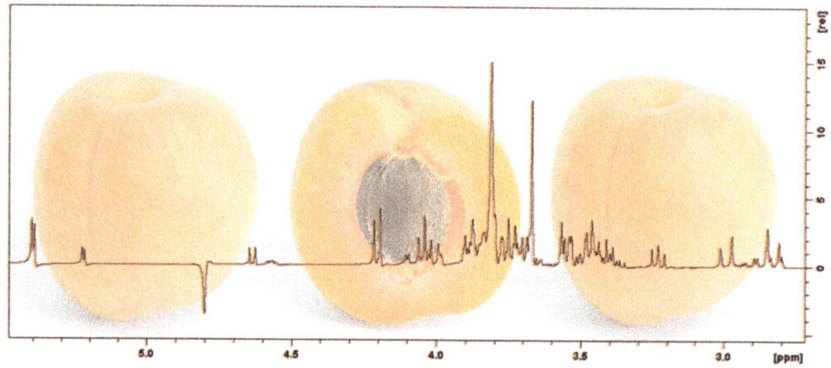

Figure 1. General ^{1}H-NMR spectra of apricot fruit juice—carbohydrates region.

Table 1. Chemical shift (δ) and assignment of metabolite resonances in the ^{1}H-NMR spectra of plum, cherry, sour cherry, and apricot juices.

Metabolites	**δ (ppm), Multiplicity (j, Hz) and Assignment**			
	Plum *Prunus domestica* **(n = 27)**	**Cherry** *Prunus avium* **(n = 22)**	**Apricot** *Prunus armeniaca* **(n = 15)**	**Sour Cherry** *Prunus cerasus* **(n = 12)**
β-d-glucose	3.23 (dd, CH), 3.40 (dd, CH), 4.63 (d, H1)	3.23 (dd, CH), 3.40 (dd, CH), 4.63 (d, H1)	3.23 (dd, CH), 3.40 (dd, CH), 4.63 (d, H1)	3.23 (dd, CH), 3.40 (dd, CH), 4.63 (d, H1)
methanol	3.36 (s, CH3)	3.36 (s, CH3)	3.36 (s, CH3)	3.36 (s, CH3)
α-d-glucose	3.43 (dd, CH), 3.50 (dd, CH), 5.22 (d, CH)	3.43 (dd, CH), 3.50 (dd, CH), 5.22 (d, CH)	3.43 (dd, CH), 3.50 (dd, CH), 5.22 (d, CH)	3.43 (dd, CH), 3.50 (dd, CH), 5.22 (d, CH)
fructose	3.60 (d, CH2) 3.99 (H5), 4.10 (d, H3, H4)	3.60 (d, CH2) 3.99 (H5), 4.10 (d, H3, H4)	3.60 (d, CH2) 3.99 (H5), 4.10 (d, H3, H4)	3.60 (d, CH2) 3.99 (H5), 4.10 (d, H3, H4)
sucrose	4.20 (d, H3), 5.39 (d, H1)		4.20 (d, H3), 5.39 (d, H1)	

3.2. Fruits Variety-Based Classification

In order to investigate the similarities between carbohydrates' metabolites in different fruit varieties, a heat map was generated (Figure 2). The heat map was built on the signals of sucrose, fructose, α-glucose, and β-glucose. As it can be seen from Figure 2, plums and sour cherries have the same content of sucrose, while cherry and sour cherry, as in the case of plum and apricot, have the same content of fructose. Moreover, similar contents of α-glucose and β-glucose in cherry and apricot compositions can be observed. The lower level of sucrose than α-glucose and β-glucose in cherry juices was also indicated in other study [28], this sugar being produced only in leaves by photosynthesis and translocated from different parts of the tree through phloem [20]. Therefore, the sucrose content is directly proportional with the photosynthetic rate. Beside this, the changes in fruit metabolism and the dilution caused by the fruit volume increase have significant effect on the whole carbohydrates content [20].

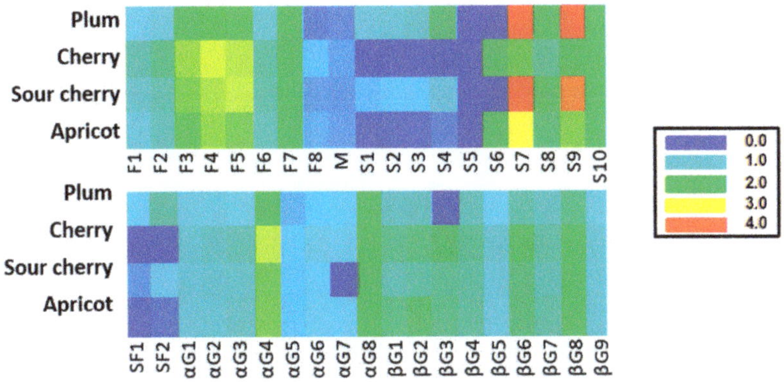

Figure 2. Heat map built on ^{1}H-NMR carbohydrates signals (F1–F8, fructose signals; M, methanol signal; S1–S10, sucrose signals; SF1 and SF2, sucrose-fructose signals; αG1–αG8, alpha glucose signals; and βG1–βG9 represent the beta glucose signals).

As previously stated, the NMR profiles reveal the existence of the same compounds but in different quantities. In order to determine the existence of latent variables linking different compounds, principal component analysis was performed to the intensity of 38 ^{1}H resonances (Table 1 and Figure 3). PCA analysis provided additional information regarding the separation of fruit juices, in term of botanical origin, as well as variable reduction, and assessment of clustering in PCA score. The four total principal components extracted, whose eigenvalues exceeded 1, explained 84.86% of the total variance. From this cumulative percentage the first two principal components accounted for 78.61% (72.81% for PC1 and 5.80% for PC2) indicating that it can be applied to obtain sample clusters in two-dimensional space. Thus, Figure 3 revealed a slight separation between the fruits with different botanical origin. The clearest separation is between apricots and the other fruit juices. The main differentiation was performed among the first principal component. Sour cherry and cherry juices samples have negative values for PC1, while apricot samples have positive values for PC1. A visible trend of separation is observed between cherries and sour cherries among PC2, the last being positively correlated with PC2. The plum samples are scattered, without a clear separation tendency.

For a closer examination regarding the relationships between the fruit variety and certain metabolites, the high dimensional data was projected to a lower dimensional subspace by means of the calculated principal components. (Figure 3b). Thus, it can be observed that the apricot juice is defined by a high content of sucrose, and cherry and sour cherry juices present a higher content of fructose, α-glucose, and β-glucose, while plums are characterized by a moderate content of sugars. Sugar accumulation in fruit is a dynamic quantitative trait that is usually influenced by environmental factors and is based on a variety of related physiological and biochemical processes. It is also determined by a number of enzymes that are correlated with the natural ecosystem and agricultural practices. Sucrose is generally accepted as the predominant sugar present in apricot fruit followed by glucose and fructose [30,31]. The mechanisms that influence the sugar profile of fruits have not been thoroughly studied. In apricots from several apricot-growing regions around the world, the proportions of the four carbohydrates varied substantially (sucrose: 18 to 82%; glucose: 5 to 28%; fructose: 2 to 17%). Individual sugar levels may vary significantly as a consequence of both genetic variability and environmental factors. According to Zhang [32], the patterns of each carbohydrate in apricots are genetically controlled, the accumulation being correlated to the sucrose-metabolizing enzyme activity [32]. Another important aspect regarding apricot sucrose content is related to the harvest maturity. Generally, the growth pattern of stone fruits is presented as a double-sigmoidal curve with three distinct growth phases: stage I characterized by a rapid growth period, stage II presented as a period of reduced growth, and stage III characterized by an even more rapid growth [33]. Overall, in a study conducted by Xi [34] regarding apricot fruit development and ripening, it was discovered that all sugars present an amplified accumulation pattern during the whole growth phase, with glucose and sucrose being the predominant identified carbohydrates. During the first stage, glucose was the most abundant sugar, but, as the process advanced, sucrose concentration increased exponentially. At the end of stage III, the sucrose concentration surpassed the total amount of glucose [34]. The investigation performed by Bae [35] presented similar results, namely sucrose content being smaller or even undetectable compared to glucose and fructose in the first stage of growth, followed by a significant increase at the full maturation stage [35]. These results indicate that the apricot carbohydrates accumulation metabolism shifts from glucose-predominated to sucrose-predominated during fruit development and ripening, presenting a balanced transition from synthesis to degradation. Furthermore, harvest maturity assessment represents a key component in determining fruit quality and customer acceptance [36]. Despite the fact that the apricot harvest ripeness period depends on the intended application, such as fresh consumption in local markets or long-term transit, realizing it too soon may have a detrimental impact among fruit sensory quality (as they will not be able to continue maturing or ripen properly), even if it makes it more resistant to postharvest handling. Apricots picked at appropriate maturity indexes are expected to present a higher customer satisfaction than harvested at commercial ripe maturity, as is the case of peaches or nectarine [37]. All of these findings suggest that apricot sucrose content could be considered a variety and maturity marker. As a result, the PCA scores plot represented a valuable tool for visualizing possible discrimination within the data set that can almost classify the juices, from Prunus genus but does not represent a perfect analysis to complete the botanical origin separation based only on carbohydrates data. In this regard, for further investigation of potential botanical origin separation of fruit juices, LDA is a technique that maximizes group separation (Figure 4). LDA proceeds by constructing discrimination functions from linear equations of variable data sets. The obtained model could be used to classify unknown observations (such as questionable or unknown samples). LDA was generated directly to the raw data set even if spectral data are known to be highly collinear.

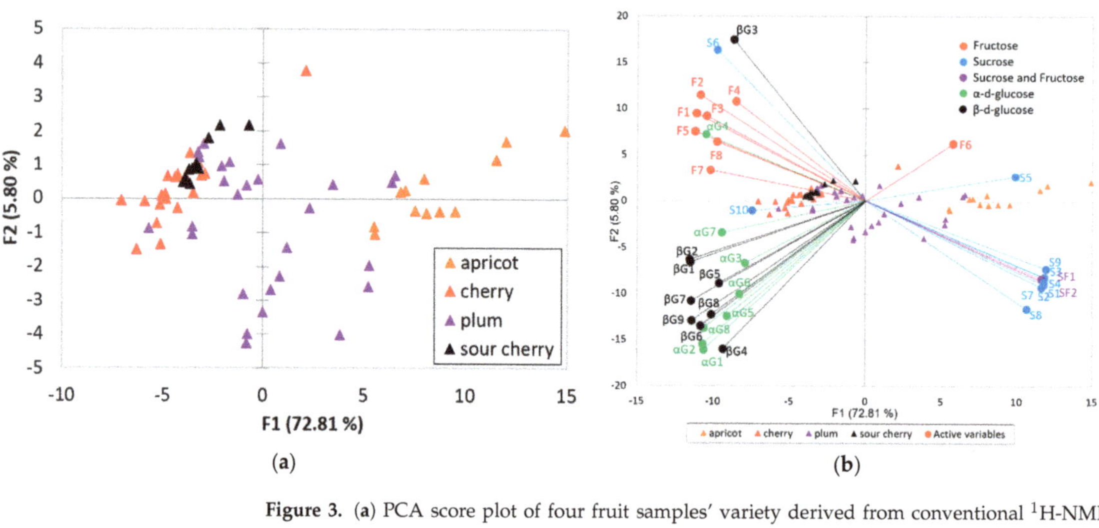

(a) (b)

Figure 3. (**a**) PCA score plot of four fruit samples' variety derived from conventional ^{1}H-NMR spectra; (**b**) correlation between the signals and factors responsible for fruits' variety separation.

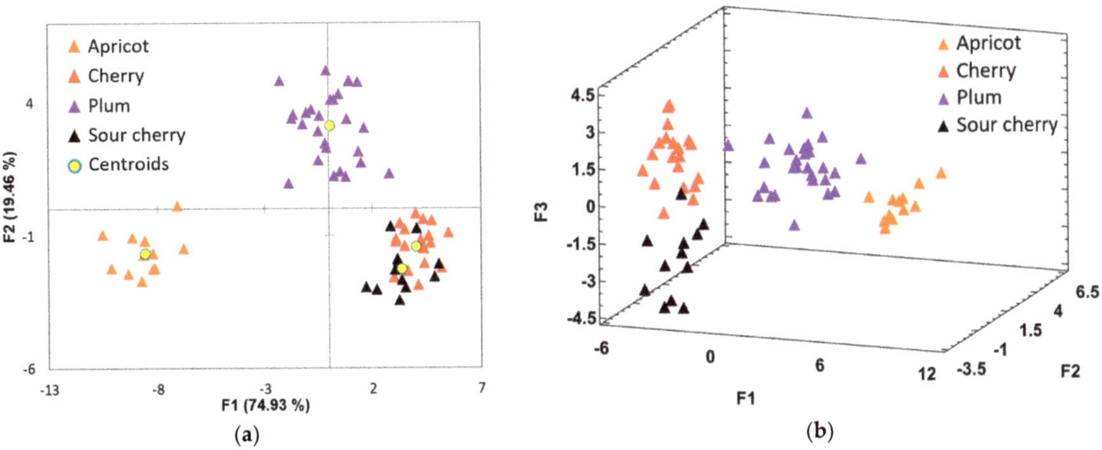

Figure 4. (**a**) 2D and (**b**) 3D plots showing the discrimination of fruit juices samples according to their botanical origin.

Three discriminant functions were obtained by LDA, which were demonstrated as suitable for correct classification. A total of 100% of the distribution was explained by this model; the first discriminant function accounted for 74.93%, the second function for about 19.46%, and the third one for 5.61%. Analyzing Figure 4a, it can be observed that the plum samples are found in the superior quadrant. The apricot samples are distributed in the third quadrant, excepting one sample. For the sour cherry and cherry, the results are not satisfactory because, as in the case of the other fruits, the separation among the two groups was not achieved due to some observations that overlapped. Function 1 provides the main separation between apricots and the other three fruit juices and was primarily correlated with sucrose, followed by fructose. Generally, the LDA results were superior and a reliable classification of fruit juices by botanical origin was achieved, except the cherry and sour cherry juices, where a slight separation was observed when using all three discriminant functions. The third discriminant function was mainly correlated with the sucrose signals, these not being identified in the cherries' and sour cherries' spectrum or being present in very low concentrations. The apricot, cherry, plum, and sour cherry juices were correctly classified with 100%, 100%, 96.30%, and 91.67%, respectively. The results obtained by the two applied multivariate techniques are similar; this fact implies that the results are reliable. More than 92% of the total fruit juices are classified correctly. However, these results could be biased to a certain extent, due to the unbalanced number of samples in each class; there two times more plum samples than apricot and sour cherry. Despite this drawback, the classifications for apricot, cherry, and plum are promising, once the classes are equilibrated.

3.3. Harvest Year-Based Classification and Climatic Condition Influence

For harvest year-based classification of fruit juices, the same procedure was followed, but the signals of carbohydrates were grouped after the three growing years (2015, 2016, and 2017) in order to obtain the statistical analysis.

The PCA results according to growing year are shown in Figure 5a. Unfortunately, the plots are spread over the all quadrants and no clusters were formed based on the harvest years.

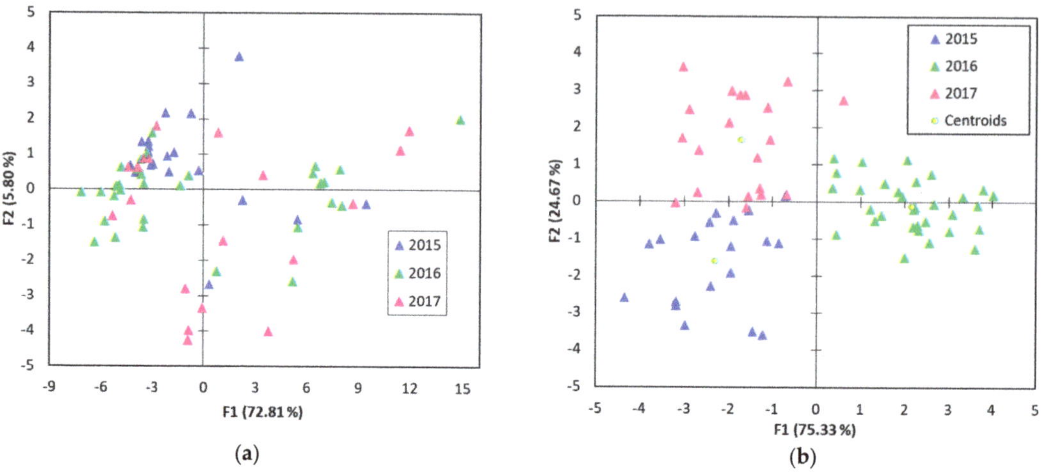

Figure 5. (**a**) PCA and (**b**) DA F1/F2 score plot showing the separation between the 3 harvest years.

Despite the fact that PCA presented a differentiation between the botanical origin of the fruits, in the case of the harvest years, the discrimination within each other with the same method was not achieved. In this respect, to come with a supplement to strengthen the results or to clear up some ambiguities related to PCA analysis, the same data were subjected to LDA analysis. The results of the LDA are superior (Figure 5b), and, according to the confusion matrix for the cross-validation results, a 96.05% was reached (two 2015 samples were classified as 2017, and one 2017 sample was classified as 2015). The first function accounted for 75.33% and the second function for 24.67%. All three groups of juice samples from fruits harvested in 2015, 2016, and 2017 are visibly separated. Taking into account the different botanical origin of the samples, a good classification of the three harvest years was done.

Due to the fact that each growing season could be different from the point of view of temperatures and precipitation, the potential influence of climatic condition on the sugar content was accessed by applying the Pearson analysis. Prior, the percentage of individual sugar from total sugars and the average temperature and precipitation recorded in the growing months of each season (June, July and August) for the studied years were calculated.

According to climatic condition data (temperature and precipitation), the growing seasons were different, presenting for the 2015 harvest year a seasonal average temperature of 20.8 °C and 58.3 mm precipitation, 20.3 °C and 86.7 mm for 2016, and a higher temperature (22.5 °C) and lower precipitation (66.7 mm) for 2017. In 2016, the average precipitation quantities were significantly higher than in 2015 and 2017, while the same year recorded the lowest average temperature, with almost 2 degrees below the average temperature recorded in 2017.

As it is shown in Table 2, the fructose and β-glucose are negatively correlated with temperature and positively with precipitation. Unfortunately, correlation between sucrose and precipitation or temperature was not found. From the same analysis, it can be remarked that fructose is highly and positively correlated with α-glucose and β-glucose. Instead, the sucrose contents significantly and negatively correlated with fructose, α-glucose, and β-glucose. Fruit juice quality is generally controlled by the growth conditions, being first dependent on the growing region climate (terroir, rainfall, humidity, hours of sun, temperature day/night, etc.) and secondly on technological influences. If the vintage year doesn't provide ideal circumstances for grape growth and quality, technology and enology can only have a limited impact. Even if a strong correlation was not observed, the temperature affects seedling development, photosynthesis, and the soluble sugar (fructose, glucose, and sucrose) content during fruit production [38,39]. The sugar content in various plant parts is known to decrease at high temperature due to plants' defense mechanisms to use photosynthetic products to support higher metabolic activities, while lower temperatures encouraged sugar accumulation. The observed differences in temperatures could have influenced the investigated fruits' soluble sugar leading to a possible explanation of the separation of the fruits according to the harvest.

Table 2. Pearson correlation coefficients between precipitation, temperature and carbohydrates.

Variables	Temperature	Precipitation	Carbohydrates	F	S	SF	αG	βG
Temperature	1	−0.564	−0.047	−0.261	0.121	0.105	−0.066	−0.173
Precipitation	−0.564	1	0.089	0.107	0.089	0.092	−0.036	0.083
Carbohydrates	−0.047	0.089	1	0.567	−0.672	−0.679	0.693	0.698
F	−0.261	0.107	0.567	1	−0.856	−0.863	0.778	0.816
S	0.121	0.089	−0.672	−0.856	1	0.993	−0.908	−0.885
SF	0.105	0.092	−0.679	−0.863	0.993	1	−0.893	−0.886
αG	−0.066	−0.036	0.693	0.778	−0.908	−0.893	1	0.903
βG	−0.173	0.083	0.698	0.816	−0.885	−0.886	0.903	1

4. Conclusions

Untargeted ^{1}H-NMR carbohydrates profiling of fruit juice proved to be a powerful tool for the classification and characterization of different *Prunus* sp. varieties. Apricot, cherry, sour cherry, and plum juices' ^{1}H-NMR spectra were observed to be dominated by sugars, which play a key role in defining the fruit's taste and flavor and make them good phytomarkers for species differentiation. Both botanical origin and harvest year could be assessed by means of multivariate analysis of the data, highlighting fruit specific sugar chemical traits. The carbohydrates profile differs among the analyzed fruit juices, and, based on their signal intensities, the PCA revealed strong correlation between sucrose and apricot juice, while the fructose and glucose were correlated with cherry and sour cherry juices. Despite these specific correlations, for a better botanical origin classification, an LDA was performed. A successful two-dimensional separation between plum and apricot juices was achieved, whereas cherry and sour cherry juices were slightly differentiated by implying all three discriminant functions. Additionally, the LDA represents a powerful technique for harvest year separation as the different botanical origin juices with different growing seasons (2015, 2016 and 2017) were clearly separated. From the point of view of climatic condition, by applying Pearson analysis, some correlations between fructose and β-glucose with temperature and precipitation were noted. However, solid associations between basic climatic factors and certain carbohydrate metabolic profiles could not be found, further studies being needed in order to obtain better correlations. Nevertheless, these results suggest the possible use of NMR-based sugar profiling for *Prunus* sp. botanical origin prediction and assessment of the possible correlation with different fruit juices, and more models can be developed for future predictions related to their quality and authenticity. In this regard, it should be noted that quality assessment was not the primary goal of the current fruit juices carbohydrate profile characterization research, which simply aimed to indicate potential metabolic profile variations across different species.

Author Contributions: Conceptualization, M.G.M. and O.R.B.; methodology, M.G.M.; software, O.R.B.; validation, M.G.M., D.C. and O.R.B.; formal analysis, M.G.M. and D.C.; investigation, M.G.M., D.C. and O.R.B.; resources, O.R.B.; data curation, O.R.B. and R.E.I.; writing—original draft preparation, M.G.M. and O.R.B.; writing—review and editing, M.G.M., R.E.I. and O.R.B.; visualization, O.R.B. and R.E.I.; supervision, O.R.B. and R.E.I.; project administration, O.R.B.; funding acquisition, O.R.B. All authors have read and agreed to the published version of the manuscript.

Funding: This work was supported by a grant of the Romanian Ministry of Education and Research, CNCS—UEFISCDI, project number PN-III-P1.1-PD-2019-0325, within PNCDI III (acronym ISO-TECH), ctr. PD 146/2020.

Institutional Review Board Statement: Not applicable.

Informed Consent Statement: Not applicable.

Data Availability Statement: The datasets generated for this study are available on request to the corresponding author.

Acknowledgments: This work was supported by a grant of the Romanian Ministry of Education and Re-search, CNCS—UEFISCDI, project number PN-III-P1.1-PD-2019-0325, within PNCDI III (acronym ISO-TECH), ctr. PD 146/2020, project no. PN 19110302 "Research on variation trends specific to stable isotopes in different tree species: deepening of fractionation mechanisms and interconnected chemical processes on the soil-water-plant chain", and Program 1—Development of the national research and development system, Subprogram 1.2—Institutional performance—Projects to finance excellence in RDI, Contract No. 19PFE/30.12.2021.

Conflicts of Interest: The authors declare no conflict of interest.

References

1. Dai, Q.; Borenstein, A.R.; Wu, Y.; Jackson, J.C.; Larson, E.B. Fruit and Vegetable Juices and Alzheimer's Disease: The Kame Project. *Am. J. Med.* **2006**, *119*, 751–759. [CrossRef]
2. Ogrinc, N.; Košir, I.J.; Spangenberg, J.E.; Kidrič, J. The Application of NMR and MS Methods for Detection of Adulteration of Wine, Fruit Juices, and Olive Oil. A Review. *Anal. Bioanal. Chem.* **2003**, *376*, 424–430. [CrossRef] [PubMed]
3. Yeganeh-Zare, S.; Farhadi, K.; Amiri, S. Rapid Detection of Apple Juice Concentrate Adulteration with Date Concentrate, Fructose and Glucose Syrup Using HPLC-RID Incorporated with Chemometric Tools. *Food Chem.* **2022**, *370*, 131015. [CrossRef] [PubMed]
4. Shojaee AliAbadi, M.H.; Karami-Osboo, R.; Kobarfard, F.; Jahani, R.; Nabi, M.; Yazdanpanah, H.; Mahboubi, A.; Nasiri, A.; Faizi, M. Detection of Lime Juice Adulteration by Simultaneous Determination of Main Organic Acids Using Liquid Chromatography-Tandem Mass Spectrometry. *J. Food Compos. Anal.* **2021**, *105*, 104223. [CrossRef]
5. Li, J.; Zhang, C.; Liu, H.; Liu, J.; Jiao, Z. Profiles of Sugar and Organic Acid of Fruit Juices: A Comparative Study and Implication for Authentication. *J. Food Qual.* **2020**, *2020*, 7236534. [CrossRef]
6. Muntean, E. Simultaneous Carbohydrate Chromatography and Unsuppressed Ion Chromatography in Detecting Fruit Juices Adulteration. *Chromatographia* **2010**, *71*, 69–74. [CrossRef]
7. Gómez-Ariza, J.L.; Villegas-Portero, M.J.; Bernal-Daza, V. Characterization and Analysis of Amino Acids in Orange Juice by HPLC-MS/MS for Authenticity Assessment. *Anal. Chim. Acta* **2005**, *540*, 221–230. [CrossRef]
8. Amaral, J.S. Target and Non-Target Approaches for Food Authenticity and Traceability. *Foods* **2021**, *10*, 172. [CrossRef]
9. Katerinopoulou, K.; Kontogeorgos, A.; Salmas, C.E.; Patakas, A.; Ladavos, A. Geographical Origin Authentication of Agri-Food Products: A Review. *Foods* **2020**, *9*, 489. [CrossRef]
10. Popescu, R.; Ionete, R.E.; Botoran, O.R.; Costinel, D.; Bucura, F.; Geana, E.I.; Alabedallat, Y.F.J.; Botu, M. 1H-NMR Profiling and Carbon Isotope Discrimination as Tools for the Comparative Assessment of Walnut (Juglans Regia L.) Cultivars with Various Geographical and Genetic Origins-a Preliminary Study. *Molecules* **2019**, *24*, 1378. [CrossRef]
11. Sobolev, A.P.; Mannina, L.; Proietti, N.; Carradori, S.; Daglia, M.; Giusti, A.M.; Antiochia, R.; Capitani, D. Untargeted NMR-based methodology in the study of fruit metabolites. *Molecules* **2015**, *20*, 4088–4108. [CrossRef]
12. Longobardi, F.; Ventrella, A.; Bianco, A.; Catucci, L.; Cafagna, I.; Gallo, V.; Mastrorilli, P.; Agostiano, A. Non-targeted 1H NMR fingerprinting and multivariate statistical analyses for the characterisation of the geographical origin of Italian sweet cherries. *Food Chem.* **2013**, *141*, 3028–3033. [CrossRef]
13. Pei, Y.; He, C.; Liu, H.; Shen, G.; Feng, J. Compositional Analysis of Four Kinds of Citrus Fruits with an NMR-Based Method for Understanding Nutritional Value and Rational Utilization: From Pericarp to Juice. *Molecules* **2022**, *27*, 2579. [CrossRef]
14. Park, S.J.; Hyun, S.-H.; Suh, H.W.; Lee, S.-Y.; Min, T.-S.; Auh, J.-H.; Lee, H.-J.; Kim, J.-H.; Cho, S.-M.; Choi, H.-K. Differentiation of black raspberry fruits according to species and geographic origins by genomic analysis and 1H-NMR-based metabolic profiling. *J. Korean Soc. Appl. Biol. Chem.* **2012**, *55*, 633–642. [CrossRef]
15. Salvino, R.A.; Colella, M.F.; Luca, G. De NMR-Based Metabolomics Analysis of Calabrian Citrus Fruit Juices and Its Application to Industrial Process Quality Control. *Food Control* **2021**, *121*, 107619. [CrossRef]
16. Tamasi, G.; Bonechi, C.; Leone, G.; Andreassi, M.; Consumi, M.; Sangiorgio, P.; Verardi, A.; Rossi, C.; Magnani, A. Varietal and Geographical Origin Characterization of Peaches and Nectarines by Combining Analytical Techniques and Statistical Approach. *Molecules* **2021**, *26*, 4128. [CrossRef]
17. Girelli, C.R.; Accogli, R.; Del Coco, L.; Angilè, F.; De Bellis, L.; Fanizzi, F.P. H-NMR-Based Metabolomic Profiles of Different Sweet Melon (Cucumis Melo L.) Salento Varieties: Analysis and Comparison. *Food Res. Int.* **2018**, *114*, 81–89. [CrossRef]
18. Vogels, J.T.W.E.; Terwel, L.; Tas, A.C.; Van Den Berg, F.; Dukel, F. Detection of Adulteration in Orange Juices by a New Screening Method Using Proton NMR Spectroscopy in Combination with Pattern Recognition Techniques. *J. Agric. Food Chem.* **1996**, *44*, 175–180. [CrossRef]
19. An, L.; Ma, J.; Wang, H.; Li, F.; Qin, D.; Wu, J.; Zhu, G. NMR-Based Global Metabolomics Approach to Decipher the Metabolic Effects of Three Plant Growth Regulators on Strawberry Maturation. *Food Chem.* **2018**, *269*, 559–566. [CrossRef]

20. Dugalic, K.; Sudar, R.; Viljevac, M.; Josipovic, M.; Cupic, T. Sorbitol and Sugar Composition in Plum Fruits Influenced by Climatic Conditions. *J. Agric. Sci. Technol.* **2014**, *16*, 1145–1155.
21. Manganaris, G.A.; Vicente, A.R.; Crisosto, C.H. Effect of Pre-Harvest and Post-Harvest Conditions and Treatments on Plum Fruit Quality. *CAB Rev. Perspect. Agric. Vet. Sci. Nutr. Nat. Resour.* **2008**, *3*, 1–9. [CrossRef]
22. Cantín, C.M.; Gogorcena, Y.; Moreno, M.Á. Analysis of Phenotypic Variation of Sugar Profile in Different Peach and Nectarine [*Prunus persica* (L.) Batsch] Breeding Progenies. *J. Sci. Food Agric.* **2009**, *89*, 1909–1917. [CrossRef]
23. Muir, J.G.; Rose, R.; Rosella, O.; Liels, K.; Barrett, J.S.; Shepherd, S.J.; Gibson, P.R. Measurement of Short-Chain Carbohydrates in Common Australian Vegetables and Fruits by High-Performance Liquid Chromatography (HPLC). *J. Agric. Food Chem.* **2009**, *57*, 554–565. [CrossRef] [PubMed]
24. Santucci, C.; Tenori, L.; Luchinat, C. NMR Fingerprinting as a Tool to Evaluate Post-Harvest Time-Related Changes of Peaches, Tomatoes and Plums. *Food Res. Int.* **2015**, *75*, 106–114. [CrossRef]
25. Girelli, C.R.; De Pascali, S.A.; Del Coco, L.; Fanizzi, F.P. Metabolic Profile Comparison of Fruit Juice from Certified Sweet Cherry Trees (*Prunus avium* L.) of Ferrovia and Giorgia Cultivars: A Preliminary Study. *Food Res. Int.* **2016**, *90*, 281–287. [CrossRef]
26. Sobolev, A.P.; Segre, A.; Lamanna, R. Proton High-Field NMR Study of Tomato Juice. *Magn. Reson. Chem.* **2003**, *41*, 237–245. [CrossRef]
27. Capitani, D.; Sobolev, A.P.; Tomassini, A.; Sciubba, F.; De Salvador, F.R.; Mannina, L.; Delfini, M. Peach Fruit: Metabolic Comparative Analysis of Two Varieties with Different Resistances to Insect Attacks by NMR Spectroscopy. *J. Agric. Food Chem.* **2013**, *61*, 1718–1726. [CrossRef]
28. Goulas, V.; Minas, I.S.; Kourdoulas, P.M.; Lazaridou, A.; Molassiotis, A.N.; Gerothanassis, I.P.; Manganaris, G.A. 1H NMR Metabolic Fingerprinting to Probe Temporal Postharvest Changes on Qualitative Attributes and Phytochemical Profile of Sweet Cherry Fruit. *Front. Plant Sci.* **2015**, *6*, 959. [CrossRef]
29. Navarro, Y.; Soengas, R.; Iglesias, M.J.; Ortiz, F.L. Use of NMR for the Analysis and Quantification of the Sugar Composition in Fresh and Store-Bought Fruit Juices. *J. Chem. Educ.* **2020**, *97*, 831–837. [CrossRef]
30. Fan, X.; Zhao, H.; Wang, X.; Cao, J.; Jiang, W. Sugar and Organic Acid Composition of Apricot and Their Contribution to Sensory Quality and Consumer Satisfaction. *Sci. Hortic.* **2017**, *225*, 553–560. [CrossRef]
31. Falchi, R.; Bonghi, C.; Drincovich, M.F.; Famiani, F.; Lara, M.v.; Walker, R.P.; Vizzotto, G. Sugar Metabolism in Stone Fruit: Source-Sink Relationships and Environmental and Agronomical Effects. *Front. Plant Sci.* **2020**, *11*, 1820. [CrossRef]
32. Zhang, Q.; Feng, C.; Li, W.; Qu, Z.; Zeng, M.; Xi, W. Transcriptional Regulatory Networks Controlling Taste and Aroma Quality of Apricot (*Prunus armeniaca* L.) Fruit during Ripening. *BMC Genom.* **2019**, *20*, 45. [CrossRef]
33. Famiani, F.; Bonghi, C.; Chen, Z.H.; Drincovich, M.F.; Farinelli, D.; Lara, M.v.; Proietti, S.; Rosati, A.; Vizzotto, G.; Walker, R.P. Stone Fruits: Growth and Nitrogen and Organic Acid Metabolism in the Fruits and Seeds—A Review. *Front. Plant Sci.* **2020**, *11*, 1427. [CrossRef]
34. Xi, W.; Zheng, H.; Zhang, Q.; Li, W. Profiling Taste and Aroma Compound Metabolism during Apricot Fruit Development and Ripening. *Int. J. Mol. Sci.* **2016**, *17*, 998. [CrossRef]
35. Bae, H.; Yun, S.K.; Jun, J.H.; Yoon, I.K.; Nam, E.Y.; Kwon, J.H. Assessment of Organic Acid and Sugar Composition in Apricot, Plumcot, Plum, and Peach during Fruit Development. *J. Appl. Bot. Food Qual.* **2014**, *87*, 24–29. [CrossRef]
36. Ceccarelli, A.; Farneti, B.; Frisina, C.; Allen, D.; Donati, I.; Cellini, A.; Costa, G.; Spinelli, F.; Stefanelli, D. Harvest Maturity Stage and Cold Storage Length Influence on Flavour Development in Peach Fruit. *Agronomy* **2018**, *9*, 10. [CrossRef]
37. Farina, V.; Lo Bianco, R.; Mazzaglia, A. Evaluation of Late-Maturing Peach and Nectarine Fruit Quality by Chemical, Physical, and Sensory Determinations. *Agriculture* **2019**, *9*, 189. [CrossRef]
38. Wu, X.Y.; Han, W.; Yang, Z.Q.; Zhang, Y.X.; Zheng, Y.J. The difference in temperature between day and night affects the strawberry soluble sugar content by influencing the photosynthesis, respiration and sucrose phosphatase synthase. *Hort. Sci.* **2021**, *48*, 174–182. [CrossRef]
39. Menzel, C.M. Effect of Temperature on Soluble Solids Content in Strawberry in Queensland, Australia. *Horticulturae* **2022**, *8*, 367. [CrossRef]

foods MDPI

Article

Chemical Authentication and Speciation of *Salvia* Botanicals: An Investigation Utilizing GC/Q-ToF and Chemometrics

Joseph Lee [1], Mei Wang [2,*], Jianping Zhao [1], Bharathi Avula [1], Amar G. Chittiboyina [1], Jing Li [3], Charles Wu [3] and Ikhlas A. Khan [1,4,*]

1 National Center for Natural Products Research, School of Pharmacy, University of Mississippi, University City, MS 38677, USA; jclee1@olemiss.edu (J.L.); jianping@olemiss.edu (J.Z.); bavula@olemiss.edu (B.A.); amar@olemiss.edu (A.G.C.)
2 Natural Products Utilization Research Unit, Agricultural Research Service, United States Department of Agriculture, University City, MS 38677, USA
3 Botanical Review Team, Office of New Drug Product, Office of Pharmaceutical Quality, Center for Drug Evaluation and Research, Food and Drug Administration, Silver Spring, MD 20993, USA; jing.li2@fda.hhs.gov (J.L.); charles.wu@fda.hhs.gov (C.W.)
4 Division of Pharmacognosy, Department of BioMolecular Sciences, School of Pharmacy, University of Mississippi, University City, MS 38677, USA
* Correspondence: mei.wang@usda.gov (M.W.); ikhan@olemiss.edu (I.A.K.); Tel.: +1-662-915-1046 (M.W.); +1-662-915-7821 (I.A.K.)

Abstract: Members of the genus *Salvia* are used as culinary herbs and are prized for their purported medicinal attributes. Since physiological effects can vary widely between species of *Salvia,* it is of great importance to accurately identify botanical material to ensure safety for consumers. In the present study, an in-depth chemical investigation is performed utilizing GC/Q-ToF combined with chemometrics. Twenty-four authentic plant samples representing five commonly used *Salvia* species, *viz. S. apiana*, *S. divinorum*, *S. mellifera*, *S. miltiorrhiza*, and *S. officinalis*, are analyzed using a GC/Q-ToF technique. High-resolution spectral data are employed to construct a sample class prediction (SCP) model followed by principal component analysis (PCA) and partial least square discriminant analysis (PLS-DA). This model demonstrates 100% accuracy for both prediction and recognition abilities. Additionally, the marker compounds present in each species are identified. Furthermore, to reduce the time required and increase the confidence level for compound identification and the classification of different *Salvia* species, a personal compound database and library (PCDL) containing marker and characteristic compounds is constructed. By combining GC/Q-ToF, chemometrics, and PCDL, the unambiguous identification of *Salvia* botanicals is achieved. This high-throughput method can be utilized for species specificity and to probe the overall quality of various *Salvia*-based products.

Keywords: *Salvia* spp.; GC/Q-ToF analysis; chemometrics; quality evaluation; chemical fingerprints

Citation: Lee, J.; Wang, M.; Zhao, J.; Avula, B.; Chittiboyina, A.G.; Li, J.; Wu, C.; Khan, I.A. Chemical Authentication and Speciation of *Salvia* Botanicals: An Investigation Utilizing GC/Q-ToF and Chemometrics. *Foods* **2022**, *11*, 2132. https://doi.org/10.3390/foods11142132

Academic Editors: Mourad Kharbach and Samuli Urpelainen

Received: 14 June 2022
Accepted: 13 July 2022
Published: 19 July 2022

1. Introduction

Members of the plant genus *Salvia* have a long and rich history of use as both culinary and medicinal herbs [1,2]. In general, the perennial shrubs have long stems which can reach heights of 50–100 cm. Although found throughout the world, most *Salvia* species grow in the Mediterranean region, Southeast Asia, and Central and South America [2]. Alluding to the importance of this plant's medicinal properties, the word "*Salvia*" is derived from the Latin word "salvere", meaning "to save" [2]. Members of the genus *Salvia* have been purported to possess a wide range of pharmacological properties, including anti-inflammatory, anti-dementia, anti-nociceptive, anti-hypertensive, anti-lipidemic, anti-mutagenic, anti-hyperglycemic, and anti-ischemic effects [1–5]. In addition to these purported properties, members of this genus have also been reported to possess anti-microbial and anti-oxidative activities [2–5]. These pharmacological properties vary among *Salvia* genus members.

Perhaps one of the most well-known members, *Salvia officinalis*, is utilized both as a culinary and a medicinal herb. This evergreen plant, native to southern Europe, is also cultivated in the United States and Central Asia [5]. As a medicinal herb, both the British Pharmacopoeia and the German Commission E have recognized its use to treat oral cavity and stomach ailments [5,6].

Another genus member, *Salvia apiana* (white sage), also has a rich history of use as a medicinal herb. This drought-resistant shrub, native to California and Baja California, can grow up to 1–3 m high. Traditionally, the herb has been used for its purported diuretic, anxiolytic, and anti-microbial properties [7]. In addition to its medicinal use, the plant is also an important part of traditional Native American religious and healing ceremonies [7].

Salvia mellifera (black sage), native to California and parts of Mexico, has also been used as a traditional healing herb. An infusion comprised of the aerial portions of the plant has traditionally been used as a drink to relieve muscle aches and pains [8].

As a popular ingredient in many traditional Chinese medicine (TCM) preparations, the red rhizomes of *Salvia miltiorrhiza* contain a unique group of compounds known as tanshinones, which have been reported to possess a broad spectrum of pharmacological activities [3,9]. The natural habitat of *S. miltiorrhiza* includes the hilly regions of China, Japan, Korea, and Mongolia; however, due to growing demand, most plant material is typically obtained from commercial farming [9]. Commonly referred to as "Danshen" or "Tanshen" in China, the rhizomes are purported to be beneficial for a number of disorders, including hyperlipidemia, vascular diseases, stroke, arthritis, and hepatitis [9].

Salvia divinorum is the only member of the *Salvia* genus to contain salvinorins, a group of neoclerodane diterpenes [10,11]. The plant, which can grow up to 1.5 m high, is native to southern Mexico. Traditionally, the Mazatec people would chew the herb or prepare an infusion using water and portions of the herb to take advantage of its psychoactive compounds. One compound, in particular, salvinorin A, possesses psychoactive properties and is a highly selective kappa-opioid receptor agonist [6,11]. Due to its high abuse potential, many local jurisdictions and countries have begun or are considering regulating the herb and/or salvinorin A as a controlled substance [6,11].

With nearly 900 species included in the genus *Salvia*, identifying plant materials and products can be a daunting task [4]. Thus, chemical fingerprint analyses represent a comprehensive approach for the quality assessment of *Salvia* botanicals and their finished products. Clearly, this is an important task given the wide range of pharmacological properties found in members of this genus. A range of methods have been developed to aid in the species identification of *Salvia* plant material [6,11–17]. Perhaps one of the most popular analytical techniques utilized for the identification and quality control of *Salvia* species is liquid chromatography/mass spectrometry (LC/MS) [6,12,13,15–19]. A brief literature search can yield numerous studies concerning this subject [6,12,13,15–19]. In addition to traditional LC/MS, techniques utilizing liquid chromatography/quadrupole time-of-flight (LC/Q-ToF) and liquid chromatography/triple quadrupole mass spectrometry (LC/TQ) have also been described [6]. While valuable to researchers, LC/MS/MS instruments are not often used by botanical industries for quality control purposes due to the cost of the instrument and the necessary technical skills required to develop and operate such tools. Due to the physical separation characteristics of LC/MS, the vast majority of previous research has focused on the non-volatile, LC-amendable polar compounds present in *Salvia* species [6,12,13,15,16,18,19]. Although polar compounds of *Salvia* are pharmacologically important, volatile constituents have also been implicated with bio-active properties and could be useful for establishing species-specific chemical fingerprinting [4,7,20,21].

DNA barcoding is another technique that has been proposed to aid in the species identification of *Salvia*. The authors of one study developed an effective DNA barcoding method to differentiate *S. miltiorrhiza* from other *Salvia* species [22]. Although differentiation was achieved, this technique involved extensive and complex sample preparation which did not lend itself to high-throughput sample analysis. In addition, the authors explained that their method was particularly developed for *S. miltiorrhiza* identification and may not be ideal for other *Salvia* species [22].

A high-performance thin-layer chromatography (HPTLC) fingerprinting method for 20 *Salvia* species was developed by Ciesla and co-workers [23]. The method utilized polar and semi-polar compounds (mostly polyphenols) for identification purposes and was successfully validated. Regarding its applicability as a high-throughput method, the authors estimated that 20 samples could be fully processed within one hour. However, one major limitation of this method was the requirement for a large amount of sample material (around 5 g), which could be problematic if the plant material is difficult to obtain, i.e., *S. divinorum* [23]. Therefore, there is a need to develop efficient and reliable methods.

Gas chromatography/mass spectrometry (GC/MS) is a well-established means of obtaining chemical fingerprints from various plants, primarily by analyzing volatile compounds. For example, this technique has been used to establish the chemical fingerprints of *Salvia* species by Rzepa and colleagues [24]. Based on the number of products sold in the U.S. market and pending botanical drug applications, five *Salvia* species, *viz. S. apiana, S. divinorum, S. mellifera, S. miltiorrhiza,* and *S. officinalis,* were selected to conduct a comprehensive and comparative study for quality evaluation and identification purposes. Currently, to the authors' knowledge, a comparative study of the five selected *Salvia* species has not been conducted. Given these five species' extensive history and current use as medicinal herbs, it is important to develop reliable and efficient identification methods for species specificity purposes and to assure the overall quality of various *Salvia*-based finished products.

With this information in mind, our goal is to develop a simple, reliable, and efficient GC method coupled with accurate mass spectrometry to establish species-specific chemical fingerprints of *Salvia*. Chemometric analysis and principal component analysis (PCA) are applied to differentiate between *Salvia* species, as well as to establish a sample class prediction model (SCP) based on partial least square discriminant analysis (PLS-DA) for the quality evaluation of commercial products. Marker and characteristic compounds present in each of the five species are identified. The integration of analytical data with statistical tools and the development of personal compound databases and libraries (PCDL) are anticipated to expedite the rapid evaluation of the quality of *Salvia*-based finished products, including raw materials used in commerce.

2. Materials and Methods

2.1. Plant Material

Twenty-four authentic plant samples from five *Salvia* species were used for this investigation. The samples included both leaf and aerial portions of the plants from *S. divinorum, S. officinalis, S. mellifera,* and *S. apiana,* with each having 3, 7, 3, and 5 individual samples, respectively. *S. divinorum* samples were procured from Trish Flaster (Botanical Liaisons, LLC, Boulder, CO, USA) (#578) and cultivated at the Medicinal Plant Garden (University, MS, USA) (#18434, #22491). *S. officinalis* samples were obtained from the Missouri Botanical Garden (St. Louis, MO, USA) (#7917, #7686, 20712), China (#16732), Richters.com (#13095), Trish Flaster (#2852), and Williams Warehouse (USA) (#1523). *S. mellifera* samples were sourced from AHP (Scotts Valley, CA, USA) (#22771, #22772), and SageRageHerb (Montclair, CA, USA) (#22506). *S. apiana* samples were obtained from AHP (#22773), Richters.com (#13096), SageRageHerb (Montclair, CA, USA) (#22502), and commercial sources (#22497, #22498). Six individual samples from the root portion of *S. miltiorrhiza* were also investigated. These samples were procured from Harvard Medical School (Boston, MA, USA) (#9729), the Medicinal Plant Garden (#11750), Beijing Yuke Botanical Development Co. Ltd. (Beijing, China) (#767), Missouri Botanical Garden (#8676, #12535), and a commercial source (#5399). The authenticity of the collected botanical samples was established based on morpho-anatomical and organoleptic properties by Dr. John Adams, a taxonomist at the National Center for Natural Products Research (NCNPR), University of Mississippi. In addition, DNA barcoding was also used for species verification purposes. Voucher samples of all the botanical material were deposited in the Botanical Repository of the NCNPR. The detailed sample information is given in Table 1.

Table 1. Analyzed authenticated *Salvia* samples.

No.	NCNPR Code	Part	Botanical Name
1	1523	Leaf	*Salvia officinalis*
2	2852	Leaf	
3	7686	Mixed Parts	
4	7917	-	
5	13095	Leaf	
6	16732	Leaf	
7	20712	-	
8	13096	Leaf	*Salvia apiana*
9	22497	Aerial	
10	22498	Aerial	
11	22502	Aerial	
12	22773	Leaf	
13	578	Leaf	*Salvia divinorum*
14	18434	Aerial	
15	22491	Leaf	
16	22506	Aerial	*Salvia mellifera*
17	22771	Leaf	
18	22772	Leaf	
19	767	Root	*Salvia miltiorrhiza*
20	5399	Root	
21	8676	Root	
22	9729	Root	
23	11750	Root	
24	12535	Root	

2.2. Chemicals

Dichloromethane was purchased from Fisher Scientific (Pittsburgh, PA, USA). Both internal standards, tridecane ($C_{13}H_{28}$) and docosane ($C_{22}H_{46}$), were obtained from Polyscience Corporation (Niles, IL, USA). The reference standards, α-pinene, β-pinene, 3-carene, eucalyptol, camphor, endo-borneol, β-caryophyllene, viridiflorol, α-bisabolol, tanshinone II, cryptotanshinone, salvinorin A, and salvinorin B, used to confirm compound identification, were purchased from Sigma-Aldrich (St. Louis, MO, USA), Agilent Technologies, Inc. (Santa Clara, CA, USA), or isolated from plant material in-house at the NCNPR.

2.3. Sample Preparation

Dry, solid plant material from each species was ground and homogenized utilizing a ball mill. Approximately 100 mg of the powdered sample material was carefully weighed and placed into a small centrifuge tube. Samples for GC/Q-ToF analysis were prepared using a two-step method. Two internal standards ($C_{13}H_{32}$ and $C_{22}H_{46}$) were selected. Each standard was combined with dichloromethane to obtain a solution with a concentration of 100 µg/mL of each internal standard. First, 340 µL dichloromethane with 80 µL of the prepared internal standard solution was added to the samples and sonicated for 1 hour. Next, the samples were centrifuged for 10 min. This procedure was repeated one more time without adding the internal standards, after which the supernatant was collected and filtered prior to the GC/Q-ToF analysis. Each sample was prepared in duplicate.

2.4. GC/Q-ToF Analysis

All prepared samples were analyzed using an Agilent 7890B (GC) instrument equipped with an RS185 PAL3 autosampler. The GC was connected to an Agilent 7250 accurate-mass Q-ToF mass spectrometer. The capillary column (30 m × 0.25 mm i.d.) was coated with a

0.25 μm film of 5% phenyl methyl siloxane (J&W, HP-5MS). Helium at a constant flow rate of 1 mL/min was used as the carrier gas. Each sample was analyzed using the following GC oven program: 50 °C, held for 2 min, then heated at 2 °C/min to 280 °C, and finally held at 280 °C for 20 min. A post-runtime period of 5 min at 300 °C was also utilized. The inlet was programmed at 280 °C, while 1 μL of each sample was injected with a split ratio of 10:1. The transfer line from the GC to the Q-ToF was held at 300 °C. Duplicate injections were made for each sample.

The Q-ToF mass spectrometer was equipped with a high-emission low-energy electron ionization source which was operated with an electron energy of 70 eV and an emission current of 5.0 μA. During the experiment, the source, quadrupole, and transfer line temperatures were 280 °C, 150 °C, and 300 °C, respectively. All mass spectra data were recorded at a rate of 5 Hz from 35 to 500 *m/z* after a 5 min solvent delay. After every second sample injection, automated ToF mass calibration was performed utilizing a keyword command in the sequence table. Data were acquired utilizing Agilent MassHunter software (version B7.06.274, Agilent Technologies, Santa Clara, CA, USA). Further data processing was accomplished using Agilent MassHunter Qualitative Analysis and Quantitative Analysis (version 10.0.10305.0, Agilent Technologies, Santa Clara, CA, USA). The NIST database (version 2.3, NIST Standard Reference Materials, Gaithersburg, MD, USA) was utilized for tentative compound identification.

2.5. Data Processing and Statistical Analysis

As a part of data processing, the GC/Q-ToF data were converted into a .cef file format utilizing Agilent MassHunter Unknown Analysis (version 10.0.7070, Agilent Technologies, Santa Clara, CA, USA). The "SureMass" peak detection and deconvolution algorithm was elected, and a peak area filter of 10,000 counts was applied. Ions with identical elution profiles and similar spectral data were extracted as entities characterized by retention time (t_R), peak intensity, and mass to charge ratio (m/z). Then, the resulting .cef file for each sample was exported into the Mass Profiler Professional software package (version B.12.05, Agilent Technologies, Santa Clara, CA, USA) which includes SCP algorithms for further data processing.

After examining various minimum abundance counts, a setting of 5000 counts was finally selected for the extraction of entities from the spectra. The alignment of retention time, with a tolerance window of 0.15 min, and the similarity of the spectral pattern were carried out and compared across the entire sample set. The internal standard docosane ($C_{22}H_{46}$) was selected to normalize the peak intensity across all spectra. A stepwise reduction of entity dimensionality was performed based on common entities found across samples to further process the data. In addition, software settings such as parameter values (filter by flags), the frequency of occurrence (filter by frequency), the abundance of respective entities in classes (filter by sample variability), and one-way analysis of variance (ANOVA) were utilized and carried out by the software to filter the raw data. After filtering the raw data, quality control of the samples was performed by PCA to further reduce the dimensionality of the GC/Q-ToF data sets, increase interpretability, and minimize information loss. Based on the PCA, an SCP model was constructed. Five algorithms, namely, partial least squares discriminant analysis (PLS-DA), support vector machines (SVM), naive Bayes (NB), decision tree (DT), and neutral network (NN), were evaluated. The PLS-DA algorithm was selected since it was particularly well-suited for the project and resulted in the best prediction accuracy when compared to other algorithms. To validate the model, a k-fold cross-validation procedure was carried out. The validation procedure had three k-folds and was repeated ten times.

2.6. Establishment of a Personal Compound Database and Library (PCDL)

A PCDL was constructed using Agilent PCDL Manager software (version B8.00). Either readily available or isolated and fully characterized in-house chemical compounds were utilized as reference standards to establish the PCDL. Data including the retention time, exact mass, and high-resolution MS fragmentation patterns were exported to the PCDL. Additional information, such as the molecular formula, compound name, and CAS number were assigned to each entry for constructing the PCDL.

3. Results and Discussion

3.1. Extraction

Although hexane is touted as an ideal extraction solvent for capturing a wide variety of volatiles in botanicals, the extraction efficiency is questionable for some of the semi-volatile polar constituents, such as salvinorins from *S. divinorum*. These limitations with hexane are alleviated by utilizing dichloromethane [25] as the solvent of choice. A simple sample extraction procedure with dichloromethane improved the overall throughput and captured a wide variety of volatile analytes for species identification.

3.2. GC/Q-ToF Analysis

After developing a satisfactory sample extraction technique and an optimized GC/Q-ToF method, the sample data were gathered (Figure 1). Upon examining the chromatograms of the investigated species, compounds were detected in the GC/Q-ToF analysis of the authentic *Salvia* plant extracts. Although there were slight variations among the concentrations of components within a particular *Salvia* species, characteristic and consistent fingerprinting patterns from the same species of *Salvia* were observed. However, distinct differences in their chemical profiles were noticed for different species, as illustrated in Figure 1.

Although approximately 200 compounds were tentatively identified from the five species, only 32 compounds which were found in the greatest abundance or were characteristic for each species were reported. The tentative identity of each analyte suggested by the NIST database was further confirmed with reference standards and the accurate mass of molecular ions when they were available for each analyte. Many early-eluting, highly volatile compounds were present in *S. officinalis*, *S. apiana*, and *S. mellifera*; however, these compounds were mostly absent in the samples of *S. divinorum* and *S miltiorrhiza*. After systematically examining the compounds present in each species, additional characteristic patterns were also established. For example, samples of *S. officinalis* contained the compounds β-thujone, viridiflorol, and verticiol, which were not detected in the other *Salvia* species. Although these compounds have been reported in other plant species, e.g., viridiflorol has been reported as a major constituent of *Allophylus edulis* [26], they were only present in *S. officinalis* among the five *Salvia* species in this study. Thus, the co-existence of β-thujone, viridiflorol, and verticiol can be used to distinguish *S. officinalis* from other *Salvia* species. This finding is also supported by a previous study comparing four *Salvia* species [27]. Likewise, samples of *S. mellifera* contained statistically significant amounts ($p < 0.05$) of camphor when compared to the other species. This is also consistent with Martino et al. report of *S. mellifera* containing approximately 12.2% camphor [28]. In addition, *S. mellifera* also contained β-amyrone, pectolinaringenin, and lupeol which were not detected in the other analyzed species. Only *S. apiana* samples contained γ-gurjunene and a statistically significant amount of isoledene. Unfortunately, due to the small amount of available literature concerning the volatile constituents of *S. apiana*, the authors were unable to confirm these findings with literature sources. *S. miltiorrhiza* samples contained the greatest amount ($p < 0.05$) of ferruginol, as well as the unique compound tanshinone II. The occurrence of tanshinone II in only *S. miltiorrhiza* samples is also supported by a review from Zhang et al. [9] In addition to being the only group that possessed the compounds salvinorin A and salvinorin B, *S. divinorum* also contained the greatest abundance ($p < 0.05$) of 8-hexadecyne. Willard and colleagues also reported the utility of salvinorin A in the identification of *S. divinorum* [25]. Utilizing these observed chemical distributions, each species' chemical fingerprint and the peak area percentage of detectable compounds can be obtained (Table 2A,B).

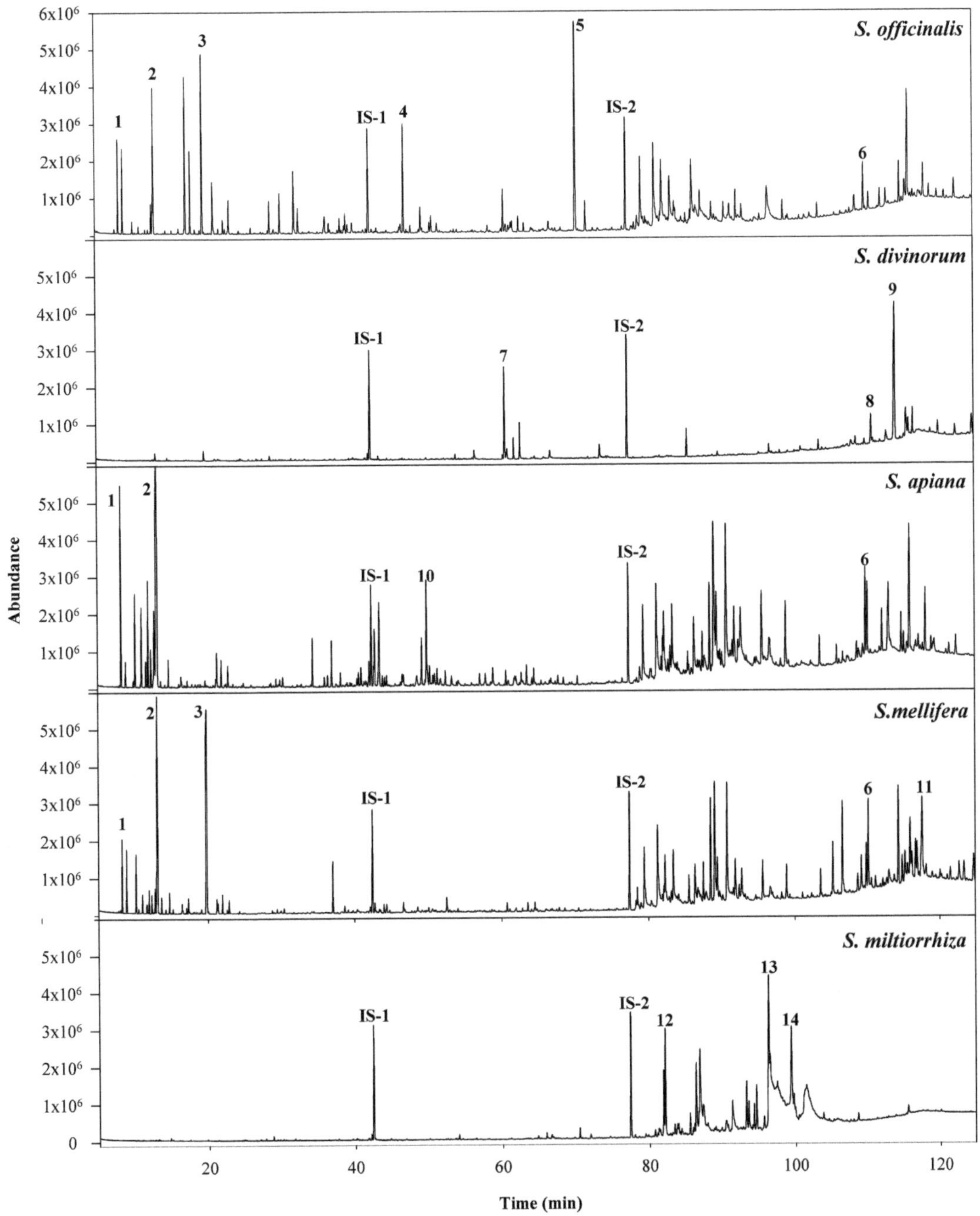

Figure 1. Representative chromatograms comparing *Salvia* species. Peak assignments: (1) α-pinene; (2) 1,8 cineole; (3) camphor; (4) viridiflorol; (5) verticiol; (6) salvigenin; (7) 8-hexadecyne; (8) salvinorin B; (9) salvinorin A; (10) γ-gurjunene; (11) lupeol; (12) ferruginol; (13) tanshinone II; (14) cryptotanshinone; (IS-1) tridecane; (IS-2) docosane.

Table 2. Tentative compound identification based on NIST library and percent (% peak area) of volatile compounds in methylene chloride extracts of (A) *S. officinalis* and *S. apiana* and (B) *S. divinorum*, *S. mellifera*, and *S. miltiorrhiza* using GC/Q-ToF analysis.

A

Compound	t_R (min)	*S. officinalis*							*S. apiana*				
		1523	2852	7686	7917	13095	16732	20712	13096	22497	22498	22502	22773
α-Pinene [a]	8.194	0.11	0.20	1.57	tr	1.91	0.54	tr	3.52	1.43	1.25	2.25	0.09
Camphene [b]	8.817	0.29	0.20	1.99	0.15	1.66	0.48	tr	0.37	1.17	0.92	0.57	0.10
β-Pinene [a,b]	10.113	tr	tr	1.99	tr	0.25	0.28	tr	1.47	1.32	1.38	1.90	0.10
3-Carene [a]	11.852	nd	nd	nd	nd	nd	nd	nd	1.88	0.57	0.38	1.23	tr
Eucalyptol [a,b]	12.983	3.14	2.11	4.34	1.36	4.41	1.16	1.47	10.90	6.20	5.89	9.59	3.04
α-Thujone	17.393	1.15	0.62	3.46	nd	nd	nd	nd	nd	nd	nd	nd	nd
β-Thujone [b]	18.086	1.53	0.73	1.51	0.30	2.19	1.33	tr	nd	nd	nd	nd	nd
Camphor [a]	19.697	3.73	5.32	6.33	3.61	6.18	3.40	4.32	tr	6.74	6.81	3.86	2.44
endo-Borneol [a]	21.160	1.50	0.74	1.78	1.84	1.66	1.23	0.47	0.12	0.30	0.21	0.2	0.21
β-Caryophyllene [a]	37.092	0.13	0.09	0.06	0.15	0.25	0.41	0.43	0.93	1.64	2.81	3.22	0.13
Isoledene [b]	43.496	tr	tr	tr	tr	0.11	tr	0.23	2.35	1.84	2.59	1.05	0.33
Viridiflorol [a]	47.199	2.73	3.14	5.38	3.93	3.54	4.14	1.94	nd	nd	nd	nd	nd
Humulenol [b]	49.514	tr	0.15	1.94	0.98	0.77	0.93	1.71	nd	nd	nd	nd	nd
γ-Gurjunene [b]	49.927	nd	nd	nd	nd	nd	nd	nd	2.73	1.46	2.06	1.01	0.23
α-Bisabolol [a]	52.517	0.23	0.36	tr	nd	tr	nd	nd	0.28	0.41	0.66	2.59	0.33
8-Hexadecyne [b]	60.731	0.11	1.32	0.59	2.86	1.22	1.63	4.15	0.28	0.31	0.41	0.43	0.75
3,7,11,15-Tetramethyl-2-hexadecen-1-ol [b]	62.814	tr	0.47	0.24	1.16	0.42	0.53	1.26	0.21	0.21	0.35	0.36	0.32
Verticiol [b]	70.524	8.19	12.17	7.81	10.59	9.75	10.17	3.84	nd	nd	nd	nd	nd
Aromandendrene [b]	71.894	0.38	0.92	1.00	0.88	0.96	1.31	0.37	nd	nd	nd	nd	nd
Ferruginol [b]	82.097	nd	nd	nd	nd	nd	nd	nd	0.36	0.07	0.09	1.15	0.48
Hexanedioic acid, mono(2-ethylhexyl)ester [b]	85.533	0.65	2.88	2.83	0.96	0.25	0.67	1.24	0.42	0.44	0.50	0.45	0.49
Unknown	86.366	5.92	4.53	4.57	4.45	2.65	5.26	0.83	1.70	0.41	1.04	1.37	1.15
Salvicanol [b]	88.476	nd	nd	nd	nd	nd	nd	nd	nd	1.41	1.19	2.05	2.45
Tanshinone II [a]	96.184	nd	nd	nd	nd	nd	nd	nd	nd	nd	nd	nd	nd
Cryptotanshinone [a]	99.365	nd	nd	nd	nd	nd	nd	nd	nd	nd	nd	nd	nd
Pectolinaringenin [b]	105.420	nd	nd	nd	nd	nd	nd	nd	nd	nd	nd	nd	nd
Heptacosane [b]	109.945	1.47	1.96	2.02	2.26	1.57	2.25	4.35	2.12	1.51	1.79	1.25	3.69
Salvigenin [b]	110.291	0.60	0.06	0.05	0.06	0.64	0.64	0.29	1.72	4.26	3.45	2.01	5.90
Salvinorin B [b]	110.841	nd	nd	nd	nd	nd	nd	nd	nd	nd	nd	nd	nd
Salvinorin A [a]	114.026	nd	nd	nd	nd	nd	nd	nd	nd	nd	nd	nd	nd
β-Amyrone [b]	116.668	nd	nd	nd	nd	nd	nd	nd	nd	nd	nd	nd	nd
Lupeol [b]	117.441	nd	nd	nd	nd	nd	nd	nd	nd	nd	nd	nd	nd

B

Compound	t_R (min)	*S. divinorum*			*S. mellifera*			*S. miltiorrhiza*					
		578	18434	22490	22506	22771	22772	767	5399	8676	9729	11750	12535
α-Pinene [a]	8.194	nd	nd	nd	1.28	0.24	0.67	nd	nd	nd	nd	nd	nd
Camphene [b]	8.817	nd	nd	nd	1.14	0.37	1.02	nd	nd	nd	nd	nd	nd
β-Pinene [a,b]	10.113	nd	nd	nd	1.09	0.26	0.69	nd	nd	nd	nd	nd	nd
3-Carene [a]	11.852	nd	nd	nd	0.38	tr	tr	nd	nd	nd	nd	nd	nd
Eucalyptol [a,b]	12.983	1.12	0.31	1.13	6.44	3.55	4.20	0.67	0.81	tr	2.12	nd	1.67
α-Thujone	17.393	nd	nd	nd	nd	nd	nd	nd	nd	nd	nd	nd	nd
β-Thujone [b]	18.086	nd	nd	nd	nd	nd	nd	nd	nd	nd	nd	nd	nd
Camphor	19.697	1.63	tr	3.71	10.21	8.75	9.73	0.40	2.13	1.95	0.90	0.05	0.84
endo-Borneol [a]	21.160	nd	nd	nd	0.33	0.52	0.52	nd	nd	nd	nd	nd	nd
β-Caryophyllene [a]	37.092	nd	nd	nd	1.28	0.73	0.82	nd	nd	nd	nd	nd	nd
Isoledene [b]	43.496	nd	nd	nd	tr	0.13	0.11	nd	nd	nd	nd	nd	nd
Viridiflorol [a]	47.199	nd	nd	nd	nd	nd	nd	nd	nd	nd	nd	nd	nd
Humulenol [b]	49.514	nd	nd	nd	nd	nd	nd	nd	nd	nd	nd	nd	nd
γ-Gurjunene [b]	49.927	nd	nd	nd	nd	nd	nd	nd	nd	nd	nd	nd	nd
α-Bisabolol [a]	52.517	nd	nd	nd	0.50	1.70	0.69	nd	nd	nd	nd	nd	nd
8-Hexadecyne [b]	60.731	12.09	10.45	9.32	0.22	0.50	0.45	nd	nd	nd	nd	nd	nd
3,7,11,15-Tetramethyl-2-hexadecen-1-ol [b]	62.814	4.70	3.11	3.20	0.10	0.11	0.15	nd	nd	nd	nd	nd	nd
Verticiol [b]	70.524	nd	nd	nd	nd	nd	nd	nd	nd	nd	nd	nd	nd
Aromandendrene [b]	71.894	nd	nd	nd	nd	nd	nd	nd	nd	nd	nd	nd	nd
Ferruginol [b]	82.097	nd	nd	nd	0.20	0.31	0.39	9.14	2.61	1.05	8.11	9.18	1.61
Hexanedioic acid, mono(2-ethylhexyl) ester [b]	85.533	4.68	4.72	2.65	0.73	0.48	1.23	4.347	24.95	35.25	19.46	1.55	14.59
Unknown	86.366	nd	nd	nd	nd	nd	nd	nd	nd	nd	nd	nd	nd
Salvicanol [b]	88.476	nd	nd	nd	3.23	2.55	2.81	nd	nd	nd	nd	nd	nd
Tanshinone II [a]	96.184	nd	nd	nd	nd	nd	nd	15.25	13.19	12.70	16.91	12.57	12.14
Cryptotanshinone [a]	99.365	nd	nd	nd	nd	nd	nd	nd	nd	nd	nd	12.16	nd
Pectolinaringenin [b]	105.420	nd	nd	nd	1.63	4.22	3.98	nd	nd	nd	nd	nd	nd
Heptacosane [b]	109.945	nd	nd	nd	2.55	1.93	2.21	nd	nd	nd	nd	nd	nd
Salvigenin [b]	110.291	nd	nd	nd	2.55	1.93	2.21	nd	nd	nd	nd	nd	nd
Salvinorin B [b]	110.841	3.60	0.75	1.12	nd	nd	nd	nd	nd	nd	nd	nd	nd
Salvinorin A [a]	114.026	33.48	20.24	22.70	nd	nd	nd	nd	nd	nd	nd	nd	nd
β-Amyrone [b]	116.668	nd	nd	nd	1.14	0.96	0.79	nd	nd	nd	nd	nd	nd
Lupeol [b]	117.441	nd	nd	nd	4.16	3.34	4.09	nd	nd	nd	nd	nd	nd

nd: not detected; tr: trace amount; [a] compound identification based on NIST library was confirmed with reference standard; [b] accurate mass was consistent with GC/Q-ToF analysis.

3.3. Chemometric Analysis

Although the GC/MS identification of *Salvia* species is a popular means of species identification, it is often time-consuming [16]. While this method is well suited for small sample sizes, it does not lend itself to high-throughput applications, such as batch processing or quality control. With the coupling of GC to a Q-ToF mass spectrometer, vast amounts of high-resolution structural data can be gathered from compounds in each sample. Utilizing this data along with chemometrics, researchers can develop an SCP model from the data obtained from species [29,30].

PCA is a useful analysis that can transform large and complex data sets into manageable information for interpretation [31]. The stepwise reduction in entity dimensionality was performed based on filtering by flags, filtering by frequency, filtering by sample variability, and the results of ANOVA. Stepwise filtering intentionally created a strong filter so that the most discriminant entities could be used to construct the prediction model. After filtering, a PCA was performed, as illustrated in Figure 2. Good separation and species-specific clustering of the different *Salvia* species was achieved. Approximately 50% of the variation among species could be attributed to component **1**. Additional variation and separation could be explained by component **2** (15%). Contributing the least, component **3** only accounted for approximately 9% of the variation observed among the species.

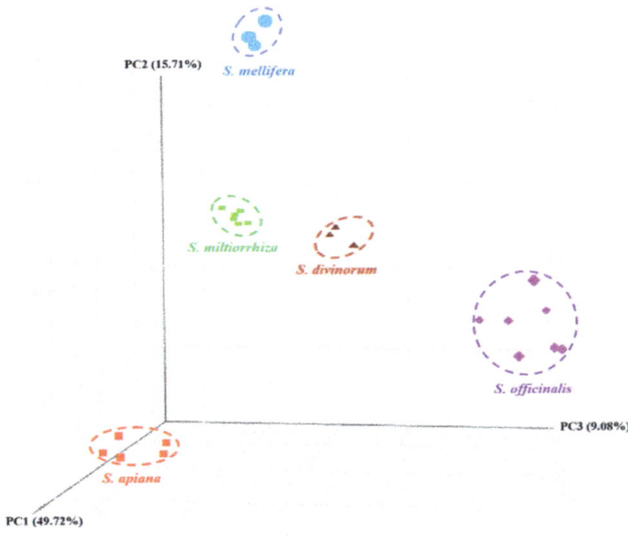

Figure 2. PCA score plot of five *Salvia* species.

Although the PCA demonstrated good separation between different *Salvia* species, it was unable to assign and predict the identity of unknown/commercial *Salvia* species sold in the U.S. market. Therefore, the GC/Q-ToF data for the authenticate samples were subjected to supervised chemometric methods. The first step in the SCP model construction process is to select the algorithm that is best suited to the project and the data set parameters. The PLS-DA [29] algorithm was found to be the best suited to construct a statistical model for *Salvia* classification and differentiation. Good separation obtained by the PLS-DA model among different *Salvia* species is shown in Figure 3. Once established, the software can use the sample characteristics and the associated algorithms to classify unknown samples. As illustrated in Figure 3, the PLS-DA successfully separated and clustered members of the authentic samples.

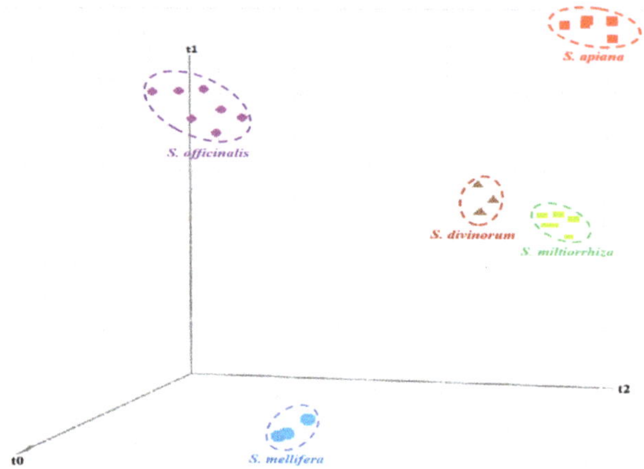

Figure 3. Score plot of the PLS-DA model constructed based on GC/Q-ToF data for the authenticated *Salvia* samples from five different species.

To validate the constructed model, the same authenticated samples used for the model training were repeatedly used due to the limited number of authenticated plant samples available. Although redundant, this is a valid statistical procedure (k-fold cross validation). Both the recognition and prediction abilities of the class prediction model were 100%, as shown in Table 3. Once the test was complete, a "confusion matrix" was generated. The test results indicated that this SCP could successfully identify and classify samples (Table 3). The construction of the SCP not only allows a large number of samples to be classified efficiently, but also in an automated manner. This allows the user to process additional samples at any point in the future.

Table 3. Summary of classification results obtained by the PLS-DA model.

	S. apiana	*S. divinorum*	*S. mellifera*	*S. miltiorrhiza*	*S. officinalis*	Accuracy (%)
Model Training						
S. apiana	5	0	0	0	0	100
S. divinorum	0	3	0	0	0	100
S. mellifera	0	0	3	0	0	100
S. miltiorrhiza	0	0	0	6	0	100
S. officinalis	0	0	0	0	7	100
Recognition ability (%)	-	-	-	-	-	100
Model validation						
S. apiana	5	0	0	0	0	100
S. divinorum	0	3	0	0	0	100
S. mellifera	0	0	3	0	0	100
S. miltiorrhiza	0	0	0	6	0	100
S. officinalis	0	0	0	0	7	100
Prediction ability (%)	-	-	-	-	-	100

3.4. Construction of a Personal Compound Database and Library (PCDL) for High-Throughput Screening

Although compound identification can be accomplished by manual inspection, this process can be both time-consuming and inefficient due to the large amount of high-resolution data obtained. With this in mind, a PCDL was constructed to facilitate the efficient throughput of samples. From the PCA loading plot (Figure S1 in the Supplementary Material), which is a visual representation of the "characteristic compounds" found in different *Salvia* species, marker compounds correlating to the separation of different species or the clustering of similar species were identified [30]. As illustrated in Table 4, each species could be distinguished by a few select compounds. Hence, the identified marker compounds that were commercially available or isolated in-house were analyzed by using the identical GC/Q-ToF method.

Table 4. Proposed marker compounds tentatively identified for the differentiation of selected *Salvia* species.

No.	Compound ID	t_R (min)	Formula	Base Peak	M+	Diff (ppm)	CAS Number
			S. officinalis				
1	β-Thujone	18.086	$C_{10}H_{16}O$	67.0542	152.1196	0.22	471-15-8
2	Viridiflorol	47.199	$C_{15}H_{26}O$	105.0697	222.1975	−1.43	552-02-3
3	Verticiol	70.524	$C_{20}H_{34}O$	95.0853	290.2598	−2.13	70000-19-0
			S. divinorum				
4	8-Hexadecyne *	60.731	$C_{16}H_{30}$	67.0542	222.2345	1.34	19781-86-3
5	Salvinorin B	110.800	$C_{21}H_{26}O_7$	94.0413	390.1679	1.53	92545-30-7
6	Salvinorin A	114.026	$C_{23}H_{28}O_8$	94.0412	432.1784	1.23	83729-01-5
			S. apiana				
7	Isoledene *	43.496	$C_{15}H_{24}$	119.0854	204.1877	1.21	95910-36-4
8	γ-Gurjunene	49.927	$C_{15}H_{24}$	105.0702	204.1875	1.21	22567-17-5
			S. mellifera				
10	Camphor *	19.697	$C_{10}H_{16}O$	95.0856	152.1195	−0.44	464-49-3
11	β-Amyrone	116.670	$C_{30}H_{48}O$	218.2034	424.3700	0.08	638-97-1
12	Lupeol	117.441	$C_{30}H_{50}O$	189.1639	426.3855	−0.28	545-47-1
13	Pectolinaringenin	105.42	$C_{17}H_{14}O_6$	271.0607	314.0785	2.58	520-12-7
			S. miltiorrhiza				
14	Ferruginol *	82.097	$C_{20}H_{30}O$	189.1275	286.2297	2.04	514-62-5
15	Tanshinone II	96.184	$C_{19}H_{18}O_3$	261.0912	294.1252	0.52	568-72-9

* Statistically significant amount detected ($p < 0.05$).

After analyzing the standards, data including the retention time, exact mass, and a curated accurate mass spectrum containing mass assignments for each spectral peak were exported to the PCDL. Utilizing the PCDL software, additional data such as the molecular formula, compound name, and CAS number were also captured. Figure 4 shows an overview of the PCDL table with the spectrum of salvinorin A, one of the marker compounds only present in *S. divinorum*.

The commercially available MassHunter Unknown Analysis software uses an algorithm called "SureMass" to find peaks in the accurate mass chromatogram and searches a mass spectral library or PCDL to identify compounds. If the library has locked retention times or index values, these can also be used as filters. If these filters are utilized, "hits" must have the correct retention time (t_R) and be similar to the database spectrum. Figure 5 illustrates the results for the identification and isotope pattern for salvinorin A in one of the *S. divinorum* samples.

The "SureMass" peak-finding algorithm uses the added information available in high-resolution accurate mass data. For instance, extracted ion chromatograms of salvinorin A are overlaid and compared in Figure 5A. In contrast, a "head-to-tail" comparison plot of the high-resolution mass spectra of the suspected target and the reference compound illustrates the matching spectra (Figure 5B). In addition, the software can generate the compound's isotope pattern if the molecular ion is detected in sufficient abundance. The compound's theoretical value is next compared to the detected isotope's m/z and relative abundance [30]. Additional confidence in the correct identification of the compound is provided when the theoretical value and detected m/z and abundance are good matches. In Figure 5C, the detected isotope pattern of salvinorin A (black vertical lines) is compared to the theoretical isotope pattern represented by red boxes. In the present study, peaks from the sample spectra of the five *Salvia* species that were identified by "SureMass" were compared to the in-house-constructed PCDL. This approach is inherently simple and data review is relatively easy. Once the PCDL is constructed, it not only allows for high sample throughput, but can be easily utilized in the future to analyze additional samples or be shared with research labs that do not have standard marker compounds.

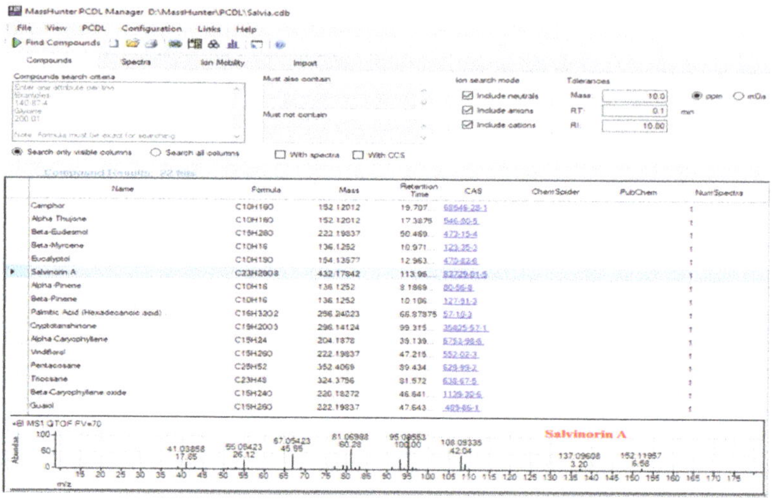

Figure 4. A section of the PCDL showing some of the content available for each entry and the accurate mass EI spectrum of salvinorin A from the PCDL.

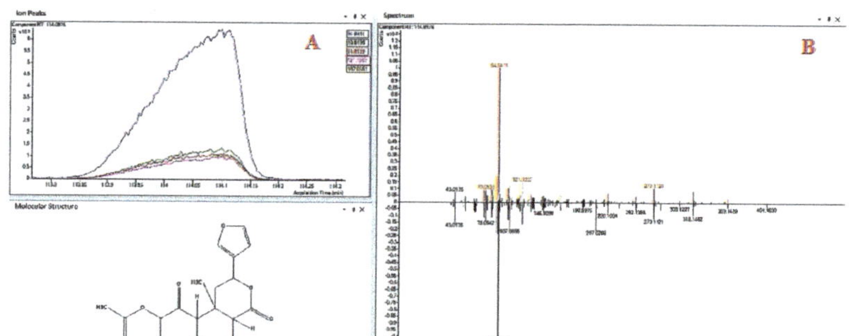

Figure 5. Identification of salvinorin A from *S. divinorum* (#22490). (**A**) Overlaid chromatograms of the five ions extracted for salvinorin A; (**B**) a "head-to-tail" comparison plot of high-resolution spectra of salvinorin A from PCDL (black) and the sample (orange); (**C**) the isotope pattern of the molecular ion (black vertical lines) compared to the theoretical pattern (red boxes).

4. Conclusions

Members of the genus *Salvia* have a rich history of both culinary and medicinal usage. With approximately 900 species included in the genus *Salvia*, the accurate species identification of processed botanical material can be a daunting task [2]. Although arduous, this task is of vast importance since the herb possesses species-specific pharmacological properties [2]. In the present study, we analyzed five species of botanically verified, medicinally important *Salvia* (*apiana, divinorum, mellifera, miltiorrhiza*, and *officinalis*) to develop a single analytical method for species differentiation purposes. Leveraging advances in software, the GC/Q-ToF of volatile organics, and the accurate mass spectral data allowed the unambiguous identification of five studied *Salvia* species. Although some of the marker compounds can be found in other plants, it is both the combination and concentration of the compounds that can aid in the species identification of *Salvia* botanical material. The implementation of chemometric analysis, *viz.* the PCA [29,30] of the *Salvia* samples, resulted in the identification of marker compounds for different *Salvia* species. Furthermore, the same PCA programs can also be expanded to build prediction models which may be utilized and modified for high-throughput sample analyses and classification purposes. To aid further, a PCDL combined with high-resolution mass spectrometry was developed with the versatility and ability to identify individual compounds present in *Salvia* samples.

In summary, by utilizing GC/Q-ToF, we obtained chemical fingerprints of each *Salvia* species being investigated. This information was further processed to construct an SCP model. By utilizing this model, future unknown samples can easily and efficiently be identified. As analytical needs change over time, the SCP model allows researchers to expand by including other economically important *Salvia* species. By leveraging advanced analytical techniques and chemometrics, the quality of closely related botanicals can be confirmed successfully, as demonstrated with a broad spectrum of biologically active *Salvia* species with complex chemistries.

Supplementary Materials: The following supporting information can be downloaded at: https://www.mdpi.com/article/10.3390/foods11142132/s1, Figure S1: PCA loading plot illustrating suggested species marker compounds.

Author Contributions: Sample preparation, gas chromatography, chemical analyses, data analysis, and manuscript preparation, J.L. (Joseph Lee) and M.W.; conceptualization, resources including funding, review, and editing, I.A.K., A.G.C., C.W. and J.L. (Jing Li); chemometric analysis, J.Z.; methodology, data validation and cross confirmation with LC-MS and B.A. All authors have read and agreed to the published version of the manuscript.

Funding: This research is supported by the "Holistic Approach for Potential Drug Interactions with Botanical Drugs" funded by the Center for Drug Evaluation and Research, United States Food and Drug Administration, grant number HHSF223201810175C and "Discovery & Development of Natural Products for Pharmaceutical & Agricultural Applications" funded by the United States Department of Agriculture, Agricultural Research Service, Specific Cooperative Agreement No. 58-6060-6-015. The content is solely the responsibility of the authors and does not necessarily represent the official views of the Food and Drug Administration and the Department of Agriculture, Agricultural Research Service.

Data Availability Statement: Data is contained within the article or supplementary material. The data presented in this study are available on request to the corresponding author.

Conflicts of Interest: The authors declare no conflict of interest.

References

1. Xu, J.; Wei, K.; Zhang, G.; Lei, L.; Yang, D.; Wang, W.; Han, Q.; Xia, Y.; Bi, Y.; Yang, M. Ethnopharmacology, phytochemistry, and pharmacology of Chinese Salvia species: A review. *J. Ethnopharmacol.* **2018**, *225*, 18–30. [CrossRef] [PubMed]
2. Kintzios, S.E. *Sage: The Genus Salvia*; CRC Press: Boca Raton, FL, USA, 2000.
3. Chun-Yan, S.; Qian-Liang, M.; Rahman, K.; Ting, H.; Lu-Ping, Q. Salvia miltiorrhiza: Traditional medicinal uses, chemistry, and pharmacology. *Chin. J. Nat. Med.* **2015**, *13*, 163–182.
4. Ghorbani, A.; Esmaeilizadeh, M. Pharmacological properties of Salvia officinalis and its components. *J. Tradit. Complementary Med.* **2017**, *7*, 433–440. [CrossRef] [PubMed]

5. Topçu, G. Bioactive triterpenoids from Salvia species. *J. Nat. Prod.* **2006**, *69*, 482–487. [CrossRef] [PubMed]
6. Avula, B.; Bae, J.-Y.; Chittiboyina, A.G.; Wang, Y.-H.; Wang, M.; Srivedavyasasri, R.; Ali, Z.; Li, J.; Wu, C.; Khan, I.A. Comparative analysis of five Salvia species using LC-DAD-QToF. *J. Pharm. Biomed. Anal.* **2022**, *209*, 114520. [CrossRef]
7. Krol, A.; Kokotkiewicz, A.; Luczkiewicz, M. White Sage (*Salvia apiana*)—A Ritual and Medicinal Plant of the Chaparral: Plant Characteristics in Comparison with Other Salvia Species. *Planta Med.* **2021**, *88*, 604–627. [CrossRef]
8. Ali, Z.; Radhakrishnan, S.; Avula, B.; Chittiboyina, A.G.; Li, J.; Wu, C.; Khan, I.A. Eupatorin 3′-O-glucopyranoside, a trimethoxyflavonoid glucoside from the aerial parts of Salvia mellifera. *Nat. Prod. Res.* **2021**, *16*, 1–8. [CrossRef]
9. Zhang, Y.; Jiang, P.; Ye, M.; Kim, S.-H.; Jiang, C.; Lü, J. Tanshinones: Sources, pharmacokinetics and anti-cancer activities. *Int. J. Mol. Sci.* **2012**, *13*, 13621–13666. [CrossRef]
10. Benoni, H. Salvinorin A—A hallucinogen from Aztec sage. *Nat. Rundsch* **2001**, *54*, 575–578.
11. Jermain, J.D.; Evans, H.K. Analyzing Salvia divinorum and its active ingredient salvinorin A utilizing thin layer chromatography and gas chromatography/mass spectrometry. *J. Forensic Sci.* **2009**, *54*, 612–616. [CrossRef]
12. Cao, J.-L.; Wei, J.-C.; Hu, Y.-J.; He, C.-W.; Chen, M.-W.; Wan, J.-B.; Li, P. Qualitative and quantitative characterization of phenolic and diterpenoid constituents in Danshen (*Salvia miltiorrhiza*) by comprehensive two-dimensional liquid chromatography coupled with hybrid linear ion trap Orbitrap mass. *J. Chromatogr. A* **2016**, *1427*, 79–89. [CrossRef] [PubMed]
13. Dal Piaz, F.; Imparato, S.; Lepore, L.; Bader, A.; de Tommasi, N. A fast and efficient LC–MS/MS method for detection, identification and quantitative analysis of bioactive sesterterpenes in Salvia dominica crude extracts. *J. Pharm. Biomed. Anal.* **2010**, *51*, 70–77. [CrossRef]
14. Gericke, S.; Lübken, T.; Wolf, D.; Kaiser, M.; Hannig, C.; Speer, K. Identification of new compounds from sage flowers (Salvia officinalis L.) as markers for quality control and the influence of the manufacturing technology on the chemical composition and antibacterial activity of sage flower extracts. *J. Agric. Food Chem.* **2018**, *66*, 1843–1853. [CrossRef]
15. Haq, F.U.; Ali, A.; Akhtar, N.; Aziz, N.; Khan, M.N.; Ahmad, M.; Musharraf, S.G. A high-throughput method for dereplication and assessment of metabolite distribution in Salvia species using LC-MS/MS. *J. Adv. Res.* **2020**, *24*, 79–90. [CrossRef]
16. Hu, P.; Liang, Q.-L.; Luo, G.-A.; Zhao, Z.-Z.; Jiang, Z.-H. Multi-component HPLC fingerprinting of Radix Salviae Miltiorrhizae and its LC-MS-MS identification. *Chem. Pharm. Bull.* **2005**, *53*, 677–683. [CrossRef]
17. Li, B.; Zhang, C.; Peng, L.; Liang, Z.; Yan, X.; Zhu, Y.; Liu, Y. Comparison of essential oil composition and phenolic acid content of selected Salvia species measured by GC–MS and HPLC methods. *Ind. Crops Prod.* **2015**, *69*, 329–334. [CrossRef]
18. Qiao, X.; Zhang, Y.-T.; Ye, M.; Wang, B.-R.; Han, J.; Guo, D.-a. Analysis of chemical constituents and taxonomic similarity of Salvia species in China using LC/MS. *Planta Med.* **2009**, *75*, 1613–1617. [CrossRef]
19. Shi, Z.; He, J.; Yao, T.; Chang, W.; Zhao, M. Simultaneous determination of cryptotanshinone, tanshinone I and tanshinone IIA in traditional Chinese medicinal preparations containing Radix salvia miltiorrhiza by HPLC. *J. Pharm. Biomed. Anal.* **2005**, *37*, 481–486. [CrossRef]
20. Kintzios, S.E. The biological/pharmacological activity of the Salvia genus. In *Sage: The Genus Salvia*; CRC Press: Boca Raton, FL, USA, 2003; pp. 161–198.
21. Lou, J.; Mao, Z.; Shan, T.; Wang, Q.; Zhou, L. Chemical composition, antibacterial and antioxidant properties of the essential oils from the roots and cultures of Salvia miltiorrhiza. *J. Essent. Oil Bear. Plants* **2014**, *17*, 380–384. [CrossRef]
22. Wang, M.; Zhao, H.-X.; Wang, L.; Wang, T.; Yang, R.-W.; Wang, X.-L.; Zhou, Y.-H.; Ding, C.-B.; Zhang, L. Potential use of DNA barcoding for the identification of Salvia based on cpDNA and nrDNA sequences. *Gene* **2013**, *528*, 206–215. [CrossRef]
23. Ciesla, Ł.; Hajnos, M.; Staszek, D.; Wojtal, Ł.; Kowalska, T.; Waksmundzka-Hajnos, M. Validated binary high-performance thin-layer chromatographic fingerprints of polyphenolics for distinguishing different Salvia species. *J. Chromatogr. Sci.* **2010**, *48*, 421–427. [CrossRef] [PubMed]
24. Rzepa, J.; Wojtal, Ł.; Staszek, D.; Grygierczyk, G.; Labe, K.; Hajnos, M.; Kowalska, T.; Waksmundzka-Hajnos, M. Fingerprint of selected Salvia species by HS-GC-MS analysis of their volatile fraction. *J. Chromatogr. Sci.* **2009**, *47*, 575–580. [CrossRef] [PubMed]
25. Willard, M.A.B.; McGuffin, V.L.; Smith, R.W. Forensic analysis of *Salvia divinorum* using multivariate statistical procedures. Part I: Discrimination from related Salvia species. *Anal. Bioanal. Chem.* **2012**, *402*, 833–842. [CrossRef] [PubMed]
26. Trevizan, L.N.F.; do Nascimento, K.F.; Santos, J.A.; Kassuya, C.A.L.; Cardoso, C.A.L.; do Carmo Vieira, M.; Moreira, F.M.F.; Croda, J.; Formagio, A.S.N. Anti-inflammatory, antioxidant and anti-Mycobacterium tuberculosis activity of viridiflorol: The major constituent of Allophylus edulis (A. St.-Hil., A. Juss. & Cambess.) Radlk. *J. Ethnopharmacol.* **2016**, *192*, 510–515.
27. Ali, A.; Tabanca, N.; Demirci, B.; Blythe, E.K.; Ali, Z.; Baser, K.H.C.; Khan, I.A. Chemical composition and biological activity of four salvia essential oils and individual compounds against two species of mosquitoes. *J. Agric. Food Chem.* **2015**, *63*, 447–456. [CrossRef]
28. Martino, L.D.; Roscigno, G.; Mancini, E.; Falco, E.D.; Feo, V.D. Chemical composition and antigerminative activity of the essential oils from five Salvia species. *Molecules* **2010**, *15*, 735–746. [CrossRef] [PubMed]
29. Agilent. *Agilent Mass Profiler Professional Operation for Chemometric Analysis, SW-MPP-3101c*; Agilent Technologies: Lexington, MA, USA, 2019.
30. Wylie, P.L.; Westland, J.; Wang, M.; Radwan, M.M.; Majumdar, C.G.; ElSohly, M.A. Screening for More than 1,000 Pesticides and Environmental Contaminants in Cannabis by GC/Q-TOF. *Med. Cannabis Cannabinoids* **2020**, *3*, 1–11. [CrossRef]
31. Worley, B.; Powers, R. Multivariate analysis in metabolomics. *Curr. Metab.* **2013**, *1*, 92–107.

foods

Article

Detection of Pesticide Residue Level in Grape Using Hyperspectral Imaging with Machine Learning

Weixin Ye [1], Tianying Yan [1], Chu Zhang [2], Long Duan [1], Wei Chen [1], Hao Song [1], Yifan Zhang [3,4], Wei Xu [3,4,*] and Pan Gao [1,*]

1 College of Information Science and Technology, Shihezi University, Shihezi 832003, China; yeweixin@stu.shzu.edu.cn (W.Y.); yantianying@163.com (T.Y.); duanlong@stu.shzu.edu.cn (L.D.); chenwei@stu.shzu.edu.cn (W.C.); songhao@stu.shzu.edu.cn (H.S.)
2 School of Information Engineering, Huzhou University, Huzhou 313000, China; chuzh@zjhu.edu.cn
3 College of Agriculture, Shihezi University, Shihezi 832061, China; zyf1005225469@163.com
4 Xinjiang Production and Construction Corps Key Laboratory of Special Fruits and Vegetables Cultivation Physiology and Germplasm Resources Utilization, Shihezi 832003, China
* Correspondence: xuwei0412@shzu.edu.cn (W.X.); gp_inf@shzu.edu.cn (P.G.)

Abstract: Rapid and accurate detection of pesticide residue levels can help to prevent the harm of pesticide residue. This study used visible/near-infrared (Vis-NIR) (376–1044 nm) and near-infrared (NIR) (915–1699 nm) hyperspectral imaging systems (HISs) to detect the level of pesticide residues. Three different varieties of grapes were sprayed with four levels of pesticides. Logistic regression (LR), support vector machine (SVM), random forest (RF), convolutional neural network (CNN), and residual neural network (ResNet) models were used to build classification models for pesticide residue levels. The saliency maps of CNN and ResNet were conducted to visualize the contribution of wavelengths. Overall, the results of NIR spectra performed better than those of Vis-NIR spectra. For Vis-NIR spectra, the best model was ResNet, with the accuracy of over 93%. For NIR spectra, LR was the best, with the accuracy of over 97%, but SVM, CNN, and ResNet also showed closed and fine results. The saliency map of CNN and ResNet presented similar and closed ranges of crucial wavelengths. Overall results indicated deep learning performed better than conventional machine learning. The study showed that the use of hyperspectral imaging technology combined with machine learning can effectively detect the level of pesticide residues in grapes.

Keywords: hyperspectral imaging; pesticide residue; table grape; deep learning; non-destructive detection

Citation: Ye, W.; Yan, T.; Zhang, C.; Duan, L.; Chen, W.; Song, H.; Zhang, Y.; Xu, W.; Gao, P. Detection of Pesticide Residue Level in Grape Using Hyperspectral Imaging with Machine Learning. *Foods* **2022**, *11*, 1609. https://doi.org/10.3390/foods11111609

Academic Editors: Samuli Urpelainen and Mourad Kharbach

Received: 9 May 2022
Accepted: 27 May 2022
Published: 30 May 2022

1. Introduction

Grapes are one of the most popular fruits due to its unique taste, multiple vitamins, and nutrients. Grapes can be eaten fresh and processed into various products, for instance, juice and wine. Thus, there exists excellent commercial potential for the grape industry. During the grape growing season, fungicides, insecticides, and herbicides are often applied to cure the stresses of the diseases and pests [1,2]. The pesticide residue in grapes has increasingly aroused the attention of consumers. Certain intake of pesticide residue content may harm consumers' health [3,4].

Various methods have been developed for the detection of pesticide residue in fruits and vegetables [5]. Generally speaking, they can be divided into conventional and rapid detection methods. Traditional detection methods for detecting pesticide residues include gas chromatography (GC) and capillary electrophoresis (CE) [6], gas chromatography-mass spectrometry (GC-MS) [7], high-performance liquid chromatography (HPLC) [8], supercritical fluid chromatography (SFC) [9], and so on. Rapid detection methods include the fast detection card method and enzyme inhibition rate method. These methods have high accuracy for the detection of pesticide residue. However, they are costly. Moreover, they require complex pre-processing and highly skilled operators.

Hyperspectral imaging (HSI) is a technology that combines spectroscopy and conventional imaging to attain the spectral and spatial information from the research object [10]. HSI has been used effectively in the non-destructive quality detection of grapes, such as total soluble solids [11–13], total phenolic compounds [12], polyphenol contents [14], amino acids [11], and PH [11,12], etc. Moreover, there have been quantitative analyses, such as discriminating geographical origin [15], the year of harvest [15,16], and the maturation stage [17], etc.

Detection of pesticide residue in agricultural products combined with HSI technology has also been used widely, due to its advantage of rapid, non-destructive, and accurate quality detection. Sun et al. used HSI technology (431–962 nm) to quantitatively identify the pesticide mixtures on lettuce leaves [18]. Jia et al. detected apple surface pesticide residue based on HSI technology (865–1712 nm) [19]. Mohite et al. used hyperspectral sensing (350–1052 nm) to detect pesticide (Cyantraniliprole) residue on grapes with no, single, and double doses [20]. Ren used HSI technology (900–170 nm) to distinguish various concentrations of pesticide residues of dimethoate on the surface of spinach leaves [21]. Sun et al. identified pesticide residues in lettuce combining chemical molecular structure and NIR hyperspectral (870–1780 nm) [22]. Jiang et al. used NIR HIS (390–1050 nm) to predict the distribution of pesticide residues on mulberry leaves and visualize the results [23]. Studies have shown that HSI technology has been widely used in the non-destructive detection of pesticide residue in agricultural products. However, research on the pesticide residue in grapes is still rare, and a single spectral region was studied for most. Therefore, it is feasible and proposed to use hyperspectral imaging technology to detect different levels of pesticide residues in grapes here.

It is a great challenge to research massive and redundant data obtained by hyperspectral imaging systems (HIS) effectively, which prevents its application. Machine learning is exceptionally crucial for predicting features and analyzing spectral information. Recently, deep learning, as a new method of machine learning, has gained remarkable results for detecting and classifying the spectral and spatio-spectral signatures in HIS. Deep learning learns features deeply and automatically, and processes large volumes of data effectively [24–26]. Thus, it can construct a network containing many neurons efficiently and quickly, and it is applied widely in spectroscopy [27–30]. Yan et al. used HIS with deep learning to detect geographical origin of Radix Glycyrrhizae [31]. Jiang et al. used HIS with AlexNet-CNN deep learning network to detect postharvest pesticide residues [32]. Dreier et al. used CNN and ResNet with HSI to identify the bulk grain [33]. Gomes et al. used deep learning CNN to predict sugar and pH levels in grapes [34]. Deep learning has decent performance, but the process is obscure and difficult to understand. The contribution of wavelength is visualized to observe crucial wavelengths, which can explain the deep learning process well and analyze data effectively.

The purpose of the study was to use hyperspectral imaging technology combined with machine learning to identify the different pesticide residue levels in grapes. The specific goals were: (1) to explore the spectral differences among different pesticide residue levels of different varieties of grape; (2) to compare the performances of hyperspectral imaging at two different spectral regions for pesticide residue level identification; (3) to compare the performances of conventional machine learning methods (LR, SVM, and RF) and deep learning (CNN and ResNet); (4) and to explore the spectral features of different models which contribute more to the identification.

2. Materials and Methods

2.1. Samples Preparation

The research was carried out in the laboratory and simulated the process of spraying pesticides. Three grape varieties were used in this study, including Munage, Cabernet Sauvignon (Cabernet), and Red grape. The fresh grapes of Munage were purchased from the Jinma Market near Shihezi University, and Cabernet and Red grape were collected from the experimental vineyard located in the School of Agriculture, Shihezi University, Xinjiang

Uygur Autonomous Region (Xinjiang), China (73°40′–96°18′ E, 34°25′–48°10′ N). Each grape variety was randomly divided into four groups, corresponding to four different concentrations of pesticide residues (corresponding to four levels mentioned later). To increase the number of samples and comply with sampling inspection in the actual production, the bunch of the grape was cut smaller, considering the cluster of 3–6 berries as a sample, as shown in Figure 1. After cutting off grape bunches, 288 clusters of Cabernet, 411 clusters of Red grape, and 372 clusters of Munage were collected. In total, 1071 small clusters of grapes were used as input samples. The sample data were randomly divided into training, validation, and test sets with a ratio of 3:1:1. The specific sample size of clusters of the grape is shown in Table 1.

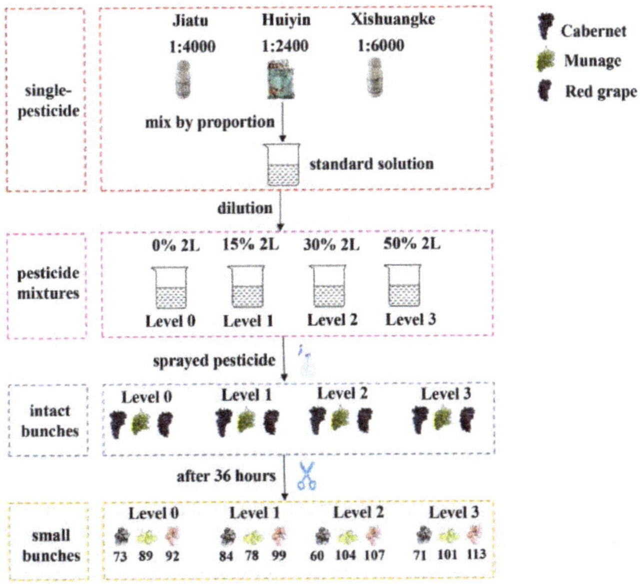

Figure 1. The flow chart of spraying pesticides and obtaining clusters of the grape.

Table 1. Number of samples after cutting intact grapes.

Category	Cabernet	Red	Munage	Total
Level 0	73	92	89	254
Level 1	84	99	78	261
Level 2	60	107	104	271
Level 3	71	113	101	285
Total	288	411	372	1071

Level 1, Level 2, and Level 3 mean the pesticide mixtures with concentrations of 10%, 15%, and 50% prepared later, and Level 0 means distilled water.

In this study, Jiatu (25% trifloxystrobin, 50% tebuconazole), Xishuangke (56% cymoxanil, 14% cyazofamid), and Huiyin (80% procymidone) were prepared, and the details are shown in Table 2. According to relevant information and instructions, these pesticide mixtures do not react chemically but only enhance the effect. Pesticide mixtures were sprayed on the grapes to simulate the pesticide residue. Different pesticides were applied to evaluate their effects on the growth of the grape. One reason for choosing these three pesticides was wide use during the ripening period of the grapes, and the other was the recommendations and suggestions of the planter. Roughly speaking, Jiatu, Xishuangke, and Huiyin are common fungicides, and they have a certain inhibitory effect on the growth of fungi.

Table 2. Information about the pesticides used in the experiment.

Category	Active Ingredients	Proportion	Efficacy
Jiatu	50% tebuconazole ($C_{16}H_{22}ClN_3O$) 25% trifloxystrobin ($C_{20}H_{19}F_3N_2O_4$)	4000	Brown spot
Huiyin	80% procymidone ($C_{13}H_{11}Cl_2NO_2$)	2400	Botrytis
Xishuangke	56% cymoxanil ($C_7H_{10}N_4O_3$) 14% cyazofamid ($C_{13}H_{13}ClN_4O_2S$)	6000	Downy mildew

There were two steps to making pesticides mixtures:

(1) Make standard pesticide mixtures. According to the instructions of each pesticide, three single-pesticide solutions (Jiatu, Huiying, and Xishaungke) were prepared with the proportion of 1:4000, 1:6000, and 1:24,000, respectively. Then, the three single-pesticide solutions were mixed together to make a 2 L pesticide mixture, as 100% standard pesticide mixtures.

(2) Make three pesticides mixtures. A beaker was used to dilute the 100% standard solution into three different pesticide mixtures. Concentrations of three pesticide mixtures were 10%, 15%, and 50% (respectively corresponding to Level 1, Level 2, and Level 3). Level 0 represented distilled water as a control group for comparing with others.

The corresponding concentration of the final configuration of each pesticide is shown in Table 3.

Table 3. Information about the concentration of each pesticide in the mixture.

Concentration	Jiatu	Xishuangke	Huiyin
Level 0 [a] (0%)	0	0	0
Level 1 [b](15%)	0.0375	0.0250	0.0625
Level 2 [c] (30%)	0.0750	0.0500	0.0125
Level 3 [d] (50%)	0.1250	0.0834	0.2085
Standard solution(100%)	0.2500	0.1667	0.4167

[a] means distilled water; [b,c,d] mean the pesticide mixtures with Level 1, 2, and 3, corresponding to concentrations of 10%, 15%, and 50%. The unit of concentrations is g/L.

With a spraying bottle, four groups of grapes were sprayed with Level 0, 1, 2, and 3 mixed pesticides, respectively. Then, the sprayed grapes were placed in a low-temperature and ventilated area for air drying for about 36 h [18,23,35–37]. When there was no more water on the grape surface, each intact bunch of grapes was cut, as shown in Figure 1.

2.2. Hyperspectral Image Acquisition and Correction

In this study, Vis-NIR and NIR HISs (SOC 710VP and SOC 710SWIR) were used in obtaining hyperspectral images. The SOC 710VP covers the spectral range of 376–1044 nm (128 bands), captures the image size of each waveband with 520 pixels × 696 pixels, and has an exposure time of 24 ms and a spectral resolution of 5 nm. The SOC 710SWIR covers the spectral range of 915–1699 nm (288 bands), captures the image size of each waveband with 512 pixels × 640 pixels, and has an exposure time of 34 ms and a spectral resolution of 2.7 nm. The distance from the sample to the imaging device was adjusted to 93.5 cm. Other information about the two HISs can be shown in Yan [31]. In the study, grapes of Level 0, Level 1, Level 2, and Level 3 were captured sequentially by the HIS. Each sample was fully photographed by shooting the front and back (randomly, one side was the front, and the other side was the back).

The raw hyperspectral images were corrected into the reflectance images by using a grayscale reference image. The correction was conducted by the following Equation (1):

$$I_r = \frac{I_{raw} - I_{dark}}{I_{white} - I_{dark}} \quad (1)$$

I_r is the reflectance image, I_{raw} is the raw image, I_{white} is the entirely white reference image, and I_{dark} is the entirely black reference image. The grayscale reference image was composed of 50% I_{dark} and 50% I_{white}.

2.3. Spectral Data Preprocessing and Extraction

The segmentation between the grape and the background was necessary to obtain accurate spectral information. In this study, ENV 5.2 (ITT Visual Information Solutions, Boulder, CO, USA) was used to crop a hyperspectral image to various hyperspectral sub-images containing a sample of 3–6 single berries. The sample in each hyperspectral sub-image was defined as a region of interest (ROI), which is a mask formed by threshold segmentation of the 804 nm Vis-NIR hyperspectral sub-image and the 1092 nm NIR hyperspectral sub-image. Further, spectra information in the ROI of the hyperspectral sub-image was extracted by Matlab R 2018b (The Math Work, Natick, MA, USA). The average spectrum of ROI was calculated as the spectral value of the sample, as shown in Figure 2. The spectral value at the beginning and the end were removed to eliminate obvious noise. The reserved wavelength range of Vis-NIR spectra was 476–890 nm (80 bands), and that of NIR spectra was 970–1594 nm (230 bands). For Vis-NIR and NIR spectral value, Savitzky–Golay (SG) [38] smoothing filter (the polynomial order was 0, the kernel size was 3) was used to improve the smoothness of the spectra and reduce noise interference. Then, the Standard Normal Variate transform (SNV) [39] was applied to avoid the impact of surface scattering, solid particle size, and the optical path change of diffuse reflection spectra.

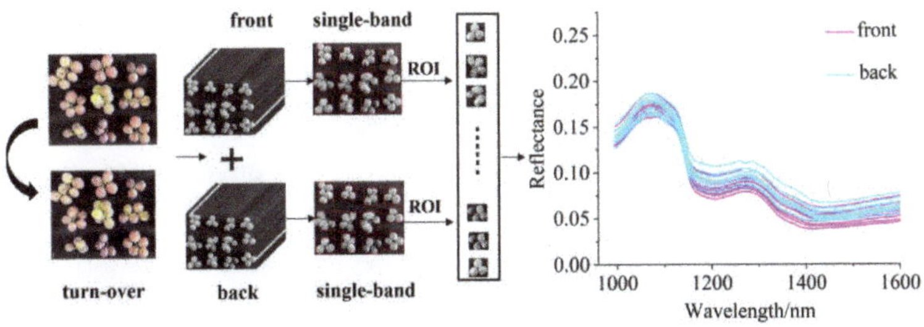

Figure 2. The flow chart of hyperspectral image data acquisition and data contact.

2.4. Data Analysis Method

2.4.1. Principal Component Analysis (PCA)

Principal component analysis (PCA) is a commonly used statistical method. A group of variables related to each other can be transformed into uncorrelated and independent ones through orthogonal transformation [40,41]. The primary purpose is to reduce the number of variables, namely dimensionality reduction. It is a linear dimensionality reduction method. The transformed variable is called the principal component (PC), and the top PCs explain most of the information of the hyperspectral image. The PCA score scatter plots for qualitative analysis of grape pesticide residues could be formed.

2.4.2. Support Vector Machine (SVM)

Support vector machine (SVM) is a supervised pattern recognition approach. SVM is a traditional classification method, and it is widely applied in classification conditions [42,43]. Moreover, SVM has excellent generalization ability, so it is widely used in spectroscopy. The kernel function is highly vital to the SVM model. In this paper, the tuning range of the kernel function was "poly, rbf, sigmoid". The kernel parameter g and penalty coefficient C were used to get optimal performance. A grid-search procedure was used to optimize g and C. The searching range of g and C were 10^{-5} to 50 and 10^{-5} to 50, respectively.

2.4.3. Logistic Regression (LR)

Logistic regression (LR) is a generalized linear regression analysis model, and it is often used in data mining, automatic disease diagnosis [44], economic forecasting [45], and other fields [46]. Linear regression is a machine learning method used to solve binary classification (0 or 1) problems, which are used to estimate the possibility of something. Adding the sigmoid active function to linear regression, LR can then be used for multiple classifications and introduced non-linear elements [47]. In this study, the optimization range of the solver was in "newton-cg", "lbfg", "liblinea", "sag", and that of C was between 10^{-5} and 10^5. The penalty was set to L2.

2.4.4. Random Forest (RF)

RF is ensemble learning, which consists of the decision tree (DT) [48]. RF shows two important traits: random sampling of training data points when building trees, and random subsets of features considered when splitting nodes [49,50]. The last result of the decision is determined by the voting method, so it has strong robustness. Random forest can process high-dimensional data without feature selection. In our study, *n_estimators* were between 100 and 1000, and *max_depth* was between 4 and 8.

2.4.5. Convolutional Neural Network (CNN)

A convolutional neural network (CNN) is a forward neural network. It usually consists of the following six layers: input layer, convolution layer, activation layer, pooling layer, fully connected layer, and output layer [31]. CNN has an excellent performance in classification. One advantage of CNN is local perception. CNN only perceives the local elements of the data and then merges local information in the higher-level network to obtain all the characterization information of the data. The second is weight sharing. By weight sharing, the number of weights of the network can be decreased, and the complexity of the network can be reduced [29]. A simple CNN architecture was designed for our study. The structure of the CNN is shown in Figure 3.

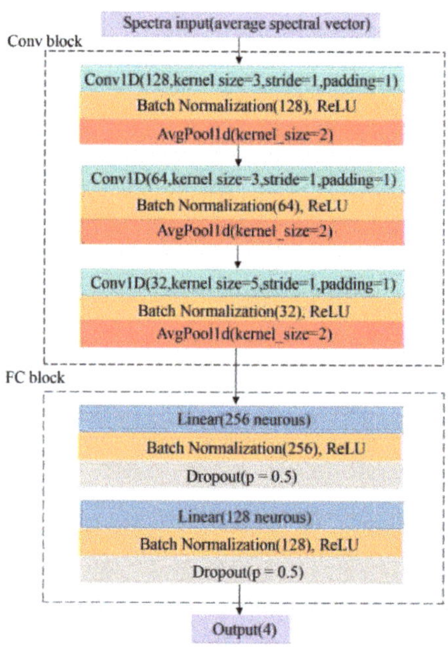

Figure 3. The proposed convolutional neural network (CNN) structure for the identification of pesticide residues in grapes.

In Figure 3, two main blocks were involved in the structure. The first block was the convolutional block (Conv Block), which consisted of three convolutional layers. Each convolutional layer was followed by a batch normalization layer (BN) and rectified linear unit (ReLU). In the end, an average pooling layer was added to alleviate the excessive sensitivity from the convolutional layer. In this process, one-dimensional (1D) spectral data were involved, and Conv1D was used, as shown in Figure 3. The second was a fully connected block (FC Block). The features extracted by the convolutional layer were learned through the fully connected layer. A linear layer was added, and BN and ReLU followed. The dropout was applied to alleviate overfitting. For the output layer, the network outputs the final result according to the probability of the four classification results. The input channels of the first three convolutional layers were 128, 64, and 32; the kernel sizes were 3, 3, and 5; the stride was 1, and the padding was 1. For the average pooling layer, kernel size was 2. The FC block included two fully connected layers, which consisted of 256 and 128 neurons, respectively. The dropout ratio was set as 0.5. Another linear layer was set for output at the end of the network. During the training process of CNN, the Adaptive Moment estimation (Adam) algorithm was used to optimize softmax cross-entropy. Weights were initialized using the Xavier algorithm.

2.4.6. Residual Neural Network (ResNet)

With the deepening of the neural network, there would be problems of overfitting, gradient explosion, and network degradation, and ResNet could effectively handle those [51]. In this study, based on the ResNet18, the ResNet was applied to identify pesticide residual levels. Figure 4a shows the structure of ResNet. The ResNet consisted of one convolutional layer and two residual blocks, the last was average pooling. The output channel of the convolutional layer was 64, kernel size was 1 × 3, and stride and padding were 1. Then a batch normalization layer (BN) and rectified linear unit (ReLU) were added. The channels of 3 residual blocks were 64, 128, and 256, kernel size was 1 × 3, and stride and padding were 1. The average pooling was followed to extract features smoothly, the last was the linear layer.

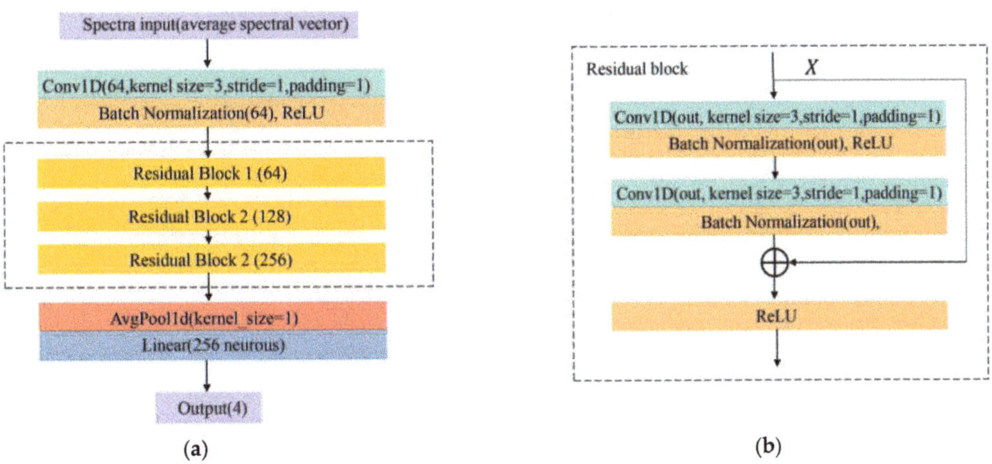

Figure 4. The proposed residual neural network (ResNet) (**a**) and residual block (**b**) structures for the identification of pesticide residues in grapes.

2.5. Saliency Map

Saliency map is a visualization technique in order to gain better insights into the decision-making of a neural network. When a sample was predicted correctly, it would be added to compute the feature importance [52]. Scale information contributions within the network could be computed [53]. Once the sample label was correctly predicted, the corresponding weights of the elements would be obtained, which represents the contribution rate (importance) of the elements. A saliency map can visualize the contribution rate of each element to intuitively see which elements play important roles in the process of CNN-based sample identification. For hyperspectral data, a saliency map could effectively visualize the importance of the wavebands.

Given the hyperspectral data X_0 with the set of the test, which was built by the classification model CNN-based, the class score function S_C (X_0) was obtained for all the wavebands [53]. When the label of this sample was correctly classified, the weight w could be calculated by the followed Equation (2).

$$w = \mathrm{abs}\left(\frac{\partial S}{\partial X} \middle| X_0 \right) \quad (2)$$

where w means the absolute value of the derivative of the score value S concerning the spectral data X_0.

In this study, test set data were used to compute the importance of all the wavelengths, when the sample label was predicted correctly.

2.6. *Software and Model Evaluation*

In this study, the areas of the samples were defined in ENVI 5.2 (ITT Visual Information Solutions, Boulder, CO, USA). The spectral data were extracted in Matlab R 2018b (The Math Work, Natick, MA, USA). The Python scripting language (version 3.8,64 bit) was applied for the numerical calculations. SVM, LR, and PLS-DA were conducted by using the machine learning library scikit learn (version 0.23.2). The 1D CNN model was built on the deep learning Pytorch framework (version 1.5.1). All data analysis procedures were implemented on a computer with a memory of 10 GB, a SSD of 238.35 GB, and a CPU of i5-7200 U.

The accuracy is used to illustrate the discrimination ability of classifier systems. The definition was the following:

$$Accuracy = \frac{TP}{All} \quad (3)$$

TP (true positive) means the number of the predicted result consistent with the actual label. *All* means the number of all samples. *Accuracy* is the index to evaluate the model.

3. Results

3.1. *Spectral Profiles*

The spectra in the range of 376–1073 and 915–1699 nm were extracted from the Vis-NIR and NIR HISs. The beginning and end of the spectra showed obvious noises. The spectral data were preprocessed by SG. The average spectra of four pesticide mixture levels and corresponding standard deviation are shown in Figure 5.

According to Figure 5, it is clear that the trend of the four average spectral curves is mostly similar. Peaks and valleys exist in the certain same positions and have no overlap (around 825, 550 and 1725 nm), which might have the potential to identify the different levels of pesticide residue in grapes due to variation of spectral reflectance in Vis-NIR and NIR regions. However, different pesticide levels and spectral ranges showed some discrepancies. In Figure 5a, the error bar overlaps at almost the entire band, and the curves of average spectra intersect at about 690 nm and 950 nm. In Figure 5b, the error bar overlaps in the spectra between 1160 nm and 1490 nm, and curves of average spectra intersect at 1310 nm. Therefore, it is impossible to directly distinguish different levels of pesticide residues in grapes clearly. It is necessary and crucial to do further research.

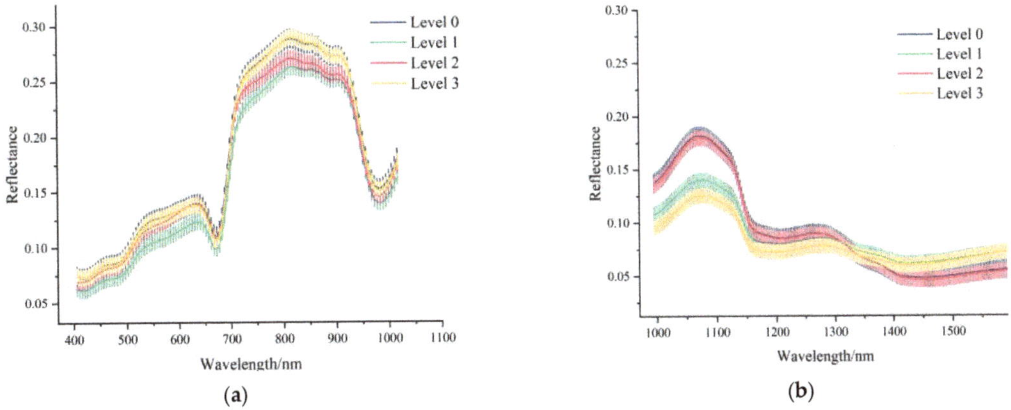

Figure 5. (**a**) Vis-NIR average (405–1016 nm) spectra with standard deviation each wavelength of different levels of pesticide residues in grape, using Vis-NIR spectrometer. (**b**) NIR average spectra (994–1641 nm) with standard deviation each wavelength of different levels of pesticide residues in grapes, using NIR spectrometer.

3.2. Principal Component Analysis (PCA)

To preliminarily explore significant differences between four levels of pesticide residues in grapes, spectral data were analyzed based on PCA. The two-dimensional PCA score plots were shown in Figure S1, with the sample's distribution of each PC. The corresponding confidence ellipse was added, with a confidence level of 0.95.

For Vis-NIR spectra, the contributions of the first three PCs of Cabernet were 48.5%, 27.5%, and 10.0%; those of Red grape were 49.4%, 26.8%, and 12.7%; those of Munage were 71.0%, 13.3%, and 4.6%. Their cumulative contributions of them were, respectively, 86.0%, 88.9%, and 88.9%, which explained most of the sample. However, the PCA score plots were clustered badly and there was serious overlap. For Cabernet, in Figure S1a–c, distributions of PC1 versus PC2, PC1 versus PC3, PC2 versus PC3 are chaotic and huddled, which means the four levels of pesticide residue are indistinguishable from each other. This phenomenon is consistent with trends of the spectral profile in Figure 5a. In addition, there is a certain similarity in Figure S1d–i.

For NIR spectra, the contributions of the first three PCs of Cabernet were 56.3%, 22.3%, 17.5%; those of Red grape were 57.4%, 31.3%, 6.6%; and those of Munage were 70.8%, 20.0%, 4.9%. The cumulative contributions of the first three PCs were 96.1%, 95.3%, and 95.7%, respectively, which also explained most of the variance information. Regarding the sample distribution, the overall clustering effect was slightly better than that of the Vis-NIR. For Cabernet, in Figure S1j–l, two major aggregating regions were shown (Level 0 and Level 2, Level 1 and Level 3), which is consistent with the phenomenon in Figure 5b. Therefore, the result comparatively illustrates the feasibility of the identification of four levels of pesticide residues in the range of NIR spectra.

In general, PCA visualizes sample distribution and provides the feasibility of classification, but it is not easy to directly distinguish the four levels of pesticide residues. Therefore, it is necessary to find other multivariate analysis methods for further research.

3.3. Classification Models

Three machine learning algorithms (SVM, LR, and RF) and two deep learning (CNN and ResNet) algorithms were conducted to analyze spectral data in this stage. The results are shown in Table 4 below.

Table 4. The classification of the accuracy of the logistic regression (LR), support vector machine (SVM), random forest (RF), convolution neural network (CNN), and residual neural network (ResNet).

Models	Categ	Parameter	Vis-NIR (%)			Parameter	NIR (%)		
			Train [a]	Val [b]	Test [c]		Train	Val	Test
SVM	0	2.0, 0.1, poly	95.9	94.8	91.4	6.6, 1.0, linear	99.4	100.0	96.6
	1	1.2, 0.1, poly	98.4	96.3	92.7	1.0, 1.0, poly	100.0	100.0	96.3
	2	1.0, 1.0, poly	1.00	88.0	93.2	1.0, 1.0, poly	100.0	100.0	95.9
LR	0	1×10^5, liblinear	100.0	89.7	93.1	100, lbfgs	99.4	93.1	98.3
	1	1×10^5, liblinear	100.0	98.8	93.9	1×10^5, liblinear	100.0	100.0	100.0
	2	1×10^4, liblinear	100.0	92.0	95.9	100, newton-cg	100.0	98.7	97.3
RF	0	8, 450	100.0	77.6	79.3	6, 750	100.0	74.1	81.0
	1	7, 500	99.6	72.3	73.2	5, 550	98.8	86.7	87.8
	2	8, 200	100.0	66.7	75.7	4, 250	99.1	98.7	93.2
CNN	0	500, 32, 0.001	99.4	98.3	93.1	500, 32, 0.001	100.0	100.0	98.3
	1	500, 32, 0.001	97.6	97.6	92.7	500, 32, 0.001	100.0	100.0	98.8
	2	500, 32, 0.001	100.0	98.7	93.2	500, 32, 0.001	99.5	100.0	98.6
ResNet	0	1000, 32, 0.005	100.0	94.8	93.1	600, 32, 0.005	100.0	93.1	86.2
	1	1000, 32, 0.005	100.0	100.0	98.8	1000, 32, 0.005	100.0	100.0	97.6
	2	1000, 32, 0.005	100.0	97.3	94.6	600, 32, 0.005	97.7	100.0	97.3

[a,b,c] represent training, validation, and test sets for the model; 0,1,2 represent Cabernet, Red grape and Munage, respectively, *Categ* mean Category of the grape. Parameters of the SVM, LR, RF, and CNN ResNet are shown. The parameters of the SVM, are (C, *gamma*, *kernel*); those of the LR are (C, *solver*); those of the RF are (*n_estimator*, *max_depth*); those of the CNN and ResNet are (epoch, batchsize, learning rate).

Vis-NIR spectra. All the models had good performances and had an average accuracy of over 90% for training, validation, and prediction sets. For Cabernet, the best models, the CNN and ResNet models, showed closed results, with the accuracy of over 99%, 94%, and 93% for train, validation, and test sets. SVM and LR models showed closed results, with the accuracy of over 91%, 89%, and 100% for training, validation, and test sets. For Red grape, all the models showed an accuracy of over 90% for training, validation, and test sets. RF showed overfitting, with the accuracy of over 100%, 77%, and 79%. For Red grape, the best model was ResNet, with the accuracy of over 100%, 100%, and 98% for training, validation, and test sets. CNN, SVM, and LR were slightly lower, with the accuracy of 97%, 96%, and 92% for training, validation, and test sets. RF still showed overfitting, with the accuracy of 99%, 72%, and 73% for training, validation, and test sets. For Munage, the best model was ResNet, with the accuracy of 100%, 97%, and 94% for training, validation, and test sets. CNN was slightly lower, with the accuracy of 100%, 98%, and 94% for training, validation and test sets. SVM performed with an accuracy of 100%, 88%, and 93.2% for training, validation, and test sets. RF was inferior to others, with the accuracy of 100%, 66%, and 75% for training, validation, and test sets. Overall, there was no significance with a different variety. ResNet performed better than other models, RF showed the overfitting, and SVM, LR, and CNN presented the fine result.

NIR spectra. Generally, all models had a slightly better result than Vis-NIR spectra, SVM, LR, CNN, and ResNet showed the average accuracy of over 90% for the validation set. For Cabernet, the CNN, LR, and SVM models presented the best and similar results, with an accuracy of close to 96% of the validation set. The following was ResNet, with the accuracy of 100%, 93%, and 86% for training, validation, and test sets. RF showed overfitting, with the accuracy of 100%, 74%, and 81% for training, validation, and test sets. For Red grape, SVM, LR, CNN, and ResNet presented closed and fine results, with the accuracy of over 100%, 100%, and 96% for training, validation, and test sets. RF showed lower results, with the accuracy of 98%, 86%, and 87.8% for training, validation, and test sets. For Munage, all models presented decent results, with the accuracy of over 93%. Overall, all the models showed fine results, and the results performed better than those of Vis-NIR. RF still showed the overfitting for Red grape and Munage. Varieties were not significant in the three grapes.

Methods. Considering different methods, there was a slight difference. For Vis-NIR spectra, overall, ResNet was the best model, with the accuracy of over 100%, 94%, and 93% for training, validation, and test sets. The following was CNN, with the accuracy of over 97%, 97%, and 92% for training, validation, and test sets. SVM and LR model were closed, with the accuracy of over 91% for the validation set. RF showed overfitting. For NIR spectra, SVM, LR, CNN, and ResNet showed closed and fine results, with an average accuracy of over 90%, but RF also showed overfitting for Cabernet. Overall, the deep learning methods (CNN, ResNet) performed better and had more stable results than those of machine learning (SVM, LR, RF).

The overall classification results showed NIR spectra performed better than Vis-NIR spectra. HIS in the NIR region was attributed to the overtone and overtone combination of molecular bonds (e.g., N-H, C-H, and O-H), and HIS in the Vis-NIR region was related to object color (e.g., chlorophyll). The results showed that spectral information on pesticide residues was related to the overtone of molecular, and more valuable information would be extracted via NIR spectra than Vis-NIR spectra regarding pesticide residue in grape. Therefore, it was more suitable to detect pesticide residues using NIR spectra. For Vis-NIR, CNN and ResNet performed best. For NIR, all results performed equally well, with the accuracy of over 95%. Overall, it shows that the deep learning method is superior to the traditional method. However, RF showed overfitting, and the reason might be the small size of the sample. The results of each grape variety showed a consistent trend. Thus, the classification accuracy did not correlate with the grape variety.

3.4. Visualization for Discovering the Wavelength Importance

Overall, deep learning (CNN and ResNet) offered finer results than machine learning, but their process of operation is hard to interpret. Therefore, CNN and ResNet were selected to visualize the wavelength importance, and saliency map was applied to analyze the model to find the critical wavelengths. The data were processed with normalization. The larger the value of the saliency map, the more critical the wavelength. The results are shown in Figure 6 for CNN and Figure 7 for ResNet.

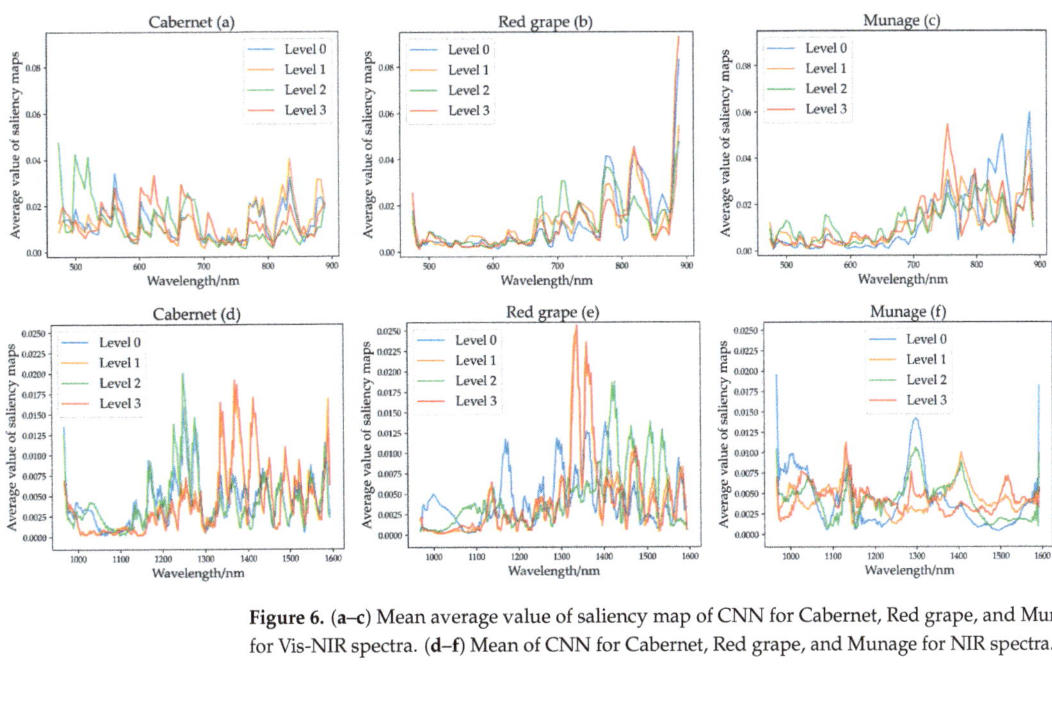

Figure 6. **(a–c)** Mean average value of saliency map of CNN for Cabernet, Red grape, and Munage for Vis-NIR spectra. **(d–f)** Mean of CNN for Cabernet, Red grape, and Munage for NIR spectra.

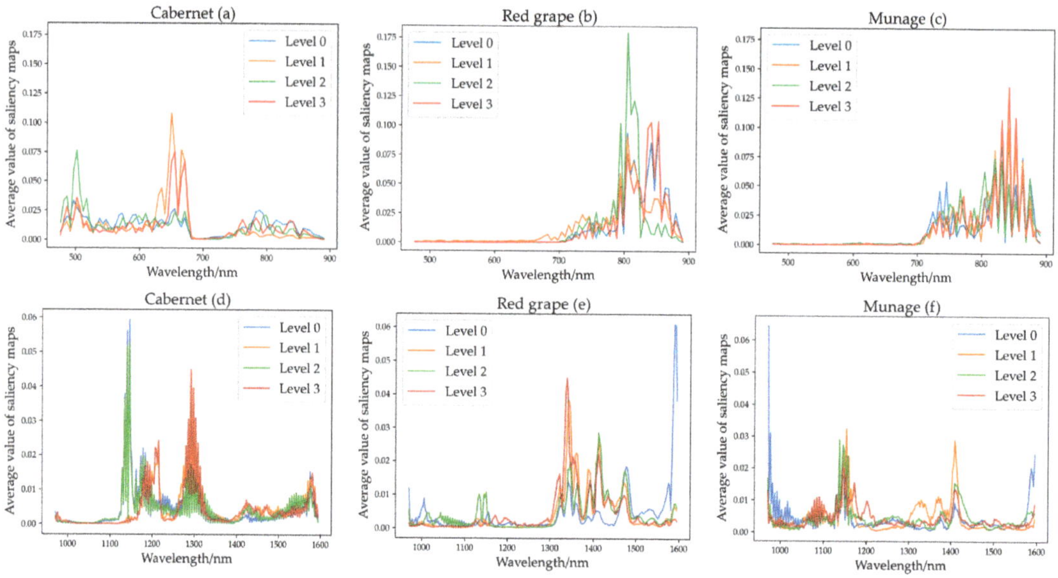

Figure 7. (**a–c**) mean average value of saliency map of ResNet for Cabernet, Red grape, and Munage for Vis-NIR spectra. (**d–f**) mean that of ResNet for Cabernet, Red grape, and Munage for NIR spectra.

Saliency map of CNN. For Vis-NIR spectra of Cabernet, approximately 500–530 nm, 550–580 nm, 600–730 nm, and 760–900 nm showed the largest contribution, and the difference between all bands was not very significant. For Vis-NIR spectra of Red grape and Munage, there were similar trends, approximately 660–900 nm contributed the most. For NIR spectra of Cabernet and Red grape, there was a consistent trend, approximately 1150–1300 nm and 1320–1600 nm contributed the most, the others showed low contribution. For NIR spectra of Red grape, regions of large contribution rate were 1290–1600 nm and 1120–1195 nm. For the NIR spectra of Munage, regions of main contribution were 960–1080 nm, 1110–1150 nm, 1280–1320 nm, 1390–1460 nm, and 1500–1550 nm.

Saliency map of ResNet. For the Vis-NIR spectra of Cabernet, the wavelengths at approximately 470–530 nm and 650–690 nm contributed the most, followed by the wavelengths at approximately 530–650 nm and 750–880 nm. For the Vis-NIR spectra of Red grape and Munage, the results presented the similarity; the wavelength at approximately 710–900 nm contributed the most. For the NIR spectra of Cabernet, the wavelengths at approximately 1120–1210 nm and 1260–1310 nm contributed the most, followed by the wavelengths at approximately 1210–1310 nm and 1420–1600 nm. For NIR spectra of Red grape, the wavelengths at approximately 1300–1500 nm and 1590 nm contributed the most, the others were low. For the NIR spectra of Munage, the wavelengths at approximately 970–980 nm, 1130–1180 nm, 1400–1420 nm, and 1580–1600 nm contributed the most, followed by the wavelengths at approximately 980–1110 nm and 1300–1440 nm.

For Vis-NIR spectra, generally, wavelengths of 380–780 nm were mainly relevant to the color variations of grape, e.g., chlorophyll [12,54]. For the rest of the NIR regions between 780 and 900 nm, those wavelengths are attributed to the third overtone stretch of O-H related to water in grapes [55]. The range of 900–980 nm was contributed to by the third overtone of C-H relevant to sugar [55]. For NIR spectra, wavelengths between 1050 nm and 1200 nm are mainly made up of the second overtone of C−H, and those between 1300 nm and 1500 nm are mainly related to the frequency of C-H [56]. The range of 1210–1450 nm is attributed to the 2nd overtone of C-H and the 1st overtone of O-H [54]. The wavelength between 975 nm and 1015 nm is mainly attributed to N-H stretch second overtone [57], and 1526 nm (N-H stretch first overtone) [58], which can reflect pesticide residue differences among different levels. Since Jiatu, Huiyin, and Xishuangke contain a large amount of C-H, O-H, and N-H as observed by their chemical molecular formula, these selected bands have a great correlation with pesticides. Overall, the saliency map of CNN and ResNet showed a similar and consistent trend, which confirmed the feasibility of visualizing the contribution of wavelengths by this method.

4. Discussion

Visible/near-infrared spectroscopy or hyperspectral imaging is a fast and non-destructive method to detect pesticide residues. Some studies have applied HSI to detect pesticide contaminants in foods [9,19–22], but the research object in those experiments was single and lacked mutual comparison between the objects. In our study, we chose three grapes to identify the difference between the varieties. Moreover, those studies used one [23] or two [18] pesticides as solvents for the research, and few studies mixed pesticides. Generally speaking, mixed pesticides can more effectively control plant diseases and insect pests without affecting the chemical properties and structure of the active ingredients. In this study, we used three pesticides together (Jiatu, Huiyin, and Xishuangke) to make the pesticide mixture and set four levels to compare the correlation among them, which was more consistent with the actual production with the use of pesticide. In addition, other studies mainly focused on assessing the pesticide residue within a single spectral range, and there is rarely a combination of Vis-NIR and NIR used on pesticide residues. In particular, to the best of our knowledge, no attempts have been made to analyze the different spectral ranges of mixed pesticides in grapes. Two spectral ranges were chosen to form a contrast and study the difference between the spectra, which increases the range of the spectrum and makes the research more comprehensive.

Due to the redundancy and high volumes of hyperspectral data, machine learning and deep learning were used to process the data and extract features. Previous studies have used SVM [22,59], DT [59], KNN [59], or RF [18] to detect pesticide residue, which showed fine results. SVM [60–62], LR [63], CNN [31,56,64], RF [60,62], and ResNet [56] have been applied widely in quality detection of hyperspectral imaging. In this study, classic machine learning and deep learning methods, CNN, ResNet, LR, SVM, and RF, were used to achieve a multivariate analysis of the detection of pesticide residue levels in grapes. More importantly, the saliency maps of CNN and ResNet were conducted to visualize the contribution rate of the wavelength, which brought us a clear understanding of the crucial wavelength information.

The two spectral ranges of the Vis-NIR (376–1044 nm) and NIR (915–1699 nm) showed great potential and decent results for detecting pesticide residue at different levels. The results (Table 4) of the NIR spectra were slightly better than those of the Vis-NIR spectra, with average accuracies of 95% and 90%, respectively. However, in this study, the main challenge was to make pesticide mixtures well-distributed in grapes. The uneven spraying of pesticides has a profound impact on the reflectance of hyperspectral images. The brightness unevenness of the hyperspectral image caused by the change in the surface curvature of the sphere also needs to be carefully corrected. The time-varying nature of spectrum acquisition deserves attention, such as drying time and acquisition sequence. The study promotes the non-destructive detection of pesticide residues in grapes, and other fruits, which accelerates the development of agro-products.

5. Conclusions

Detection of pesticide residuals in agro-products is of significant importance for food safety. This study successfully identified pesticide residual levels of grapes using hyperspectral images at two different spectral ranges. The results showed that it was feasible to detect different residual levels treated by the mixtures of different pesticides which were in accordance with the real-world pesticide usage of grapes. Furthermore, to validate the performances of the HSI technology, three different varieties of grapes were studied, and all of them showed good performance. The comparison between conventional machine learning methods and deep learning illustrated the effectiveness of deep learning in pesticide residual level identification by HSI. More importantly, the wavelengths contributing more to the identification were identified by saliency maps of deep learning models, which was of great help to understand the spectral responses to the pesticides. This study illustrated that HSI can be used for pesticide residual levels identification. The non-destructive approach of HSI can be conducted in a contactless, rapid, and accurate manner, which improves the detection efficiency and reduces the costs and the use of chemical reagents. HSI can further be studied for on-line pesticide residual level identification. In future studies, a larger number of samples and more varieties of grapes should be studied to establish more robust models for real-world application. The optimization of deep learning models should be studied. Deep transfer learning can be used to improve the generalization ability of the established deep learning models. Furthermore, in addition to qualitative analysis, the quantification of pesticide residual content and the limit of detection (LOD) should be determined by HSI with deep learning methods. The mechanism of the active ingredients of pesticides on the spectral responses of grapes should also be studied.

Supplementary Materials: The following supporting information can be downloaded at: https://www.mdpi.com/article/10.3390/foods11111609/s1, Figure S1: PCA score plot for the Cabernet, Red grape, and Munage spectral images photographed by using the Vis-NIR and NIR spectrometers.

Author Contributions: Conceptualization, W.Y. and T.Y.; data curation, W.Y., C.Z., L.D., W.C., H.S. and Y.Z.; formal analysis, C.Z., P.G. and W.X.; funding acquisition, P.G. and W.X.; methodology, W.Y., C.Z, T.Y. and L.D.; project administration, W.Y., T.Y. and C.Z.; writing—original draft, W.Y.; writing—review and editing, W.Y., Y.Z. and C.Z. All authors have read and agreed to the published version of the manuscript.

Funding: This research was funded by the National Natural Science Foundation of China, grant numbers 61965014 and 32060685, and the Scientific and Technological Research Projects in Key Areas of Xinjiang Production and Construction Corps (XPCC) grant number 2020AB005.

Data Availability Statement: The data presented in this study are available on request from the corresponding author.

Acknowledgments: The authors want to thank Yan Jiangkun, a student at the College of Information Science and Technology of Shihezi University in China, for providing help with this experiment.

Conflicts of Interest: The authors declare that the research was conducted without any commercial or financial relationships that could be construed as a potential conflict of interest.

References

1. Moeder, M.; Bauer, C.; Popp, P.; Van Pinxteren, M.; Reemtsma, T. Determination of pesticide residues in wine by membrane-assisted solvent extraction and high-performance liquid chromatography-tandem mass spectrometry. *Anal. Bioanal. Chem.* **2012**, *403*, 1731–1741. [CrossRef] [PubMed]
2. Popp, J.; Peto, K.; Nagy, J. Pesticide productivity and food security. A review. *Agron. Sustain. Dev.* **2013**, *33*, 243–255. [CrossRef]
3. Grewal, A.S.; Singla, A.; Kamboj, P.; Dua, J.S. Pesticide Residues in Food Grains, Vegetables and Fruits: A Hazard to Human Health. *J. Med. Chem. Toxicol.* **2017**, *2*, 1–7. [CrossRef]
4. Bouagga, A.; Chaabane, H.; Toumi, K.; Hamdane, A.M.; Nasraoui, B.; Joly, L. Pesticide residues in Tunisian table grapes and associated risk for consumer's health. *Food Addit. Contam. Part B-Surveill.* **2019**, *12*, 135–144. [CrossRef]
5. Narenderan, S.T.; Meyyanathan, S.N.; Babu, B. Review of pesticide residue analysis in fruits and vegetables. Pre-treatment, extraction and detection techniques. *Food Res. Int.* **2020**, *133*, 109141. [CrossRef]
6. Simonetti, G.; Castellani, F.; Di Filippo, P.; Riccardi, C.; Pomata, D.; Risoluti, R.; Buiarelli, F.; Sonego, E. Determination of Mancozeb, a Pesticide Used Worldwide in Agriculture: Comparison among GC, LC, and CE. *Curr. Anal. Chem.* **2020**, *16*, 1041–1053. [CrossRef]
7. Grimalt, S.; Dehouck, P. Review of analytical methods for the determination, of pesticide residues in grapes. *J. Chromatogr. A Incl. Electrophor. Other Sep. Methods* **2016**, *1433*, 1–23. [CrossRef]
8. Perez-Navarro, J.; Izquierdo-Canas, P.M.; Mena-Morales, A.; Martinez-Gascuena, J.; Chacon-Vozmediano, J.L.; Garcia-Romero, E.; Hermosin-Gutierrez, I.; Gomez-Alonso, S. Phenolic compounds profile of different berry parts from novel Vitis vinifera L. red grape genotypes and Tempranillo using HPLC-DAD-ESI-MS/MS: A varietal differentiation tool. *Food Chem.* **2019**, *295*, 350–360. [CrossRef]

9. Zhang, X.Z.; Zhao, Y.C.; Cui, X.Y.; Wang, X.R.; Shen, H.S.; Chen, Z.M.; Huang, C.; Meruva, N.; Zhou, L.; Wang, F.; et al. Application and enantiomeric residue determination of diniconazole in tea and grape and apple by supercritical fluid chromatography coupled with quadrupole-time-of-flight mass spectrometry. *J. Chromatogr. A* **2018**, *1581*, 144–155. [CrossRef]
10. Gowen, A.A.; O'donnell, C.P.; Cullen, P.J.; Downey, G.; Frias, J.M. Hyperspectral imaging-an emerging process analytical tool for food quality and safety control. *Trends Food Sci. Technol.* **2007**, *18*, 590–598. [CrossRef]
11. Fernandez-Novales, J.; Garde-Cerdan, T.; Tardaguila, J.; Gutierrez-Gamboa, G.; Perez-Alvarez, E.P.; Diago, M.P. Assessment of amino acids and total soluble solids in intact grape berries using contactless Vis and NIR spectroscopy during ripening. *Talanta* **2019**, *199*, 244–253. [CrossRef] [PubMed]
12. Xiao, H.; Li, F.; Song, D.J.; Tu, K.; Peng, J.; Pan, L. Grading and sorting of grape berries using visible-near infrared spectroscopy on the basis of multiple inner quality parameters. *Sensors* **2019**, *19*, 2600. [CrossRef] [PubMed]
13. Kanchanomai, C.; Ohashi, S.; Naphrom, D.; Nemoto, W.; Nakano, K. Non-destructive analysis of japanese table grape qualities using near-infrared spectroscopy. *Hortic. Environ. Biotechnol.* **2020**, *61*, 725–733. [CrossRef]
14. Baca-Bocanegra, B.; Hernandez-Hierro, J.M.; Nogales-Bueno, J.; Heredia, F.J. Feasibility study on the use of a portable micro near infrared spectroscopy device for the "in vineyard" screening of extractable polyphenols in red grape skins. *Talanta* **2019**, *192*, 353–359. [CrossRef]
15. Lemos, A.M.; Machado, N.; Egea-Cortines, M.; Barros, A.I. ATR-MIR spectroscopy as a tool to assist 'Tempranillo' clonal selection process: Geographical origin and year of harvest discrimination and oenological parameters prediction. *Food Chem.* **2020**, *325*, 126938. [CrossRef]
16. Fernández-Novales, J.; Tardáguila, J.; Gutiérrez, S.; Diago, P.M. On-the-go vis + sw nir spectroscopy as a reliable monitoring tool for grape composition within the vineyard. *Molecules* **2019**, *24*, 2795. [CrossRef]
17. Costa, D.D.; Mesa, N.F.O.; Freire, M.S.; Ramos, R.P.; Mederos, B.J.T. Development of predictive models for quality and maturation stage attributes of wine grapes using vis-nir reflectance spectroscopy. *Postharvest Biol. Technol.* **2019**, *150*, 166–178. [CrossRef]
18. Sun, J.; Cong, S.L.; Mao, H.P.; Wu, X.H.; Yang, N. Quantitative detection of mixed pesticide residue of lettuce leaves based on hyperspectral technique. *J. Food Process Eng.* **2018**, *41*, e12654. [CrossRef]
19. Jia, Y.G.; He, J.R.; Fu, H.F.; Shao, X.T.; Li, Z.K. Apple Surface Pesticide Residue Detection Method Based on Hyperspectral Imaging. *Intell. Sci. Big Data Eng.* **2018**, *11266*, 539–556. [CrossRef]
20. Mohite, J.; Karale, Y.; Pappula, S.; Shabeer, T.P.A.; Sawant, S.D.; Hingmire, S. Detection of pesticide (Cyantraniliprole) residue on grapes using hyperspectral sensing. In *Sensing for Agriculture and Food Quality and Safety IX*; SPIE: Bellingham, WA, USA, 2017; Volume 10217, pp. 192–199. [CrossRef]
21. Ren, Z.Q.; Rao, Z.H.; Ji, H.Y. Identification of Different Concentrations Pesticide Residues of Dimethoate on Spinach Leaves by Hyperspectral Image Technology. *Ifac Pap. Online* **2018**, *51*, 758–763. [CrossRef]
22. Sun, J.; Zhou, X.; Mao, H.; Wu, X.; Zhang, X.; Li, Q. Discrimination of pesticide residues in lettuce based on chemical molecular structure coupled with wavelet transform and near infrared hyperspectral. *J. Food Process Eng.* **2017**, *40*, e12509. [CrossRef]
23. Jiang, S.; Sun, J.; Xin, Z.; Mao, H.; Wu, X.; Li, Q. Visualizing distribution of pesticide residues in mulberry leaves using NIR hyperspectral imaging. *J. Food Process Eng.* **2017**, *40*, e12510.1–e12510.6. [CrossRef]
24. Petersson, H.; Gustafsson, D.; Bergstrom, D. Hyperspectral image analysis using deep learning—A review. In Proceedings of the 2016 Sixth International Conference on Image Processing Theory, Tools and Applications (IPTA), Oulu, Finland, 12–15 December 2016; pp. 1–6. [CrossRef]
25. Wang, C.; Liu, B.; Liu, L.; Zhu, Y.; Hou, J.; Liu, P.; Li, X. A review of deep learning used in the hyperspectral image analysis for agriculture. *Artif. Intell. Rev.* **2021**, *54*, 5205–5253. [CrossRef]
26. Signoroni, A.; Savardi, M.; Baronio, A.; Benini, S. Deep learning meets hyperspectral image analysis: A multidisciplinary review. *J. Imaging* **2019**, *5*, 52. [CrossRef]
27. Li, S.; Song, W.; Fang, L.; Chen, Y.; Benediktsson, J.A. Deep learning for hyperspectral image classification: An overview. *IEEE Trans. Geosci. Remote Sens.* **2019**, *57*, 6690–6709. [CrossRef]
28. Yamashita, R.; Nishio, M.; Do, R.K.G.; Togashi, K. Convolutional neural networks: An overview and application in radiology. *Insights Imaging* **2018**, *9*, 611–629. [CrossRef]
29. Xie, L.; Chen, X.; Bi, K.; Wei, L.; Xu, Y.; Chen, Z.; Xiao, A.; Chang, J.; Zhang, X.; Tian, Q. Weight-sharing neural architecture search: A battle to shrink the optimization gap. *ACM Comput. Surv.* **2020**, *54*, 1–24. [CrossRef]
30. Paoletti, M.E.; Haut, J.M.; Plaza, J.; Plaza, A. Deep learning classifiers for hyperspectral imaging: A review. *ISPRS J. Photogramm. Remote Sens.* **2019**, *158*, 279–317. [CrossRef]
31. Yan, T.Y.; Duan, L.; Chen, X.P.; Gao, P.; Xu, W. Application and interpretation of deep learning methods for the geographical origin identification of Radix Glycyrrhizae using hyperspectral imaging. *RSC Adv.* **2020**, *10*, 41936–41945. [CrossRef]
32. Jiang, B.; He, J.; Yang, S.; Fu, H.; Li, T.; Song, H.; He, D. Fusion of machine vision technology and alexnet-cnns deep learning network for the detection of postharvest apple pesticide residues. *Artif. Intell. Agric.* **2019**, *1*, 1–8. [CrossRef]
33. Dreier, E.S.; Sorensen, K.M.; Lund-Hansen, T.; Jespersen, B.M.; Pedersen, K.S. Hyperspectral imaging for classification of bulk grain samples with deep convolutional neural networks. *J. Near Infrared Spectrosc.* **2022**, *30*, 107–121. [CrossRef]
34. Gomes, V.; Mendes-Ferreira, A.; Melo-Pinto, P. Application of Hyperspectral Imaging and Deep Learning for Robust Prediction of Sugar and pH Levels in Wine Grape Berries. *Sensors* **2021**, *21*, 3459. [CrossRef]

35. Li, M.; Lu, L.; Zhang, X. Qualitative Determination of Pesticide Residues in Purple Cabbage Based on Near Infrared Spectroscopy. *J. Phys. Conf. Ser.* **2021**, *1884*, 012015. [CrossRef]
36. Soltani Nazarloo, A.; Rasooli Sharabiani, V.; Abbaspour Gilandeh, Y.; Taghinezhad, E.; Szymanek, M.; Sprawka, M. Feasibility of using VIS/NIR spectroscopy and multivariate analysis for pesticide residue detection in tomatoes. *Processes* **2021**, *9*, 196. [CrossRef]
37. Özbey, A.; Karagöz, Ş.; Cingöz, A. Effect of Drying Process on Pesticide Residues in Grapes. *Gida J. Food* **2017**, *42*, 204–209. [CrossRef]
38. Gorry, P.A. General least-squares smoothing and differentiation by the convolution (Savitzky-Golay) method. *Anal. Chem.* **1990**, *62*, 570–573. [CrossRef]
39. Barnes, R.J.; Dhanoa, M.S.; Lister, S.J. Standard normal variate transformation and de-trending of near-infrared diffuse reflectance spectra. *Appl. Spectrosc.* **1989**, *43*, 772–777. [CrossRef]
40. Ringnér, M. What is principal component analysis? *Nat. Biotechnol.* **2008**, *26*, 303–304. [CrossRef]
41. Bro, R.; Smilde, A.K. Principal component analysis. *Anal. Methods* **2014**, *6*, 2812–2831. [CrossRef]
42. Cao, X.H.; Wang, D.; Wang, X.Z.; Zhao, J.; Jiao, L.C. Hyperspectral imagery classification with cascaded support vector machines and multi-scale superpixel segmentation. *Int. J. Remote Sens.* **2020**, *41*, 4528–4548. [CrossRef]
43. Kuo, B.C.; Ho, H.H.; Li, C.H.; Hung, C.C.; Taur, J.S. A Kernel-Based Feature Selection Method for SVM With RBF Kernel for Hyperspectral Image Classification. *IEEE J. Sel. Top. Appl. Earth Obs. Remote Sens.* **2014**, *7*, 317–326. [CrossRef]
44. Divya, R.; Kumari, R.S.S.; Initia, A.D.N. Genetic algorithm with logistic regression feature selection for Alzheimer's disease classification. *Neural Comput. Appl.* **2021**, *33*, 8435–8444. [CrossRef]
45. Schauberger, G.; Tutz, G. Regularization Methods in Economic Forecasting. *Springer Int. Publ.* **2015**, *48*, 61–68. [CrossRef]
46. Cheng, Q.; Varshney, P.K.; Arora, M.K. Logistic regression for feature selection and soft classification of remote sensing data. *IEEE Geosci. Remote Sens. Lett.* **2006**, *3*, 491–494. [CrossRef]
47. Stephan, D.; Lucila, O.M. Logistic regression and artificial neural network classification models: a methodology review. *J. Biomed. Inform.* **2002**, *35*, 352–359. [CrossRef]
48. Witten, I.H.; Frank, E.; Hall, M.A.; Pal, C.J. Ensemble learning. In *Data Mining*, 4th ed.; Elsevier: Amsterdam, The Netherlands, 2017; pp. 479–501. [CrossRef]
49. Liaw, A.; Wiener, M. Classification and regression by randomForest. *R News* **2002**, *23*, 18–22.
50. Cutler, A.; Cutler, D.R.; Stevens, J.R. Random forests. *Mach. Learn.* **2004**, *45*, 157–176. [CrossRef]
51. He, K.; Zhang, X.; Ren, S.; Sun, J. Deep residual learning for image recognition. In Proceedings of the 2016 IEEE Conference on Computer Vision and Pattern Recognition (CVPR), Las Vegas, NV, USA, 27–30 June 2016; pp. 770–778. [CrossRef]
52. Li, Z. A saliency map in primary visual cortex. *Trends Cogn. Sci.* **2002**, *6*, 9–16. [CrossRef]
53. Simonyan, K.; Vedaldi, A.; Zisserman, A. Deep Inside Convolutional Networks: Visualising Image Classification Models and Saliency Maps. *arXiv* **2014**, arXiv:1312.6034. [CrossRef]
54. Wang, X.; Li, Z.; Wang, W.; Wang, J. Chlorophyll content for millet leaf using hyperspectral imaging and an attention-convolutional neural network. *Ciência Rural.* **2020**, *50*, e20190713. [CrossRef]
55. Yu, X.; Lu, H.; Wu, D. Development of deep learning method for predicting firmness and soluble solid content of postharvest korla fragrant pear using vis/nir hyperspectral reflectance imaging. *Postharvest Biol. Technol.* **2018**, *141*, 39–49. [CrossRef]
56. Jin, B.C.; Zhang, C.; Jia, L.; Tang, Q.; Gao, L.; Zhao, G.; Qi, H. Identification of Rice Seed Varieties Based on Near-Infrared Hyperspectral Imaging Technology Combined with Deep Learning. *ACS Omega* **2022**, *7*, 4735–4749. [CrossRef] [PubMed]
57. Xu, Z.; Fan, S.; Liu, J.; Liu, B.; Wu, Y. A calibration transfer optimized single kernel near-infrared spectroscopic method. *Spectrochim. Acta Part A Mol. Biomol. Spectrosc.* **2019**, *220*, 117098–117105. [CrossRef]
58. Jerry, W. *The Handbook of Organic Compounds. NIR, IR, Raman, and UV-Vis Spectra Featuring Polymers and Surfactants*; Elsevier: Amsterdam, The Netherlands, 2001.
59. Zhang, L.; Rao, Z.; Ji, H. Nir hyperspectral imaging technology combined with multivariate methods to study the residues of different concentrations of omethoate on wheat grain surface. *Sensors* **2019**, *19*, 3147. [CrossRef]
60. Cheng, Y.L.; Yang, S.Q.; Liu, X.; Zhang, E.Y.; Song, Z.S. Identification of wine grape varieties based on near-infrared hyperspectral imageing. *Asabe Res.* **2019**, *35*, 959–967. [CrossRef]
61. Salvador, G.; Juan, F.N.; Diago, M.P.; Javier, T. On-the-go hyperspectral imaging under field conditions and machine learning for the classification of grapevine varieties. *Front. Plant Ence* **2018**, *9*, 1102. [CrossRef]
62. Mohite, J.; Sawant, S.; Borah, K.; Pappula, S. Temporal Detection of Pesticide Residues in Tea Leaves Using Hyperspectral Sensing. In *IGARSS 2019-2019 IEEE International Geoscience and Remote Sensing Symposium, Yokohama, Japan, 28 July–2 August 2019*; IEEE: Piscataway, NI, USA, 2019.
63. Liu, W.; Zeng, S.; Wu, G.; Li, H.; Chen, F. Rice Seed Purity Identification Technology Using Hyperspectral Image with LASSO Logistic Regression Model. *Sensors* **2021**, *21*, 4384. [CrossRef]
64. Li, X.; Jiang, H.; Jiang, X.; Shi, M. Identification of geographical origin of chinese chestnuts using hyperspectral imaging with 1d-cnn algorithm. *Agriculture* **2021**, *11*, 1274. [CrossRef]

foods

MDPI

Review

Current Application of Advancing Spectroscopy Techniques in Food Analysis: Data Handling with Chemometric Approaches

Mourad Kharbach [1,2,*], Mohammed Alaoui Mansouri [3,4], Mohammed Taabouz [5] and Huiwen Yu [6,7,*]

1 Department of Food and Nutrition, University of Helsinki, 00014 Helsinki, Finland
2 Department of Computer Sciences, University of Helsinki, 00560 Helsinki, Finland
3 Nano and Molecular Systems Research Unit, University of Oulu, 90014 Oulu, Finland
4 Research Unit of Mathematical Sciences, University of Oulu, 90014 Oulu, Finland
5 Biopharmaceutical and Toxicological Analysis Research Team, Laboratory of Pharmacology and Toxicology, Faculty of Medicine and Pharmacy, University Mohammed V in Rabat, Rabat BP 6203, Morocco
6 Shenzhen Hospital, Southern Medical University, Shenzhen 518005, China
7 Chemometrics group, Faculty of Science, University of Copenhagen, Rolighedsvej 26, 1958 Frederiksberg, Denmark
* Correspondence: mourad1kharbach@gmail.com (M.K.); huiwen.yu@food.ku.dk (H.Y.)

Abstract: In today's era of increased food consumption, consumers have become more demanding in terms of safety and the quality of products they consume. As a result, food authorities are closely monitoring the food industry to ensure that products meet the required standards of quality. The analysis of food properties encompasses various aspects, including chemical and physical descriptions, sensory assessments, authenticity, traceability, processing, crop production, storage conditions, and microbial and contaminant levels. Traditionally, the analysis of food properties has relied on conventional analytical techniques. However, these methods often involve destructive processes, which are laborious, time-consuming, expensive, and environmentally harmful. In contrast, advanced spectroscopic techniques offer a promising alternative. Spectroscopic methods such as hyperspectral and multispectral imaging, NMR, Raman, IR, UV, visible, fluorescence, and X-ray-based methods provide rapid, non-destructive, cost-effective, and environmentally friendly means of food analysis. Nevertheless, interpreting spectroscopy data, whether in the form of signals (fingerprints) or images, can be complex without the assistance of statistical and innovative chemometric approaches. These approaches involve various steps such as pre-processing, exploratory analysis, variable selection, regression, classification, and data integration. They are essential for extracting relevant information and effectively handling the complexity of spectroscopic data. This review aims to address, discuss, and examine recent studies on advanced spectroscopic techniques and chemometric tools in the context of food product applications and analysis trends. Furthermore, it focuses on the practical aspects of spectral data handling, model construction, data interpretation, and the general utilization of statistical and chemometric methods for both qualitative and quantitative analysis. By exploring the advancements in spectroscopic techniques and their integration with chemometric tools, this review provides valuable insights into the potential applications and future directions of these analytical approaches in the food industry. It emphasizes the importance of efficient data handling, model development, and practical implementation of statistical and chemometric methods in the field of food analysis.

Keywords: food analysis; food authenticity; food chemicals; spectroscopy techniques; chemometrics; multivariate analysis

Citation: Kharbach, M.; Alaoui Mansouri, M.; Taabouz, M.; Yu, H. Current Application of Advancing Spectroscopy Techniques in Food Analysis: Data Handling with Chemometric Approaches. *Foods* **2023**, *12*, 2753. https://doi.org/10.3390/foods12142753

Academic Editors: Ricard Boqué, Antonello Santini and Clara Cecília Santana Sousa

Received: 13 February 2023
Revised: 30 June 2023
Accepted: 18 July 2023
Published: 19 July 2023

1. Introduction

The growing world population is increasing the demand for food in multiple ways, which is leading to a higher demand for safety and quality control of commercialized products. Food can become contaminated by chemicals and physical substances through accidental or intentional means. In recent years, there have been several major cases of food adulteration, highlighting the critical need for controlling product authentication [1]. In one instance, wheat gluten samples were infused with melamine to improve their protein content. In 2008, China experienced a milk scandal where milk was found to be adulterated with melamine, and in 2012, India had a similar scandal where milk was found to be adulterated with detergent, urea, and other substances [2]. Several noteworthy incidents occurred in 2005, including the adulteration of chili powder with dye and the contaminated chili powder in Indian Worcestershire sauce [3]. Gelatin-like chemicals were used in the aquaculture market recently to increase weight in many instances in China. Spices are frequently adulterated with ground material worldwide, particularly in Europe and India [4,5]. These incidents demonstrate the importance of product authentication and quality control in the food industry to protect consumers from harmful and potentially dangerous adulterated products. To gain economic benefit, consumers are at risk of being exposed to serious health threats if food products are adulterated with cheap or chemical materials. This was seen in the Chinese milk scandal, where six infants passed and several thousand were hospitalized. A further case is the contamination of paprika with lead oxide to give it a reddish color, which caused over sixty hospitalized. In recent years, both controlling agencies and the public have grown increasingly concerned about the application of the phthalate plasticizer di-2-ethylhexyl phthalate as a clouding agent in beverages and food [6,7]. In the past decade, both food and feed products have been found to be adulterated. Some of the most adulterated agro-food commodities include honey, edible oils, and spices. Additionally, it has been reported that food products such as milk products, fruit extract, flour, coffee, alcohol, and meats, are being adulterated more frequently [8]. All these examples illustrate a significant global problem that poses a threat to consumers and has prompted food authorities to increase their scrutiny and inspection of the food production chain from farming to consumption.

Food adulteration can occur for various reasons, including complex production processes and long supply chains. Therefore, authentication is crucial for both labeling organizations and industries that must test raw materials and finished products to ensure compliance with specifications [9]. Additionally, confirming the authenticity of the food is essential for maintaining quality and preventing economic fraud. To address authentication challenges and ensure product quality, fast, reliable, and competent analytical methods are needed. The detection of composition properties and contaminants in food and agricultural commodities can be accomplished using a wide variety of physical and chemical methods. These properties, such as density, texture, color, acidity, and solubility, are typically measured by physical or chemical methods. The chemical composition of the samples determines the different chemical techniques involved and used in the identification of components and contamination detection of food commodities [10]. Despite being powerful analytical techniques, separation techniques including liquid-chromatography or gas-chromatography are not always suitable when the workflow is rapid and the experiments are costly, as the samples may be damaged during the study process. Therefore, advanced spectroscopic techniques, along with chemometrics, are now being used for the quality analysis and authentication of a wide range of food products. The main advantages of these techniques are that they are non-damaging, rapid, ecologically friendly, and economical. Spectroscopy data are complex and are typically handled by chemometric approaches for supervised and unsupervised pattern recognition, such as hierarchical cluster analysis (HCA), principal component analysis (PCA), linear discriminant analysis (LDA), soft independent modeling of class analogy (SIMCA), and partial least squares-discriminant analysis (PLS-DA) among others. These tools are mainly used to assess classes, such as attributing samples as either adulterated/unadulterated or authentic/not authentic. In addition to the qualitative chemometric approaches, there are also quantitative approaches including multivariate calibration tools, including principal component regression (PCR), and partial least squares (PLS) [11,12]. These tools are mainly used for quantitative parameters to quantify the number of adulterants and fatty acids content, based on data generated from spectroscopic techniques.

The focus of this review is to provide an overview of recent studies on advanced spectroscopy techniques and chemometric methods that are widely utilized in food evaluation, safety assessment, quality analysis, and manufacturing processes. The review aims to emphasize the advantages of these techniques over traditional analytical methods and underscore the importance of efficient data handling, model construction, and practical implementation of statistical and chemometric approaches in the field of food analysis. Additionally, the review aims to discuss the potential applications and future directions of these techniques in the food industry, while addressing the challenges associated with traditional analytical methods, including their destructiveness, laborious nature, time consumption, cost, and negative environmental impact.

2. Chemometric Approaches in Spectroscopy Data

The advancement of modern instruments, represented by spectroscopy techniques, has accelerated the development of the food industry and food research in recent decades. As a result, the data available to food analysts has become increasingly complex. Not only does the amount of data tend to be large, but the dimensionality of the data can also increase dramatically. Effectively analyzing and managing large amounts of spectroscopy data from food production, food processing, and food research is both a practical and theoretical issue. Chemometrics, which combines the power of statistics and mathematics for use by chemists, provides a valuable solution to the challenging analytical issues in food spectroscopy analysis. The general framework and pipeline of advanced spectroscopic techniques coupled with chemometric approaches applied in food analysis is illustrated in Figure 1. Advanced spectroscopic techniques combined with chemometric approaches are of great significance in food analysis. These techniques provide fast, precise, and non-destructive measurements of diverse food properties (Figure 1). Their integration offers multiple advantages, such as rapid analysis, non-destructive measurement, multivariate data analysis, quality control, process optimization, and allergen detection. These advancements contribute significantly to enhancing food safety, ensuring quality assurance, and safeguarding consumer protection. On the other hand, several chemometric tools have been developed and validated to be powerful in terms of information extraction, multivariate relationship analysis, prediction, and discrimination analysis, among others, in food spectroscopy data analytics. A roadmap workflow example in Figure 2 for using chemometric tools to handle spectral data in different scenarios for either qualitative or quantitative purposes. However, the interpretation of spectroscopy data, be it in the form of signals or images, can be intricate without the aid of statistical and innovative chemometric approaches. These approaches encompass crucial steps, including pre-processing, exploratory analysis, variable selection, regression, classification, and data integration. By employing these methods, researchers can extract pertinent information and effectively manage the complexities inherent in spectroscopic data.

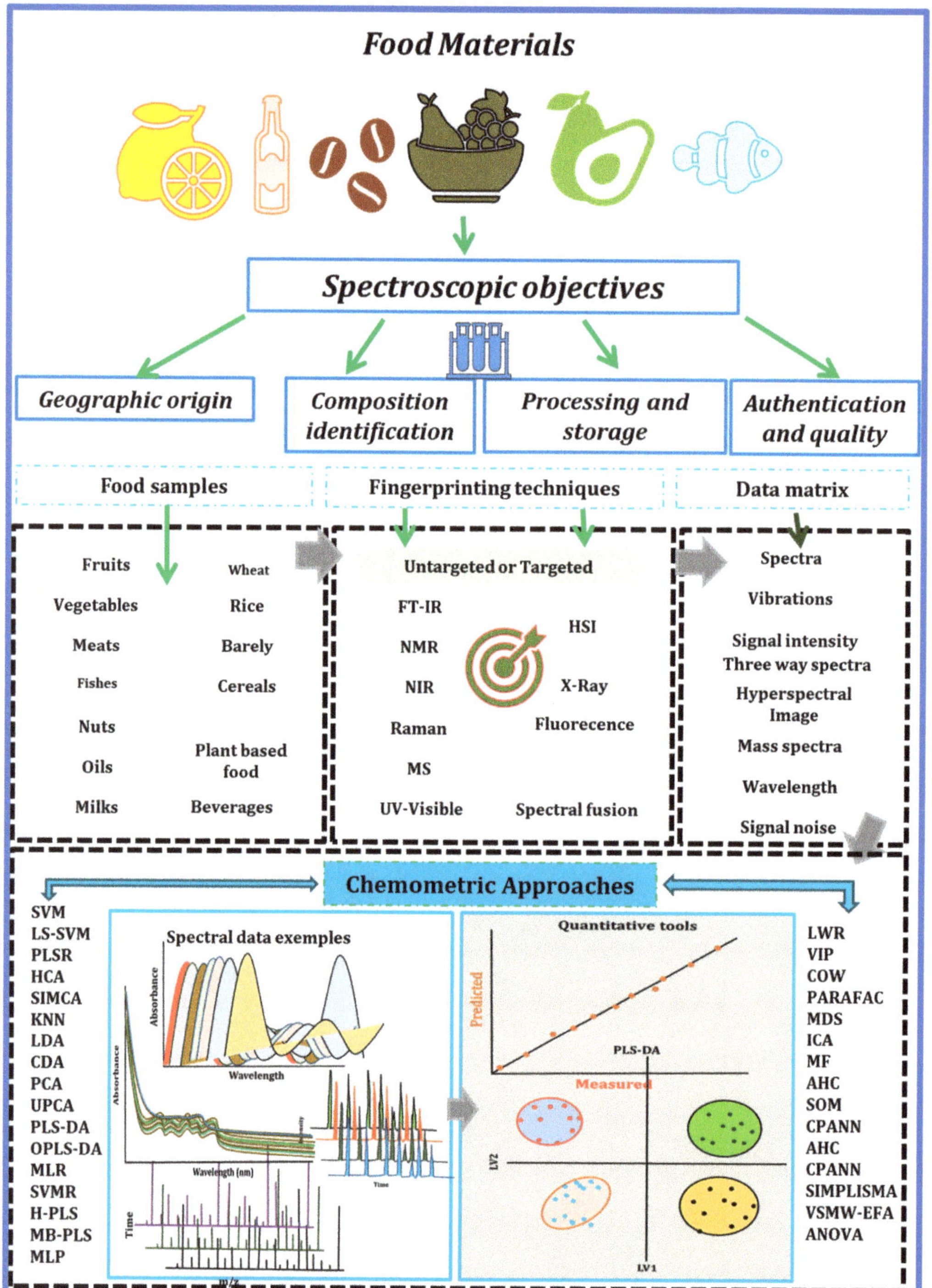

Figure 1. A general workflow of advanced spectroscopic techniques combined with chemometric approaches for food analysis.

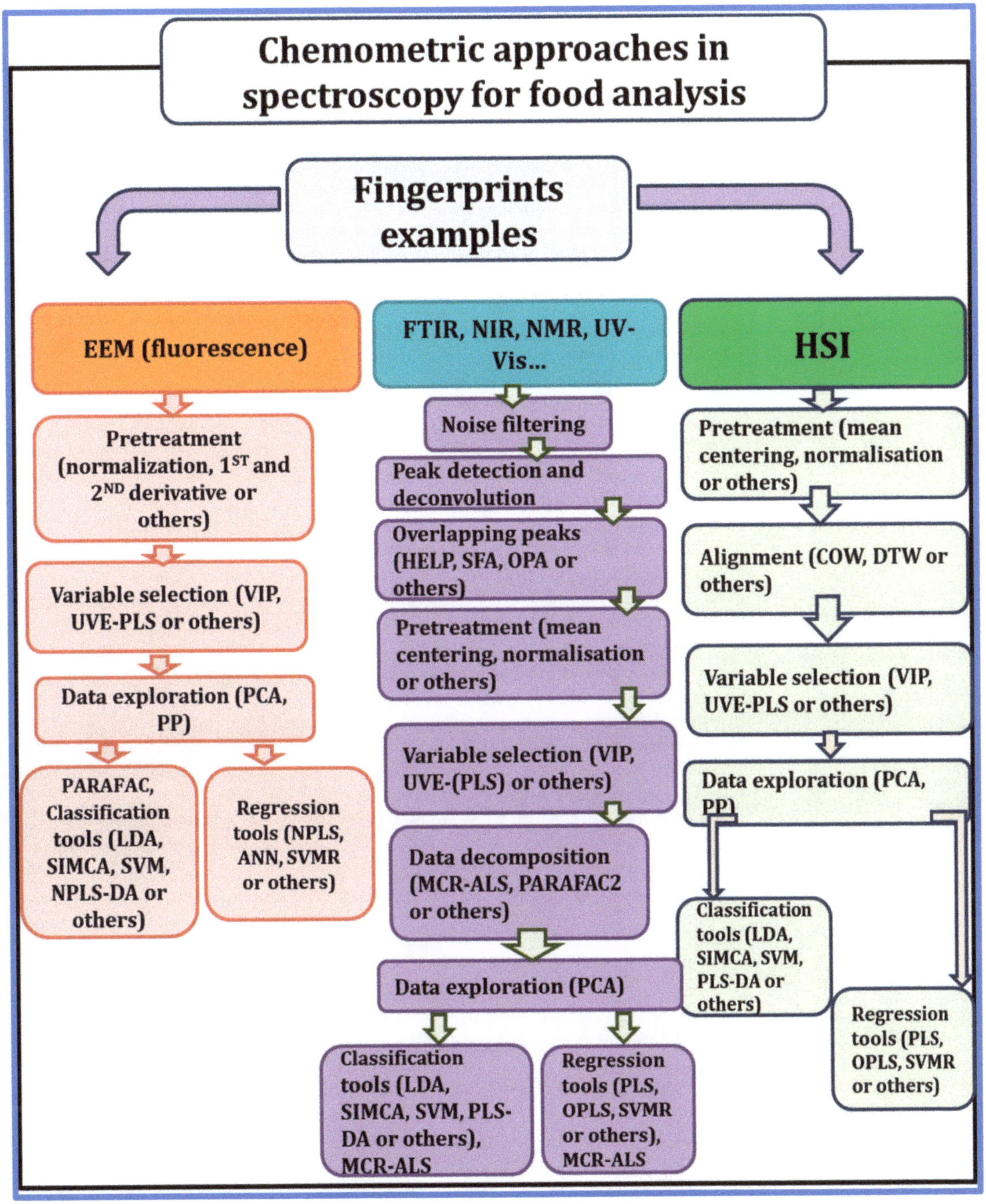

Figure 2. A roadmap example for chemometric approaches applied to different spectroscopic data for both quantitative and qualitative food analysis purposes.

2.1. *Pre-Processing Techniques*

Pre-processing plays a crucial role in spectral data analysis as it prepares the data for further analysis and modeling. Various pre-processing techniques are employed to address common challenges such as scattering correction, baseline correction, peak shift alignment, denoising, and handling missing values. Effective pre-processing is essential for enhancing the performance of models by eliminating artifacts from the data and reducing fitting errors. By applying appropriate pre-processing methods, the spectral data becomes more reliable and conducive to accurate analysis and interpretation. In the context of missing data, several chemometric tools have been developed to address this issue [13]. For instance, maximum likelihood PCA-based imputation [14], as well as data regression methods like KDR-PLSR (kernel density-ratio-based partial least squares regression) and KDR-PCR (kernel density-ratio-based principal components regression) [14], have recently emerged as highly effective approaches for handling missing data. For multi-way data, such as fluorescence data, practical solutions for missing data imputation include alternating least squares with single imputation, Parallel Factor (PARAFAC) analysis, and the Levenberg-Marquardt method [15,16]. These techniques offer reliable and efficient means to impute missing data, enabling more comprehensive and accurate analyses of complex datasets. In the context of scatter correction, Multiplicative Scatter Correction (MSC) [17], Standard Normal Variate (SNV) [18], and other normalization methods are widely employed chemometrics techniques. These methods effectively address the issue of scatter in spectral data. Additionally, advancements in scatter correction tools have led to the development of new versions and improved approaches [17]. To tackle baseline drift problems in spectral data, adaptive reweighing schemes for polynomial fitting and penalized least squares [19], as well as Tikhonov regularization [20], have proven successful in removing unwanted baseline variations. These techniques offer reliable means to address baseline drift and enhance the accuracy of subsequent analysis.

In order to mitigate peak shift problems, several methods have been reported to be useful. Automatic time shift alignment (ATSA) [21], coherent point drift peak alignment [22], Global peak alignment with point matching algorithm [23], and PARAFAC Applied to Shift Invariant Amplitude Spectra (PARASIAS) [24] are among the techniques that have demonstrated effectiveness in aligning peaks accurately despite shifts or distortions. Derivative calculation of spectral data is a promising solution to some different artefacts problems especially when it is combined with other techniques, e.g., combining the first derivative and simple spectral ratio can correct both the additive effects and multiplicative effects [25].

Overall, the utilization of these advanced techniques in scatter correction, baseline drift correction, and peak shift alignment significantly improve the quality and reliability of spectral data, facilitating more robust and accurate analyses in various chemometric applications.

2.2. *Variable Selection Tools*

Variable selection is a valuable technique in spectral data analysis as it enhances model performance, provides better interpretations, and reduces measurement costs [26]. Several popular methods are commonly used in chemometrics for spectral data analysis, including model factors and assessment, model-based feature importance statistics, interval partial least squares regression (iPLS) [27], and genetic algorithm (GA) [28]. When considering model parameters, variables with lower loadings and regression coefficients may not be as important as those with higher values. In the case of model-based variable importance statistics, variables such as variables important for projection (VIP) [29], which measures the contribution of a variable in describing the data, and selectivity ratios [30], which evaluate the predictive performance of variables, are commonly used in chemometrics analysis of spectral data. These statistics help identify the most influential variables in the model. The iPLS is a variable selection method that operates by selecting windows of variables, making the selection of window size a critical aspect of iPLS. On the other hand, GA performs variable selection by simulating the process of natural selection, estimating models involving patterns in variable generation [26]. However, GA requires more complex parameter settings compared to other variable selection methods. Regardless of the specific variable selection method employed, it is important to remember that validation is always necessary to avoid erroneous conclusions resulting from overfitting. Validation procedures help ensure the robustness and reliability of the selected variables and the overall model performance.

In summary, the utilization of variable selection techniques in spectral data analysis offers improved model performance, interpretability, and cost-effectiveness. Methods such as iPLS, GA, and model-based feature importance statistics contribute to selecting relevant variables and enhancing the accuracy and reliability of chemometric analyses. Validation procedures are crucial to validate the selected variables and mitigate the risk of overfitting.

2.3. Exploratory and Clustering Tools

Exploratory data analysis is a critical component of food spectroscopy data analysis, serving as a means to understand descriptive statistical characteristics and gain multivariate insights from complex datasets. A range of exploratory analysis tools, including both graphical techniques and quantitative dimensionality reduction techniques, are widely utilized in the analysis of food spectroscopy data. Graphical techniques such as box plots, histograms, and scatter plots are valuable for visualizing the distribution, variation, and presence of missing values in samples. These tools are commonly employed in the analysis of metabolite data obtained from mass spectroscopy [31]. By utilizing graphical tools, researchers can effectively explore the properties and patterns present within the data.

Principal Component Analysis (PCA) is a representative dimensionality reduction technique and one of the most popular tools employed in food spectroscopy data analysis. The earliest invention of PCA dates back to the early 20th century [32]. However, this first invention focused more on the modeling property and explained the variation of PCA. However, it was later expanded upon by Hotelling, who introduced the concept of PCA as a linear combination of variables [33]. Nowadays, PCA is now widely recognized for its ability to reveal complex relationships within multivariate data, making it a powerful tool for obtaining an overview of complex datasets. It is frequently used to explore relationships between samples and variables, identify outliers, discover and determine patterns (groups), as well as generating new hypotheses [34]. Moreover, PCA can also be utilized for conducting clustering analysis. By examining the relationships between samples, PCA can effectively divide samples with different statistical characteristics into distinct groups. In addition to PCA, Hierarchical Cluster Analysis (HCA) [35] and k-means clustering analysis [36] are widely used as clustering analysis tools in food spectroscopy data analysis. In the format of a dendrogram, HCA constructs a dendrogram that hierarchically divides samples into groups based on their similarities, facilitating the identification of distinct clusters. On the other hand, K-means clustering analysis partitions objects into k-clusters, with each object belonging to the group with the closest average.

Overall, exploratory data analysis, comprising graphical techniques, dimensionality reduction methods like PCA, and clustering analysis tools such as HCA and k-means, enables researchers to gain insights into complex food spectroscopy datasets. These approaches facilitate the understanding of data properties, identification of patterns, and generation of hypotheses, ultimately advancing the knowledge and interpretation of food analysis.

2.4. Regression and Prediction Tools

Prediction models are essential in food spectroscopy data analysis, as they enable the successful prediction of chemical and physical properties, supporting green food research and the sustainable food industry. Partial least squares (PLS) regression is a standard chemometric method used for prediction analysis of spectral data [37]. By projecting the predictor variables and response variables onto a new space, PLS seeks to uncover the underlying relationship between X and Y matrices by modeling the covariance structures in this new space [38]. PLS has the added advantage of effectively handling the collinearity problem often encountered in spectral data [39]. In addition to PLS, N-way PLS [40] is a powerful tool for handling multi-way spectral data and the N-way PLS model is advantageous in well modeling performance, robustness to noise, stabilized solution, and improved prediction capabilities [41] specifically for multi-way spectral data. Various other regression methods, such as support vector machine (SVM) regression artificial neural network (ANN) and multiple linear regression (MLR), are also employed in spectral data analysis [42]. MLR is a linear regression method that establishes the relationship between independent variables and dependent variables by fitting a linear model. However, MLR has strict assumptions that need to be met, including constant variance of the residuals, multivariate normality, and linear assumption [43], which can limit its applications in certain cases. In contrast, SVM regression and ANN are often utilized for non-linear prediction analysis in spectral data. For instance, ANN can capture highly non-linear relationships between inputs and outputs, enabling the prediction of an output variable based on input data [44]. The non-linear models such as SVM and ANN can offer attractive predictive performance in many applications [42], but the issue of overfitting must be carefully addressed. Therefore, validation is a crucial step in non-linear prediction analysis to ensure the reliability and generalization of the models when applied to spectral data.

In short, prediction models, including PLS regression, N-way PLS, SVM regression, ANN, and MLR, are valuable tools in food spectroscopy data analysis. These models enable the prediction of chemical and physical properties, and their selection depends on the nature of the data, linearity assumptions, and the need for handling collinearity or capturing non-linear relationships. Validation procedures are essential to evaluate and validate the performance of these models, ensuring their accuracy and robustness in spectral data analysis.

2.5. Classification Tools

Classification is an essential task in food analysis, aimed at identifying and assigning categories to samples in order to detect pattern differences and enable subsequent analysis. Various chemometrics techniques are employed for the classification analysis in food analysis, including tree-based methods, regression-based methods, discriminant analysis, and neural networks, among others. One popularly used classification tool in food analysis is partial least squares-discriminant analysis (PLS-DA) [45]. PLS-DA is a discrimination method based on PLS regression but incorporates an additional classification step based on the thresholding of predicted y-values. Another method, the soft independent modelling class analogy (SIMCA) method combines the concept of PCA in performing classification analysis [46]. SIMCA utilizes the residuals from disjoint PCA models to assign samples to one or several classes, with the critical distance being based on the F-distribution [47]. In addition to PLS-DA and SIMCA models, Linear Discriminant Analysis (LDA) [48] and SVM [49] are also frequently used for classification analysis in food analysis. LDA models the differences between groups by finding a linear combination of features and projecting them from a higher dimensional space into a lower dimensional space, effectively separating them into distinct classes. On the other hand, SVM can also be used for non-linear classification analysis. In the case of non-linear classification, SVM employs a kernel function to transform the data from a non-linear space to a linear space, enabling the classification task in a high-dimensional space. However, it's important to note that performing SVM classification on large spectroscopy datasets may be time-consuming due to the training and kernel computation requirements.

In brief, classification is a fundamental data analysis task in food analysis, and several chemometric techniques such as PLS-DA, SIMCA, LDA, and SVM are commonly utilized for this purpose. These methods enable the identification of patterns and the assignment of samples to different categories, facilitating subsequent analysis and decision-making in the field of food analysis. It's important to consider the nature of the data, linear or non-linear relationships, and computational considerations when selecting the appropriate classification method for a given analysis.

2.6. Mixture Analysis Tools

Mixture analysis is a critical area of focus in food analysis, driven by advancements in omics technologies and the demand for improved production processes in the food industry. A wide range of chemometric tools is employed to analyze mixtures in food, encompassing both two-way analytical methods and multi-way analytical methods.

Multivariate curve resolution-alternating least squares (MCR-ALS) is one of the representative two-way chemometrics tools used for food mixture analysis. It allows for the extraction of chemically meaningful bilinear models from a data matrix that includes mixed measurements, with an additive model structure [50]. Basically, MCR-ALS decomposes the mixture data matrix into two matrices and a residual matrix. For instance, in the case of high-performance liquid chromatography with diode array detection (HPLC-DAD) data, MCR-ALS separates the data into a matrix containing elution profiles of all components, another matrix containing the corresponding pure spectra, and a residual matrix of the same dimensions as the raw data, capturing unexplained variations. Although MCR-ALS encounters challenges such as permutation ambiguity, intensity ambiguity, and rotational ambiguity, these issues can be partially addressed through specific strategies [51]. Parallel Factor Analysis (PARAFAC) [52] and Parallel Factor Analysis2 (PARAFAC2) [53] models are widely used multi-way models for complex mixture analysis. These models generalize the bilinear models to handle multi-way data. Instead of generating a set of bilinear components, PARAFAC and PARAFAC2 decompose the high-order tensor data into a set of trilinear components (in the case of three-way data) in which each vector represents the information from each mode. Due to the unique advantages of PARAFAC and PARAFAC2 models [54], they are powerful for decomposing the pure chemical from the multi-way fluorescence and mass spectroscopy data. The main difference between PARAFAC and PARAFAC2 models is that the strict multilinear assumption in the PARAFAC model is relaxed in PARAFAC2 model, meaning that the profiles for each slab in the multi-way data is not required to be the same in PARAFAC2 model if the cross products of the components keep the same [55].

In outline, mixture analysis holds significant importance in food analysis, and a range of chemometric tools are applied for this purpose. MCR-ALS is a powerful two-way method for mixture analysis, while PARAFAC and PARAFAC2 models are widely used in multi-way analysis of complex mixtures. These methods enable the extraction of chemically meaningful information from mixture data, aiding in the identification and characterization of individual components within complex food matrices. While certain challenges exist, strategies have been developed to address them and enhance the reliability and applicability of these chemometric tools in mixture analysis.

In this section, a general overview was provided regarding the various chemometric tools utilized for handling spectroscopy data. These tools encompass pre-processing techniques, exploratory analysis methods, variable selection approaches, regression models, classification algorithms, and mixture analysis methodologies. However, the specific methodologies and algorithms underlying each technique were not thoroughly explored, nor were the advantages or disadvantages of individual approaches discussed in detail.

3. Advanced Spectroscopy Techniques with Chemometrics in Food Analysis

Advanced spectroscopy techniques paired with chemometric tools are crucial in analyzing food by providing a fast, non-destructive, and efficient means of obtaining detailed information about food samples. This information can be used to improve the quality, safety, and authenticity of food products. Table 1 summarizes recent applications of advanced spectroscopic techniques and chemometric approaches for quantitative analysis in food whereas a summary of recent applications of advanced spectroscopic techniques linked to chemometric approaches for qualitative analysis in food can be found in Table 2. In this section, the focus was on discussing the most notable studies conducted on advanced spectroscopic techniques, including X-ray-based methods, hyperspectral and multispectral imaging, NMR, Raman, IR, UV, visible, fluorescence, and portable techniques. These studies were examined with respect to their applications, both qualitative and quantitative, and their overall utility in the field of food analysis.

Table 1. Advanced spectroscopic techniques combined with chemometric approaches for quantitative analysis in food application. Table was ordered regarding used techniques.

Quantitative Analysis						
Authors	**Food Materials**	**Fingerprinting Technique**	**Analysed Components**	**Objectives**	**Chemometric Analysis**	**Ref.**
Gamela et al.	Cocoa beans	EDXRF	Cu, K, Sr and Zn	Cocoa beans	PLS	[56]
Sperança et al.	Bananas	X-ray fluorescence	Zn	Determination Zn content in Banans	PLS	[57]
Priyashantha et al.	Cheese	Near-infrared hyperspectral (NIR-HS) imaging	-	Explain the relationship between average spectra and cheese maturity	PLS	[58]
Darnay et al.	Semi-hard cheese	HSI	Transglutaminase	Detection of the enzyme of transglutaminase in the cheese	PLS	[59]
Lu et al.	Potatoes	Fluorescence HSI	Solanine	Predict the solanine content in potatoes	SVR	[60]
Xiao et al.	Fresh-cut potato	HSI	Color parameters (bruising index) and water content	Assess the quality of potatoes	LS-SVM and PLS	[61]
Tian et al.	Purple sweet potato	Vis-NIR HSI	Moisture and anthocyanins	Predict the critical indexes of moisture and anthocyanins in purple sweet potato	PLSR	[62]
Li et al.	Plum	VNI-HSI	Nan	Predict the soluble solid contents and the color of two plums cultivars	PLS	[63]
He et al.	Wheat flour	HSI	Talcum powder	Detection of talcum powder adulterated in wheat flour	SNV-CARS-PLS	[64]
Kim et al.	Wheat flour	SWIR-HSI	Benzoyl peroxide	Detecting the bleaching agent of benzoyle peroxide in wheat flour	PLS	[65]
Sun et al.	Melons	HSI	Nan	Predict the sugariness and hardness of melons	PLS, SVM and ANN	[66]
Wang et al.	Chinese steamed bread	FT-NIR	Potato flour content	Predicting potato flour in Chinese steamed bread	PLS-R	[67]
Tu et al.	Wheat flour	FT-NIR	Talcum powder	Quantitation of low content of talcum powder in wheat flour	Gradient-boosted decision tree (GBDT)	[68]
Kandpal	Tuber flour	NIR and MIR	Chemical components:amylose, starch, protein, glucose, cellulose, and moisture contents	Prediction of quality traits in tuber traits by mean of Data fusion of FT-IR and FT-NIR	SOPLS	[69]
Kamboj et al.	Wheat	FT-NIR	Crude protein and carbohydrate	Compare chemometrics for predicting the quality parameters of wheat	PLS, MLR, SVM	[70]
Liang et al.	Potatoes	FT-NIR	Sugar content	Detection of zebra chip disease (ZC) in potatoes	PLS	[71]
Jiang et al.	Wheat flour	Portable NIR	Fatty acid	Quantitation of fatty acids in wheat	Variable combination population analysis (VCPA), extreme learning machine (ELM)	[72]

Table 1. *Cont.*

Quantitative Analysis						
Authors	**Food Materials**	**Fingerprinting Technique**	**Analysed Components**	**Objectives**	**Chemometric Analysis**	**Ref.**
Ning et al.	Wheat grains	FT-NIR	Zearalenone	Detection of zearalenone in wheat	SVM	[73]
Cámara et al.	Clove and pomegranate	IR	Antioxidant activity	Estimation of antioxidant activity in clove and pomegranate	MCR-ALS and PLS	[74]
Castro et al.	Peanut oil	NIR and Raman	Adulterants (corn oil and vegetable oil)	Assessment of vibrational spectroscopy with chemometrics	MCR-ALS and PLS regression	[75]
Castro et al.	Saffron	FT-NIR	Saffron adulterants (onion, calendula, pomegranate, and turmeric)	Detection of Saffron adulterants	MCR-ALS and PLS regression	[76]
Li et al.	Saffron	FT-NIR	Saffron adulterants (lotus stamens and corn stigmas)	Detection of saffron adulterants	Synergistic interval PLS (SI-PLS), competitive adaptive reweighted sampling PLS (CARS-PLS)	[77]
Li et al.	Saffron	FT-NIR	Corcin	Determination of corcin content in Saffron	PLS	[78]
Liu et al.	Panax notoginseng	FT-NIR	Adulterants (hizoma curcumae, *Curcuma longa* and rhizoma alpiniae offcinarum)	Quantification of Panax notoginseng with its adulterants	PCR, PLS, ELM and SVR	[79]
Liu et al.	Vegetable oils	FT-NIR	Phytosterols	Determination of phytosterols in vegetable oils	Pls	[80]
Joshi et al.	Eggs	FTIR	Constituents of eggs (yolk and albumen)	Detection of fabricated eggs	PLS-DA and SVM	[81]
Mazivila et al.	Milk	FT-NIR	Melamine and sucrose	Estimation the adulterant contents in the milk	MCR-ALS	[82]
Novianty et al.	Palm fruit	FT-NIR	Oil content	Quantitation of oil content in palm fruit	EMD-ANN	[83]
Basar et al.	Honey	FTIR	Adulterant (beet sugar and corn syrup)	Determination of honey adulteration	Genetic-algorithm-based inverse least squares (GILS) and (PLS)	[84]
Qin et al.	Wheat flour	Raman chemical imaging	Benzoyl peroxide	Detection of benzoyle peroxide	PLS	[85]
Yuan et al.	Duck meat	Surface-enhanced Raman	Testosterone propionate and nandrolone residues	Quantitation of residues in the duck meat	LS-SVR	[86]
Nakajima et al.	Banana	Raman	Starch	Quantification of starch in banana	PLS	[87]
Hara et al.	Tomatoes	Raman	Carotenoids	Determination of carotenoids in tomatoes	PLS	[88]
De Olieveira mendes et al.	Raw milk	Raman	Whey	Quantitation of whey in raw milk	PLS	[89]
Czaja et al.	Youghurts	Raman	Fat, lactose, and protein	Determination of nutritional parameters of yoghurts	PCA and PLS	[90]

Table 1. *Cont.*

Quantitative Analysis						
Authors	**Food Materials**	**Fingerprinting Technique**	**Analysed Components**	**Objectives**	**Chemometric Analysis**	**Ref.**
Tian et al.	Milk	Raman	Adulterants (maltodextrin, sodium carbonate, and whey)	Prediction of adulterants in raw milk	PLS	[91]
Berzins et al.	Breast milk	Raman and FTIR	Macronutrients (protein, fat, and carbohydrate)	Determination of macronutrients in the breast milk	PLS	[92]
De sa oliveira et al.	Spreadable cheese	Raman	Starch	Quantitation of starch in adulterated spreadable cheese	PLS	[93]
Liu et al.	Edible oils	Raman and FT-IR data fusion	Peroxide values and acid values	Determination of chemical quality indices of edible oils during thermal oxidation	PLS	[89]
Puertas et al.	Egg yolk	Data fusion of FTIR and UV-Vis	Cholesterol	Prediction of cholesterol in egg yolk	PLS and PCR	[94]
Wang et al.	Infant formula	Vis-NIR and Raman data fusion	-	Assessment of infant formula storage temperature and time	SVM	[95]
Valinger et al.	Honey	UV-Vis and NIR data fusion	Sugar syrups	Detection of honey adulteration	PLS and ANN	[96]
Wang et al.	Camelia oil	Excitation-emission matrix fluorescence	Vegetable oils	Quantitation of adulterant in camelia oil	N-PLS and PARAFAC	[97]
Baretto et al.	Milk	Fluorescence	Melamine	Determination of melamine in milk	PARAFAC and UPLS	[98]
Gu et al.	Rapessed oil in water	Fluorescence	Lipid	Quantitative assessment of lipid oxidation in a rapeseed oil-in-water	GA-SVR	[99]
Tarhan	Extra virgin olive oil (EVOO)	FTIR, UV–Vis and fluorescence	Squalene	Quantification of squalene in extra virgin olive oils	PLS	[100]
Wu et al.	Edible blend oil	UV-Vis	Adulterant (vegetable oil)	Quantification of vegetable oils in edible blend oil	Weighted multiscale SVR	[101]
Zhang et al.	Edible oils	UV-Vis	Acid value	Impact of heating on edible oils	PLS and PCR	[102]
Rios-Reina et al.	Wine and balsamic vinegar	UV-Vis	Grape-must caramel (E-150d caramel)	Quantitation of grape-must caramel in wine and balsamic vinegars	PLS	[103]
Cavdaroglu et al.	Vinegar	UV-Vis and MIR	Phenolic components, p-coumaric and syringic acids, citric and acetic acids,	Predict quality and chemical parameters of vinegar	PLS and OPLS	[104]
Santos et al.	Milk	NMR	Adulterants (Whey, urea, hydrogen peroxide, synthetic urine and synthetic milk)	Quantification of milk adulteration	PLS	[105]
Liu et al.	Cream	NMR	Artificial bright blue pigment	Detecting additives content in cream	PLS and MLR	[106]
Sun et al.	Carrot, banana and pleurotus eryngii	NMR	Moisture	Monitor water states of typical fruits and vegetables during microwave vacuum drying	PLS, SVM and BP-ANN	[107]

Table 1. *Cont.*

Quantitative Analysis						
Authors	**Food Materials**	**Fingerprinting Technique**	**Analysed Components**	**Objectives**	**Chemometric Analysis**	**Ref.**
Hajjar et al.	Hen egg	NMR	Fatty acids	Quantification of fatty acids in hen eggs	PLS	[108]
Galvan et al.	Edible oils	NMR	Fatty acids and iodine value	Analysis of edible oils	PLS and SVR	[109]
Haddad et al.	Cheese	NMR	Fatty acids	Quantitation of individual fatty acids	PLS	[110]
Jiang et al.	Rice	Surface-enhanced Raman scattering	Chlorpyrifos residue	Quantify chlorpyrifos residues in rice samples	GA-PLS, UVE-PLS, VCPA-PLS and CARS-PLS	[111]
Richardson et al.	Coconut water	Raman	Sugars	Detection of adulteration in Coconut water	PLS	[112]

Table 2. Advanced spectroscopic techniques combined with chemometric approaches for qualitative analysis in food application. Table was ordered regarding used techniques.

Qualitative Analysis						
Authors	**Food Materials**	**Fingerprinting Technique**	**Analysed Components**	**Objectives**	**Chemometric Analysis**	**Ref.**
Galvan et al.	Tomato and sweet paper	EDXRF	-	Discrimination of tomato or sweet pepper samples effectively according to the agronomic production mode or geographical origin	PLS-DA	[113]
Scatigno et al.	EVOO	EDXRF	Ni, Fe and Ti	Discrimination of EVOO	PCA	[114]
Panebianco et al.	Tomato fruit	XRF	-	Establish an assessment procedure for the origin and quality assessment of Sicilian tomato fruits	PCA	[115]
Allegretta	Beans	TXRF	-	Clustering of the seeds of beans according to their geographical origin	PCA and PLS-DA	[116]
Vitali et al.	Croatian wines	TXRF	Contents of metals (K, Ca, Fe, Cu, Zn, Mn, Sr, Rb, Ba, Pb, Ni, Cr and V)	Classification of origin and type of Croatian wines	PCA and cluster analysis	[117]
Li et al.	Peaches	Short-wave near-infrared (SW-NIR) and long-wave near-infrared (LW-NIR) hyperspectral imaging	-	Detection bruises in peaches	PCA	[118]
He et al.	Flour	Vis-NIR HSI	Mites *Tyrophagus putrescentiae* and *Cheyletus eruditus*	Detection of mites *Tyrophagus putrescentiae* and *Cheyletus eruditus* in flour	Random forest and PCA-ANN	[119]
Al-Sarayreh et al.	Meat	NIR-Vis HSI	-	Deep learning approach for red-meat classification by combining the spectral and spatial features of HSI data	CNN	[120]
Pan et al.	Peaches	Hyperspectral reflectance imaging	-	Detection of cold injury in peaches	ANN	[121]
Sun et al.	Peaches	Hyperspectral reflectance imaging	-	Characterization of chilling injury in peaches	PLS-DA, ANN and SVM	[122]

Table 2. *Cont.*

Qualitative Analysis						
Authors	**Food Materials**	**Fingerprinting Technique**	**Analysed Components**	**Objectives**	**Chemometric Analysis**	**Ref.**
Babellahi et al.	Green bell peppers	HSI	-	Detection of chilling injury in green bell peppers	PLS-DA	[123]
Cen et al.	Cucumber fruit	HSI	-	Detection of chilling injury in cucumber fruit	SVM and KNN	[124]
Carreiro Soares et al.	Cotton seeds	HSI	-	Discrimination of different varieties of seeds	PLS-DA	[125]
Fan et al.	Blueberry	HSI	-	Detection of blueberry internal bruising over time	LS-SVM	[126]
Sun et al.	Tomatoes	HSI	-	Characterization of bruised tomatoes	PLS-DA	[127]
Susic et al.	Tomatoes	HSI	-	Discrimination between abiotic and biotic drought stress in tomatoes	PLS-DA PLS-SVM	[128]
Zhao et al.	Wheat seeds	HSI	-	Characterization the purity of wheat seeds	CNN	[129]
Zhao et al.	Maize seeds	HSI	-	Classification of maize seeds	Neural network	[130]
Tsouvaltzis et al.	Eggplant fruit	FT-NIR and NIR-HSI	-	Evaluating the temperature effect on chilling injury of eggplant	PLS-DA, SVM and KNN	[131]
Liang et al.	Potatoes	FT-NIR	Sucrose, glucose fructose	Detection of zebra chip disease (ZC) in potatoes	Canonical discriminant analysis	[71]
Huang et al.	Honey	NIR and FTIR	Syrup adulterant	Distinguish the normal honey from adulterant one	SVM	[132]
De Girolamo	Wheat	FT-MIR and FT-NIR	Ochratoxin A	Assessment of the adulteration of wheat by ochratoxin	PLS-DA and PC-LDA	[133]
Chen et al.	Eggs	FT-NIR	-	Verifying the authenticity of native eggs	Data-driven-based class-modeling (DDCM), PCA	[134]
Liu et al.	Panax notoginseng	FT-NIR	Adulterants (rhizoma curcuma, *Curcuma longa* and rhizoma alpiniae offcinarum)	Identification of panax notoginseng with its adulterants	PLS-DA and SVM	[79]
Marquetti et al.	Arabica Coffee	FT-NIR	-	Evaluation of geographic and genotypic origin of arabica coffee	PLS-DA	[135]
Mazivila et al.	Milk	FT-NIR	Melamine and sucrose	Discrimination of pure milk from the adulterant one	DD-SIMCA	[82]
Miao et al.	Rice	FT-NIR	-	Classification of rice based on storage time	PCA, KNN and PLS-DA	[136]
Rovira et al.	Cashew nuts	FT-NIR	Adulterants (peanuts)	Characterization of the adulterant cashew nuts by other nuts	SIMCA	[137]
Visconti et al.	Cheese	FT-NIR	Cellulose and silicon dioxide	Determination of additives in the grated hard cheese	PLS-DA	[138]

Table 2. *Cont.*

Qualitative Analysis						
Authors	**Food Materials**	**Fingerprinting Technique**	**Analysed Components**	**Objectives**	**Chemometric Analysis**	**Ref.**
Xie et al.	Waxy rice	FT-NIR	Amylose and amylopectin	Determination of quality parameters by FT-NIR	Modified PLS (MPLS)	[139]
Ziegler et al.	Kernels and flours	FT-NIR	-	Differentiation of flours and kernels of costly ancient species from less expensive bread wheat	PLS-DA	[140]
Joshi et al.	Eggs	FTIR	Constituents of eggs (yolk and albumen)	Detection of fabricated eggs	PLS-DA and SVM	[81]
Rozali et al.	Crude palm oil	FTIR	-	Authentication of different geographical and temporal origins of crude palm oils	OPLS-DA	[141]
Li et al.	Hazelnuts	FT-Raman and NIR data fusion	Almonds adulterant	Discriminate the unadulterated hazelnuts from the adulterated hazelnuts with almonds	SIMCA	[142]
Yuan et al.	Duck meat	Surface-enhanced Raman	Testosterone propionate and nandrolone residues	Classification of duck meat based on residues	Particle swarm optimization–support vector classification (PSO-SVC)	[86]
Unuvar et al.	Durum wheat flour	Raman spectroscopy, FT-NIR, synchronous fluorescence spec-troscopy (SFS), (ATR-FTIR)	-	Distinguishing common and durum wheat flour samples with different genotypes	PCA, PLS-DA	[143]
Amjad et al.	Milk	Raman	Proteins, milk fats, lactose	Differentiation between milk samples of different species	Random forest classifier (RF), PCA	[144]
De Oliveira et al.	Enriched eggs	Raman	Omega-3 fatty acids	Discrimination between conventional and omega-3-fatty acids enriched eggs	PLS-DA	[145]
De sa oliveira et al.	Spreadable cheese	Raman	Starch	Classify spreadable cheese as adulterated or without starch	PLS-DA	[93]
Nieuwoudt et al.	Milk	Raman spectroscopy	Nitrogen-rich molecules and sucrose	Detecting adulteration of milk	PLS-DA	[146]
Ning et al.	Duck meat	Raman	Sulfadimidine and Sulphapyridine	Classification of duck meat based on Sulfadimidine and Sulphapyridine	SVM and PCA	[87]
Robert et al.	Meat	Raman	-	Discrimination between different species of meat (intact beef, venison, and lamb meat)	PLS-DA	[147]
Tian et al.	Milk	Raman spectroscopy	Adulterants ofaltodextrin, sodium carbonate, and whey	Distinguishing raw milk from the adulterated one	PLS-DA	[91]
Tian et al.	Rice	Raman spectroscopy	-	Distinguishing rice based on producing areas	PCA-KNN, SPA-KNN, PCA-LS-SVM and SPA-LS-SVM	[148]

Table 2. *Cont.*

Qualitative Analysis						
Authors	**Food Materials**	**Fingerprinting Technique**	**Analysed Components**	**Objectives**	**Chemometric Analysis**	**Ref.**
Wu et al.	Honey	Raman spectroscopy	Adulterants (fructose corn syrup, rice syrup, maltose syrup, blended syrup)	Characterization of adulterant honey	CNN	[149]
Wang et al.	Infant formula	Vis-NIR and Raman data fusion	-	Assessment of infant formula storage temperature and time	SVM	[95]
Yao et al.	Boletus mushrooms	Data fusion of FT-IR and UV	-	Discrimination of different geographical origins of Boletus mushrooms	PLS-DA and SVM	[150]
Antonio et al.	Honey	Spectrofluorimetry	Adulterants (corn syrup, sugar cane molasses and polyfloral honey)	Detection of adulterations in a valuable Brazilian honey	Multilinear PLS-DA (NPLS-DA), unfolded PLS-DA (UPLS-DA), PARAFAC	[151]
Fang et al.	Chinese lager beers	Excitation-emission matrix fluorescence	-	Characterization and classification of Chinese pale lager beers produced by different manufacturers	PARAFAC-KNN	[152]
Jiménez-Carvelo et al.	Extra virgin olive oils	Fluorescence and NIR	Adulterant (vegetable oil)	Authenticate the geographic origin of Argentinean EVOO samples	NPLS–DA	[153]
Meng et al.	Olive oil	Excitation-emission matrix fluorescence	Adulterant (soybean)	Detection of adulteration of olive oil with soybean oil	Multiway-PCA (MPCA), ANN, PLS-DA	[154]
Yuan et al.	Edible vegetable oils	Infrared, NIR and fluorescence	-	Identification of different vegetable oils	MPCA, NPLS-DA	[155]
Uncu et al.	Fresh olive oils	Mid-infrared, UV–Vis and fluorescence	Adulterant (old olive oil)	Detection of adulteration of olive oil	OPLS-DA	[156]
Gonçalves et al.	Monovarietal Extra Virgin Olive Oils	UV-Vis	Phenolic compounds	Monitor the behavior of autoxidative processes through the storage time in two packaging systems of different EVOO	MCR-ALS	[157]
Suhandy et al.	Peaberry coffee	UV-Vis	-	Classify coffee samples as either pure peaberry or pure normal coffee	SIMCA and PLS-DA	[158]
Torrecilla	Vinegar	UV-Vis	-	Characterization of vinegars produced from six different raw materials	PLS-DA and ANN	[159]
Cavdaroglu et al.	Vinegar	UV-vis and FTIR	Adulterant (spirit vinegar and acetic acid)	Discrimination of non-adulterated vinegar from the adulterated	ANN	[160]
Kucharska-Ambrożej et al.	Mint	UV-Vis and FTIR		Distinguish between two species of mint (peppermint or spearmint)	PLS-DA and SVM	[161]
Botoran et al.	Fruits	NMR	Amino acid	Differentiation of the fruit samples in varietal origin	PCA and LDA	[162]

Table 2. *Cont.*

Qualitative Analysis						
Authors	**Food Materials**	**Fingerprinting Technique**	**Analysed Components**	**Objectives**	**Chemometric Analysis**	**Ref.**
Consonni et al.	Coffee	NMR	Fatty acids, β-(1-3)-d-galactopyranose, quinic acid and its cyclic ester)	Characterizing organic roasted coffee from the conventional roasted coffee	OPLS-DA	[163]
De Moura Ribeiro et al.	Roasted coffee	NMR	Adulterants (corn, coffee husks, barley, and soybean)	Investigating the authenticity of the roasted coffee	PCA	[164]
Da Silva et al.	Larger beer	NMR	Carbohydrates	Discriminate lager beer samples from two different classes, according to their style and information provided on the label	PCA, PLS-DA	[165]
Gougeon et al.	Wines	NMR	-	Classifying wines of different geographical origins	OSC-PLS-DA	[166]
Marseglia et al.	Cocoa beans	NMR	Amino acids, polyalcohols, organic acids, sugars, methylxanthines, lipids	Assess the geographical origin of cocoa beans	OSC-PCA, OPLS-DA	[167]
Milani	Ground coffee	NMR	Adulterants	Authentication of roasted and ground coffee based on adulterants	PCA, SIMCA	[168]
Rachineni et al.	Honey	NMR	Adulterants (brown rice syrup, corn syrup, and jaggery syrup)	Identifying type of sugar adulterants in honey	Deep learning-based neural network	[169]
Santos et al.	Milk	NMR	Adulterants (Whey, urea, hydrogen peroxide, synthetic urine and synthetic milk)	Detection of adulterated milk	SIMCA, KNN	[105]
Shi et al.	Camelia oils	NMR	Adulterants (cheap vegetable oils)	Detection of adulteration in camellia oils	PCA, OPLS-DA	[170]
Zhang et al.	Edible oils	NMR	Fatty acids	Distinguishing plant origin of edible oils	PCA, OPLS-DA	[171]

3.1. X-ray-Fluorescence-Based Methods

Energy dispersive X-ray spectroscopy (EDXRF) is a technique commonly used for determining mineral content in food samples. Additionally, its association with the unsupervised and supervised data analysis tools demonstrated its efficiency to deal with the challenges of food analysis. The scope of this section is to discuss the application and usefulness of X-ray-based spectroscopic techniques in combination with chemometric tools for qualitative and quantitative analysis of various food samples.

For example, [57] EDXRF has been applied with PLS Regression for analyzing the micronutrient zinc in biofortified banana samples. This method showed good results mainly on low limits of detection (LOD) and quantification (LOQ). Another research work, conducted by Gamela et al. [56] used the same combination of EDXRF and PLS to determine not only zinc but also the contents of copper and strontium in cocoa bean samples. The study proved satisfactory results through the evaluation of developed PLS models in terms of the same criteria. Additionally, this study has been extended and proved the ability of EDXRF to be fused with Laser Induced Breakdown Spectroscopy (LIBS) to determine the micronutrient potassium in cocoa beans using the supervised technique of multivariate calibration. This fusion showed besides the satisfactory results an advantage to minimize the matrix effect induced by samples of cocoa beans.

The combination of Energy-Dispersive X-ray Fluorescence (EDXRF) and chemometric tools have been used for qualitative purposes too. Galvan et al. [113] carried out an analysis by EDXRF under two measurement conditions, to classify the geographical area of two food species and also according to the production mode by PLS-DA. These food species were tomato and sweet pepper samples Based on the good results of classification EDXRF was considered an excellent technique for authentication of plant-based food products based on the mineral elements K, Ca, Mn, and Fe. Another study for the same qualitative purpose [115] carried out by the association of X-ray Fluorescence (XRF) to (PCA) permitted to identify elements like Cl, K, Ca, Fe, Br, Cl, Rb and Sr which establish a clear fingerprint pattern of the tomato. Similarly, other work applied several chemometrics tools for the discrimination of Italian Extra Virgin Olive Oil (EVOO) geo-markers through the analysis of mineral constituents using EDXRF and associated with PCA and SIMCA [114]. Besides EDXRF and XRF, Total-Reflection X-ray Fluorescence was also employed for food screening [172]. For example, different wine samples from two different geographical regions of Croatia were discriminated against based on the analysis of thirteen metal contents through the association of TXRF with PCA, cluster analysis, and Linear Discriminant Analysis (LDA). Thanks mainly to PCA, elements such as K, Mn, Ba, and Ni were determined as the most relevant to characterize between different origins of wines [117]. TXRF has already been associated with both PCA to obtain the clustering of the bean seeds according to their geographical origin, then it was coupled to PLS-DA for classification purposes [116]. Specific studies that utilize EDXRF in combination with chemometric tools for qualitative and quantitative analysis of various food samples are highlighted. The end points emphasize EDXRF's efficiency in determining mineral content, addressing challenges in food analysis, and its application in food authentication and geographical classification.

XRF-based methods are commonly used in food quality control and analysis due to their non-invasive and time-efficient nature. They can simultaneously detect and quantify trace elements and contaminants in food. However, they have limitations such as limited sensitivity, making them unsuitable for some applications, and being a surface analysis technique, they may not provide information on deeper layers of the sample. The presence of other compounds in the food matrix may also interfere with the analysis, necessitating calibration and standardization to minimize such effects.

3.2. *Hyperspectral and Multispectral Imaging*

In contrast to traditional spectroscopy, hyperspectral imaging affords continuous and high-resolution narrow-band spectral data linked to both physical and chemical sample composition [173]. With The HSI, an object's spectral and spatial information can be retrieved simultaneously by integrating spectroscopic and imaging techniques. This technique has immense potential and has been reported in the detection of various food adulteration, especially when it is associated with chemometric approaches for quantitative purposes. The scope of the proposed paragraph is to discuss the application and advantages of hyperspectral and multispectral imaging combined with chemometric tools in detecting food adulteration, assessing food composition, monitoring food quality, and classifying different food products.

Various studies have been conducted recently on wheat flour to estimate its different contents [174]. Unlike conventional methods, HSI is a reagent-free, non-invasive [175]. One of the special HSI characteristics is to exhibit metabolic transformations, making it useful to assess food composition. A significant amount of recent work has been focused on the application of HSI to various food and agricultural products and animal products. For instance, hyperspectral imaging was used within 400–800 nm to develop a method for analyzing impurities of mites in wheat flour through the supervised chemometric tool of ANN [119]. Benzoyl peroxide, which can also be found as a bleaching agent in wheat [85], was investigated using shortwave infrared (SWIR) HSI and PLS regression [65]. The estimation of talcum content has also been done using hyperspectral imaging and the SNV-PLS model, which proved to estimate adequately the talcum content [64]. In terms of food analyzed by HSI, research by Al-Sarayreh et al. investigated the efficiency of hyperspectral imaging systems to detect meat adulteration, depending on its storage conditions. This analysis proved efficiently the advantage of CNN compared to SVM for this analysis purpose [120]. In addition to wheat and meat, HSI has also been applied to other food samples such as cheese. Priyashantha et al. developed and evaluated a predictive model based on coupling the NIR-HS imaging technique and PLS for determining the maturity state of cheese. The model was then applied on a pixel-wise basis, producing prediction images, and allowing for the determination of how and where the maturity spread in the cheese [58]. On the same food product, PLS and Monte Carlo Cross Validation (MCCV) were applied to HSI to detect the main wavelengths of fat and microbial transglutaminase (mTG), which is responsible for the color and yield of the cheese. Additionally, this study mentioned the possibility of using HSI to inspect the cheese remotely through its transparent foil [59]. Potato is another food sample that has recently started to be useful for monitoring its quality with HSI. Lu et al. assessed the impact of storage times on the evolution of solanine content in potatoes by using HSI in the spectral region of 500–1000 nm and support vector regression (SVR) and then allowed estimating the edibility of the potatoes [60]. Besides solanine content, the color is another indicator, that is also considered as a parameter to judge the quality of potatoes. Xiao et al. combined HSI in the region of 477–947 nm and used and compare two supervised tools: (PLS) and (LS-SVM) for predicting this last parameter [61]. The sweet potato has been subjected to HSI analysis. Tian et al. investigated how the moisture and total anthocyanin contents of potato samples under various drying conditions, by using HSI in the region of 400–1000 nm and PLS. The obtained results of this method showed a low prediction error and a high R^2p [62]. The use of HSI has been shown to be useful in predicting specific parameters that characterize food products. Recently, Li et al. analyzed the quality of plum fruit based on color and soluble solid content using HSI and PLSR and showed how short-wave infrared (SWIR) hyperspectral imaging can predict soluble solid content, [63]. Sun et al. monitored the quality of melon through its indicators as sweetness and hardness by associating NIR hyperspectral imaging system to PLSR [66].

Besides the quantitative advantages of HSI that have been cited, HSI has also been widely used for qualitative purposes. HSI in the spectral domain has been applied to detect chilling injury in agri-food products, which could not be achieved without subjecting HSI to multivariate data analysis. For example, Cen et al. employed HSI in reflectance (500–675 nm) and transmittance (675–1000 nm) modes with supervised classification tools for the detection of chilling injury in cucumbers [124]. Tsouvaltzis et al. evaluated the chilling injury in eggplant fruit by coupling visible and Near-Infrared (NIR) HSI to classification tools such as PLS-DA, SVM, and KNN to classify eggplant fruit according to storage temperature [131]. Recently, Babellahi et al. demonstrated the convenience of HSI with PLS-DA to discriminate between cold-stored green peppers that can be impacted by chilling injury and fresh ones [123]. Another example applied to fruits; peaches might have chill damage during cold storage which Pan et al. associated HSI to Artificial Neural Network (ANN) to differentiate normal peaches from chill-damaged ones [121]. Sun et al. assessed the classification of peaches based on the chilling injury by PLS-DA, SVM, and ANN with Spectral Angle Mapper (SAM) which achieved the best classification performances [122]. Related to the application of HSI on peaches, Li et al. investigated and compared Long-Wavelength-Near-Infrared (LW-NIR) and Short-Wavelength-Near-Infrared (SW-NIR) hyperspectral imaging by associating them with PCA and the approach of watershed segmentation for discriminating bruised from healthy peaches [118]. This study clearly proved the advantage of SW-NIR in detecting early bruises in peaches. Moreover, bruises have also been identified in blueberries samples by SWIR hyperspectral image and the developed models based on two approaches: considering Least Squares-Support Vector Machine (LS-SVM) with full spectra and optimum selected wavelengths by (CARS) (CARS-LS-SVM model) [126]. There are many factors that can lead to bruising food items. For example, Hyperspectral imaging technology applied in the region 400–1000 nm, was used with PLS-DA to classify bruised tomatoes that were caused by falling damage, detection times, falling heights or fruit sizes [127]. Besides fruits and vegetables, HSI with chemometric tools has shown its convenience in the qualitative analysis of wheat. Zhao et al. developed an approach based on a hybrid CNN model with hyperspectral imaging technology to classify different varieties of wheat seed [129,176]. In addition to wheat seeds, maize seeds were classified through the association of HSI with the chemometric approach of Radial Basis Function Neural Network (RBFNN) [130]. Soares et al. also presented a new strategy for fast and non-destructive classification of cotton, based mainly on coupling NIR-HSI images to PLS-DA. The results of this method showed good accuracy in the classification of test samples, with correct classification rates [125]. The endpoints of this section are to emphasize the potential of hyperspectral and multispectral imaging in providing continuous and high-resolution spectral data linked to physical and chemical composition, their non-invasive and reagent-free nature, and their ability to analyze various food samples.

Hyperspectral and multispectral imaging are valuable tools for food quality control and analysis, with advantages and limitations to consider. These imaging techniques offer non-destructive analysis and high spatial resolution for detailed surface analysis, simultaneous detection of multiple analytes, and real-time analysis for efficient quality control. However, hyperspectral, and multispectral imaging equipment can be expensive, have limited penetration for internal composition analysis, may lack sensitivity for detecting low analyte levels, and require complex image processing and specialized expertise for data analysis.

3.3. *Infrared Spectroscopy*

Infrared including (NIR) and (MIR) are ones of the conventional spectroscopy that have been usually used with many multivariate data analysis tools in food quantitative analysis. The scope of the proposed paragraph is to discuss the application and benefits of infrared spectroscopy with chemometrics, in quantitative and qualitative analysis of food components, such as carbohydrates, proteins, fats, and moisture content, as well as the determination of functional groups, carbon, and nitrogen.

MIR is responsible mainly for detecting functional groups as well as carbon and nitrogen, whereas NIR is used for determining carbohydrates, proteins, fats, and moisture content in various foods [177]. However, the method is not sensitive enough for samples containing just small amounts of target components. The fundamental vibrations occur when absorbed in the NIR [178]. Wang et al. combined NIR PLSR to estimate the content of potato flour in steamed bread [67]. Recently, the same association of NIR with PLS regression was applied to wheat flour samples to estimate the quantity of low-content talcum. In this study, several chemometrics were applied with PLS together to select the effective feature as genetic algorithm and elastic net, thus improving the capacity of the PLS model [68]. In addition to talcum, zearalenone might have an impact on the quality and safety of wheat grains. Recently, a study was carried out to determine zearalenone in wheat by NIR spectroscopy and (SVM) model. The results were significantly improved after the application of a variable selection approach called least absolute shrinkage and selection operator (LASSO) to extract useful spectral regions of NIR. In contrast to the contents that can have an impact on wheat, the determination of valuable contents has been featured in many recent research works [73]. Kamboj et al. predicted quality parameters mainly protein and carbohydrate of wheat content that has been stored at different temperatures using NIR Spectroscopy (NIRS) with PLS, MLR, and SVM [70]. Additionally, the fatty acid value is also considered an important indicator of the quality of wheat flour, particularly during storage. Therefore, Jiang et al. demonstrated the feasibility of using portable NIR spectroscopy in conjunction with appropriate chemometric methods to achieve quantitative determination of fatty acid values in wheat flour during storage. Jiang et al. used a method called variable combination population analysis (VCPA) in addition to PLS to improve NIR spectral characteristic wavelengths [72]. In addition to the chemometric tools that were cited and used for quantitative purposes, MCR-ALS is one of the chemometric tools that is combined with FT-NIR spectroscopy to estimate certain characteristics of food samples. For instance, an assessment of the combination of multivariate tools, including PLS regression and MCR-ALS, was used to predict antioxidant activity from clove and pomegranate extracts. The results showed that MCR-ALS with FT-NIR stood out among PLS with high R^2 and low RMSEP [74]. Another application of PLS and MCR associated with FT-NIR was used successfully to estimate peanut oil adulterants, [75]. Castro et al. also proved the efficiency of coupling FT-NIR with MCR-ALS for the quantitative purpose of four adulterants at low levels in a complex mixture of saffron, including onion, calendula, pomegranate, and turmeric [76]. In terms of determining adulterants in saffron, PLS-R was applied to FT-NIR data of saffron to estimate lotus stamens and corn stigmas. This study proved the efficiency of combining PLS with the variable selection approach of competitive adaptive reweighted sampling (CARS) showing good results [77]. Additionally, crocin I and II were analyzed using near-infrared spectroscopy and chemometrics. Crocin I and II are considered the most important indicators of the quality and commercial value of saffron [179]. Le et al. used FT-NIR and PLS to determine these two contents in saffron with low RMSECV [78].

Many studies have shown how MIR and NIR spectroscopy are efficient for the qualitative analysis of different food varieties comprised for example identification, classification, and authentication, based on, for example, country of origin. For instance, Liang et al. used NIR spectroscopy appropriately for the detection of zebra chip disease using Canonical Discriminant Analysis with a low classification error rate [71]. Discriminative analysis was applied to durum wheat to determine if they were contaminated by ochratoxin by combining FT-IR and FT-NIR with PLS-DA and PCA-LDA. In this study, FT-IR and FT-NIR were convenient spectroscopic techniques for discrimination purposes [133]. PLS-DA was compared to other classification tools such as HCA, SVM, and ANN to identify and classify Panax notoginseng with its adulterants. The classification purpose of this work was achieved by both PLS-DA and SVM with 100% classification accuracy [79]. PLS-DA showed its efficiency in detecting the freshness of rice based on storage time using FT-NIR with an accuracy of 96%, whereas the application of KNN achieved an accuracy of 100% [136]. In relation to the analysis of rice by NIR, L.-H. Xie et al. led a discrimination of two kinds of rice, waxy, which contains very low apparent amylose content, and non-waxy rice. The developed PLS-DA model allowed the recognition of these two types of rice with 100% accuracy [139]. Detecting fake eggs from authentic ones is another example that proves the efficiency of FT-IR and chemometrics in this field of food analysis. Joshi et al. showed how PLS-DA and SVM achieved a good classification of 100% [81]. The authenticity of the native was subjected to FT-NIR analysis by Chen et al. who proved the efficiency of using Data-Driven Class Modeling (DD-SIMCA) as an alternative tool for this classification [134]. The end points of the paragraph are to emphasize the effectiveness of infrared spectroscopy in estimating the content of target components in food samples, such as potato flour, talcum, zearalenone, protein, carbohydrate, fatty acid values, and antioxidant activity. The paragraph also mentioned the successful application of chemometric approaches in enhancing the accuracy and reliability of quantitative analysis using NIR and MIR spectroscopy.

Overall, infrared spectroscopy is a powerful and versatile tool for food quality control and analysis. However, limitations such as limited penetration, sample homogeneity requirements, calibration requirements, complexity of data analysis, and interference from other components should be taken into consideration when using this technique.

3.4. Raman Spectroscopy

Raman spectroscopy is a vibration spectroscopy technique that is based on monochromatic light diffusion. It involves the excitation of a sample by collisions with photons, which causes the sample to reach an unstable state of virtual energy. The scope of the proposed paragraph is to discuss the application of Raman spectroscopy in food analysis, both qualitatively and quantitatively.

Raman spectroscopy was carried out for the determination of fat content in various food samples such as milk and meat. Heterogeneous foods have recently been detected chemically with Raman microscopy [180]. This combination is used qualitatively and quantitatively to evaluate food value. Many organic components are detected and identified based on the absorption curves. Microscopic food species can be analyzed too. Raman microscopy has been carried out to determine the main composition of wheat and to detect protein content changes during milling [181,182]. Raman spectroscopy detects changes in protein secondary structure, conformational changes in lipid-binding proteins.

Based on the Raman spectrum, it is possible to estimate the relative concentration of food contents. For instance, a recent study aimed to determine starch using Raman spectroscopy and a linear regression model for a specific band, and PLS regression for a specific spectral region [183], which confirmed the efficiency of association of FT-Raman to PLS for the estimation of gluten content in flour [184]. Carotenoids have been determined in tomatoes by Raman spectroscopy and PLS regression and proved low prediction error [88]. The main characteristic of Raman spectroscopy is that it can directly measure aqueous solutions because of the low effect of water, and even the sample preparation of liquids for Raman analysis is simple, which can be considered an advantage to estimating the contents in food liquids such as milk [185]. Whey is one of the contents that has been quantified accurately in the milk [89]. In addition to whey, macronutrients such as fat, lactose, and protein have been successfully quantified in commercial yoghurt samples using FT-Raman spectroscopy and PLS models [90]. In a recent study, milk adulterants such as sodium bicarbonate, maltodextrin, and whey were also analyzed using Raman spectroscopy and the PLS chemometric tool, with a low detection limit [91]. A handheld Raman spectrometer has also been applied to quantify lard in adulterated butter, another milk derivative, through PLS [186]. Richardson et al. demonstrated how Raman spectroscopy is able with PLS to detect coconut water adulteration [112]. In a recent study, various variable selection approaches were tested on surface-enhanced Raman scattering spectra of rice, used with PLS for quantifying the target residue analyte of chlorpyrifos [109].

As it has shown its relevance for quantitation, Raman spectroscopy has proven its efficiency with chemometric tools in many recent studies. For example, Robert et al. built a classification model using SVM and PLS-DA to discriminate lamb meat from beef meat despite the similar chemical composition of these two species [147]. Hai Chao et al. classified duck meat according to the residues of testosterone propionate and testosterone nandrolone using Raman Spectroscopy and Support Vector Classification (SVC) which shows a classification rate of 100% for the test set [86]. Other residues that have an impact on duck meat and have been subjected to analysis by means of Raman spectroscopy and chemometrics are sulfonamides, comprised of sulfadimidine and sulphapyridine [187]. A recent research work used a support vector classification on Raman Spectroscopic data to classify duck meat into four groups, which are as follows: samples free of residues, samples containing one of the two residues, and samples containing both residues [87]. Another mode of Raman spectroscopy called Spatially Offset Raman Spectroscopy, which allows to measure the chemical compounds under the surface of meat tissues [188]. Besides that, a study has also proven that the use of Raman spectroscopy in combination with the SVM method can discriminate rice samples according to their regions with high-rate accuracy [148]. Raman was applied to differentiate four categories of milk species of cow, buffalo, goat, and human. Thus, Principal Component Analysis (PCA) besides Random Forest (RF) was applied on Raman data to highlight and characterize the Raman spectra of different milk samples with high accuracy of 93.7% [144]. In addition to benchtop Raman, Handheld Raman spectroscopic devices have shown their efficiency using SIMCA to classify milk samples from adulterated ones [189]. In addition to milk and its derivative samples, PLS-DA was employed with Raman to accurately classify a milk derivative of cheese whether it was adulterated by starch or not [93]. Related to handheld Raman, Aykas, et al. succeeded in seeking to characterize commercial honey by combining handheld Raman equipment and SIMCA [190]. A recent research work monitors according to a new method the adulteration in cassava starch, by means of Raman spectroscopy and supervised tool One-Class Modelling (OC-SVM) which proved its higher accuracy compared to the SIMCA, allowing for the discrimination of samples [191]. Discriminant analysis by PLS-DA of coffee genotypes by Raman spectroscopy based on two main contents, kahweol and fatty acids, has shown how Raman with chemometrics was more effective compared to sensorial analysis [192]. Sha et al. combined Raman with PCA, HCA, and SVM for feature extraction to improve the efficiency of identification of rice varieties [176]. For oil samples, Jiménez-Carvelo et al. used chemometrics for the classification and characterization of pure olive oil from adulterated using Raman spectroscopy in addition to NIR by employing classification models. While PCA was used to reduce the features, other supervised techniques were applied to for the discrimination goal [193]. Raman analysis was also employed to discriminate waste cooking oil from edible vegetable oil. Thanks to PCA, signals at 869, 969, 1302, and 1080 cm^{-1} were found to be the most important features to differentiate between these two types of oils. In addition, PCA demonstrated its ability to separate adulterated from pure oils when the adulteration proportions reached 10% and 20% [194]. The endpoints of this part include the successful application of Raman spectroscopy in estimating the content of specific food components and the detection and classification of various residues and adulterants in food samples.

Finally, Raman spectroscopy is a valuable method for analyzing and controlling the quality of food, offering several advantages such as non-destructiveness, molecular specificity, sensitivity, minimal sample preparation, and high spatial resolution. However, when using this technique, some limitations must be considered, including its limited penetration depth, susceptibility to fluorescence interference, equipment cost, complexity of data analysis, and sensitivity to water.

3.5. NMR Analysis

Nuclear magnetic resonance (NMR) is a spectroscopic technique used to determine the molecular structure and physical properties of substances, and efficiently used to ensure the quality of different varieties of food samples [195]. The scope of the proposed paragraph is to discuss the application of NMR spectroscopy in food analysis, both qualitatively and quantitatively. It focuses on the use of NMR spectroscopy and chemometric tools for food identification, discrimination, and characterization purposes.

For example, the combination of low and high-field NMR and chemometrics, including PLS-R and SVR, has proved its ability to accurately estimate essential quality parameters of edible oils, especially to detect potential adulteration. The results summarized in statistical parameters indicate that all developed models, whether of PLS-R or SVR on the three different fields of NMR, were similar. In addition to oil applications, Haddad et al. have carried out a quantitative analysis of fatty acids based on ^{1}H-NMR variables as predictors and relative mass percentages of fatty acids as targets, including caproic, caprylic, capric, oleic, palmitic, and margaric [110]. Fatty acids have been accurately estimated in hen egg samples by ^{1}H-NMR and PLS regression [108]. Proton nuclear magnetic resonance (^{1}H-NMR) associated with chemometrics were combined to investigate the camellia oil adulterants with other vegetable oils [170]. In addition to ^{1}H-NMR, ^{1}H TD-NMR was combined efficiently with PLS regression to detect the percentage of adulterants such as water and whey in milk products varied from 5% to 50% through milk package and without sample preparation [105]. Besides PLS regression, Sun et al. successfully set up a model to detect moisture content through the association of low-field NMR and ANN with a low RMSE [107].

As previously shown, NMR spectroscopy supported by multivariate data analysis tools has been applied for various qualitative purposes in different foods. For example, Milani et al. successfully explored the versatility of 1H NMR with pattern recognition of PCA and SIMCA for identification and discrimination of pure Brazilian coffee from adulterated ones by corn, barley, or even coffee husks. The built SIMCA model ensured its high classification accuracy [168]. In relation to these quality analysis of coffee, 1H NMR data of roasted coffee samples were analyzed qualitatively by OPLS-DA to characterize organic roasted coffee from conventional coffee. The orthogonal signal correction (OSC) allowed for the extraction of the main features of each coffee category and thus improved the PLS-DA model discrimination. While fatty acids, β-(1-3)-d-galactopyranose, quinic acid, and its cyclic ester were the major metabolites characterizing organic roasted coffee, conventional coffee was characterized mainly by trigonelline and chlorogenic acid isomers [163]. The OSC filter was used with PCA (OSC-PCA) and applied to HR MAS 1H NMR data of cocoa beans to discriminate them based on their origin, whether they were American or African, based on the fatty acids, acetate, and saccharides components [167]. In another research work, both 1H NMR and 13C NMR were employed to analyze refined edible oils from different sources. By applying PCA on the 1H NMR or on 13C NMR, it was possible to identify and characterize these plants based on their fatty acids [171]. Amino acids were analyzed by NMR and explored by chemometric tools in fruits, since they are considered essential metabolites in cell function and enable distinction between plants of the same fruit. For example, Botoran et al. identified ten kinds of amino acids that allowed for the observation of differences and distinctions of different varieties of juice using PCA and LDA, which accurately classified juices from different plant sources [162]. In the honey adulteration problem, Rachineni et.al analyzed successfully honey by associating 1H NMR with supervised machine learning (neural network) for the characterization purpose of authentic honey from the adulterated whether by sugar, brown rice syrup or jaggery syrup [169]. The endpoints of the paragraph include the successful application of NMR spectroscopy combined with chemometrics, and machine learning, in accurately estimating and detecting various quality parameters and adulterants in food samples.

Overall, NMR spectroscopy is a valuable tool for food quality control and analysis, offering numerous advantages such as non-destructiveness, molecular specificity, sensitivity, and versatility. Additionally, it allows for quantitative analysis, making it particularly useful for determining the concentration of specific compounds in food products. However, the technique also has limitations that should be considered, including equipment cost, limited penetration, sample preparation requirements, and sensitivity to sample properties.

3.6. *UV-Visible*

UV-visible spectroscopy is known as one of the most sensitive techniques for determining less concentrated contents in food samples. Its association with multivariate tools offers an added advantage for such quantitative analysis. This technique uses electromagnetic radiation between 200 and 800 nm and detects two different aspects: color and fat oxidation [104]. The scope of the proposed paragraph is to discuss the application of UV-visible spectroscopy in food analysis, particularly for quantitative and qualitative purposes.

In addition to other analytical techniques, a recent research work for the same food product employed UV-Vis with PLS regression to accurately determine squalene in Extra virgin olive oils (EVOO) [100]. Wu et al. integrated empirical mode decomposition with SVR (EMD-SVR) to evaluate the quality of edible blend oils samples, concluding that EMD-SVR was more accurate for the quantitative analysis of ternary edible blend oil [101]. Zhang et al. developed models by coupling UV-Vis to Partial least squares regression (PLS) and principal component regression (PCR) for the quantitation of acid value in various oils. The PLS models performed well compared to PCR models [102]. In addition to different oil analyses, UV-Vis spectroscopy was proved to be more efficient as a method associated with PLS instead of univariate tools for the quantitation of grape-must caramel in Balsamic vinegars of different varieties of wine vinegars [103].

The UV-Vis spectroscopy has been combined with multivariate techniques for various qualitative purposes in food analysis. For instance, UV-Vis combined with MCR-ALS is a suitable tool to pursue the autoxidation of edible oils and to monitor the quality of extra virgin olive oil (EVOO) in different packaging systems [157]. In addition to olive oil samples, multivariate discrimination tools using UV-Vis spectroscopy, such as PLS-DA and SVM, have been used to distinguish between two specific mint species, such as spearmint and peppermint, while SIMCA has been used to detect outlier samples other than the two species [161]. Similarly, coffee has been analyzed by UV-Vis spectroscopy and SIMCA to accurately classify and discriminate between Peaberry and normal coffee [158]. In addition to SIMCA and PLS-DA, artificial neural networks (ANN) have been applied to UV-Vis spectroscopy to discriminate between vinegars produced from different raw materials, showing the discrimination efficiency compared to PLS-DA [159]. Another study used UV-Vis spectroscopy and ANN to discriminate between vinegars adulterated with spirit vinegar or acetic acid [160]. The end points of the paragraph include the successful application of UV-visible spectroscopy combined with multivariate techniques for food analysis adulteration, improved discrimination, and classification purposes.

Generally, UV-Visible spectroscopy is a valuable tool for food quality control and analysis, with several advantages such as simplicity, non-destructiveness, versatility, and cost-effectiveness. However, it has limitations in sensitivity, interference, and surface analysis that should be considered while using this technique.

3.7. Fluorescence Spectroscopy

Fluorescence spectroscopy is a technique that focuses mainly on the molecular level. It refers to the process in which a specific wavelength of light is irradiated in a solution, and the fluorescent substance in the solution absorbs the released energy. The scope of the proposed paragraph is to discuss the application of fluorescence spectroscopy in food analysis, both for quantitative and qualitative purposes. It highlights the molecular-level focus of fluorescence spectroscopy and its ability to detect and analyze various elements in food samples.

In recent years, fluorescence spectroscopy has been applied for the analysis of various elements of food. For example, a recent research study exhibited the application of front-face fluorescence mode spectroscopy and supervised PLS to estimate cow milk adulteration with other milk kind [196]. Another study used excitation-emission matrix (EEM) fluorescence spectroscopy and second-order calibration ways, like (PARAFAC) and (U-PLS), to detect and estimate the content of melamine in milk [98]. Additionally, fluorescence spectroscopy has been used to detect and quantify adulteration in olive oils [156]. Three-dimensional fluorescence spectra were subjected to analyze the same analysis purpose using the supervised approach of GA-SVR [99].

Many studies have shown the potential of fluorescence spectroscopy combined with multivariate data analysis tools for the qualitative analysis of various food samples. For example, Yuan et al. [155] conducted a comparative study using excitation-emission matrix fluorescence, FTIR, and vis-NIR on different types of vegetable oils for discrimination purposes using advanced chemometric tools (PCA, multiway-PCA, PLS-DA, and unfold-PLS-DA). The study found that FTIR and Vis-NIR, were more suitable compared to EEM for the identification of vegetable oil species. This is because most chemical components in vegetable oil produce FTIR and NIR absorption, while only a small number of fluorophores produce fluorescence [155]. Another study proved the same classification results of these techniques for detecting olive oil adulteration. This highlights the importance of combining analytical techniques with the appropriate chemometric tool [154]. Fluorescence spectroscopy has been combined with chemometrics to distinguish pure Aroeira honey from adulterated. The advanced chemometric methods (PARAFAC, PLS-DA, unfolded PLS-DA (UPLS-DA), and N-way PLS-DA (NPLS-DA)) were used to decompose the spectral data and build classification models. This qualitative analysis has proven the convenience of fluorescence spectroscopy with UPLS-DA for this kind of honey analysis [151]. It can be noted from previous research that multi-way chemometric techniques are often applied conveniently to EEMs data, whether on edible oils, honey, or beverages, as demonstrated by Fang et al. for the classification of Chinese lager beers made by different manufacturers [152]. The end points of the paragraph include the successful application of fluorescence spectroscopy combined with multivariate data analysis tools for quantitative and qualitative analysis of various food samples.

Fluorescence spectroscopy is a powerful tool for food quality control and analysis, offering advantages such as high sensitivity, specificity, non-destructiveness, and rapid analysis. However, there are certain limitations that should be considered when using this technique, including the complexity of sample preparation, potential interference from other compounds, limited penetration depth, and high instrumentation costs.

3.8. Fusion of Spectroscopic Techniques

In recent years, the strategy of data fusion combined with multivariate statistical analysis, that has been widely used to ensure the safety of food and to extract more information for both qualitative and quantitative purposes. The scope of the proposed paragraph is to discuss the application of data fusion combined with multivariate statistical analysis in food analysis for both qualitative and quantitative purposes. It highlights the use of various spectroscopic techniques such as UV-VIS, NIR, Raman, FT-IR, FT-Raman, and MID, and their fusion with multivariate statistical models for food analysis.

A recent research work developed PLS and ANN models for the quantification of adulteration in honey, using data fusion of non-pre-processed UV-VIS and NIR spectra [96]. Vis-NIR and Raman have also been merged and applied to predict the storage time of infant formula between 0–12 months [95]. UV-Vis-NIR was combined with PLS to accurately quantify cholesterol in egg yolk, whether in the shell or in pasteurized form [94]. Additionally, the combination of PLS regression with the data fusion of FT-IR with Raman spectroscopy allowed the determination of peroxide values and acid values in oils [197]. A study was elaborated to test merging mid-infrared (MIR) with Raman spectroscopy for the fructose syrup determination in honey samples. The PLS model was used to estimate the adulterant, and the results were improved after the data fusion compared to the results obtained by each of the two spectroscopic techniques [198]. The same conclusion was achieved by a recent research work that evaluated the data fusion of NIR and MIR, combined with the sequential orthogonalized partial least square regression (SOPLS), to estimate different quality traits of tubers and root flours. These traits included different chemical compounds including for example amylose and protein [69].

The efficiency of data fusion methodology for qualitative purposes has been proved by Yao et al., who established a method based mainly on Fourier transform infrared (FT-IR) and ultraviolet (UV) spectroscopies associated with data fusion to distinguish different regions of mushroom samples [150]. A synergistic strategy of FT-Raman and NIR for the classification of two classes of hazelnut: unadulterated and adulterated with almonds using SIMCA was also demonstrated. The obtained results proved that merging the two techniques can be more effective than using each technique alone, based on sensitivity and specificity [142]. NIR and MID were also used with the SVM model to discriminate natural honey from syrup-adulterated one. In this case study, Huang et al. showed two levels of data fusion. A low level, in which redundant and irrelevant variables were introduced, and an intermediate level, where PCA was applied to extract the feature variables. The results acquired from this study had a significant increase in SVM model parameters of accuracy, precision, and sensitivity using the intermediate-level data fusion [132]. The endpoints of this section include the successful application of data fusion methodologies with chemometrics for quantification of adulteration, for estimating different chemicals, as well as for qualitative purposes and classifying unadulterated and adulterated food samples. However, there are limitations to consider when using data fusion methodologies in food analysis. The success of data fusion relies on the compatibility and complementarity of the combined techniques and the availability of appropriate statistical models. Proper calibration and validation procedures are necessary to ensure the reliability and robustness of the fused data. Furthermore, data fusion may introduce additional complexity and computational requirements, requiring careful data preprocessing and analysis. It is also important to consider the specific requirements and limitations of each spectroscopic technique and statistical model being used for data fusion.

The fusion of spectroscopic techniques provides an advanced tool for food quality control and analysis, offering advantages such as enhanced accuracy, complementary information, improved sensitivity, and non-destructiveness. However, it should be noted that this technique has some limitations, such as complexity, high equipment cost, sample preparation requirements, and limited penetration depth. These factors should be considered before implementing this technique for food analysis.

3.9. Portable Spectroscopic Techniques

The food industry is constantly seeking faster, more accurate ways to assess the safety and quality of their products while also detecting possible adulteration. Portable spectroscopic equipment, such as Raman, NIR and HSI among other techniques, have become increasingly popular due to their portability, accuracy, and ability to control food products [199]. The scope of the proposed paragraph is to discuss the application of portable spectroscopy techniques in the food industry for assessing food safety and quality, detecting adulteration, and enabling in-process monitoring. It highlights the advantages of portable spectroscopy equipment, such as Raman, NIR, and HSI, including their portability, accuracy, low sample preparation requirements, and cost-effectiveness.

Portable spectroscopies equipment in general requires less sample preparation and less hazardous consumption than traditional laboratory-based processes, granting fast results on food quality and safety. Furthermore, these technologies provide the food industry the low cost-effective analysis and product safety [200,201]. Portable spectroscopy techniques are highly effective in detecting food fraud and contaminants, such as pesticides, heavy metals, and pathogens. Previous developments in portable fluorescence and Raman spectroscopy have enabled water detection in milk and honey products with less expensive syrups, respectively [202,203]. Moreover, they are involved to analyze food quality attributes, for instance, mid-infrared spectroscopy used for fatty acid profile and fat content in lamb meat [202], and the amount of fat in meat and its degree of tenderness [204]. The NIR-HSI as a portable technique coupled with chemometric tools was showed excellent application in different food products [205,206].

Portable spectroscopy techniques were demonstrated to be a valuable solution for food manufacturers and processors, as they can be used for in-process monitoring of food quality and safety. The technology can also be used for post-harvest processing, such as the detection of mold and fungal infections in food products [207]. In addition, they have demonstrated to be versatile in identifying various food tampering and contaminants, including pesticides, heavy metals, and biological contaminants. They are also beneficial with chemometrics for quality control testing and authentication for rapid food chain analysis aimed at a perfect digital traceability system [201]. For a deep understanding, a recent review article provides an overview of how miniaturized NIR spectroscopy can be applied to address a range of issues in food-related settings [208]. They provide a comprehensive summary of the latest research trends, highlighting key factors driving the development of the micro-NIR analytical framework for modern food analysis, quality control, and safety risk monitoring. Emphasis is placed on the significance of combining complementary tools with the NIR analytical method, which enhances its precision, dependability, and versatility for food applicability. The endpoints of this section include the successful use of portable spectroscopy techniques and the use of chemometric tools for detecting food fraud and contaminants, analyzing food quality attributes, monitoring in-process food quality, safety, and rapid food chain analysis.

Finally, to highlight portable spectroscopy tools have several advantages in food applications, including non-destructive sample analysis, rapid analysis where timely decisions are needed, in situ analysis which is particularly useful in food-based applications. However, there are also some limitations to the use of portable spectroscopy techniques in food applications, such as limited accuracy particularly for complex samples, limited range of wavelengths which may not be suitable for all applications, regular calibration requirements which can be time-consuming, sensitivity to environmental conditions such as temperature and humidity, and often more cost-effective than traditional laboratory-based techniques [201]. These limitations should be taken into account when implementing portable spectroscopy in food analysis and quality control processes.

4. Importance of Integrating Advanced Spectroscopy Techniques in Food Analysis

The use of advanced spectroscopy techniques in food analysis is crucial for several reasons. The scope of the proposed paragraph is to discuss the significance of advanced spectroscopy techniques in food analysis, including their non-destructive nature, ability to provide detailed information about food composition, and the use of chemometric tools for data analysis. Firstly, techniques such as NIR, Raman, FTIR, and UV-Vis offer a non-destructive and non-intrusive way of analyzing food samples, making them ideal for large-scale analysis in the food industry without affecting the quality or safety of the samples. Secondly, they provide detailed information about the chemical composition and structure of food, which can be used to classify and identify different food types, detect contamination, and monitor food quality changes. Lastly, the implementation of advanced chemometric tools enables effective analysis of complex spectral data for food quality assessment. The food industry faces a significant challenge in ensuring consistent quality and overall food safety information throughout the entire supply chain, from production to distribution, to meet consumer demands and expand the market. To address this challenge, advanced non-invasive technologies are increasingly being used to monitor the nutritional and hygienic properties of raw and end food products. The initial perception of a food's quality is often based on its texture, while the nanostructures in food play a role in determining its color, shape, and sensory appeal (Figure 3).

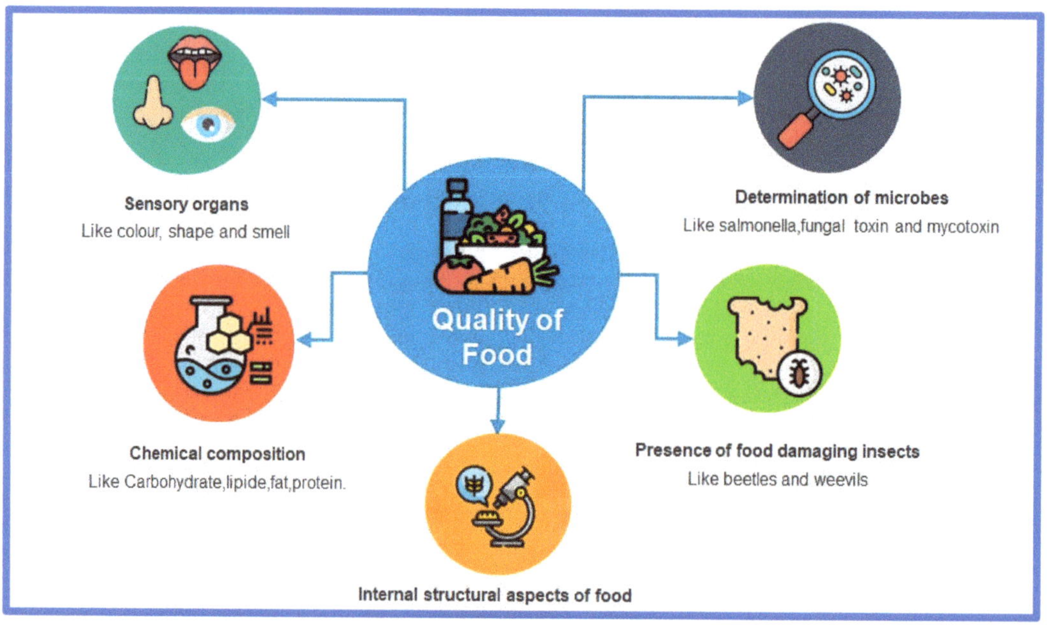

Figure 3. Factors that affect the quality of food.

Spectroscopy techniques can help to determine the presence and quantity of key components like flavonoids and antioxidants in food, which play a vital role in assessing its quality. The effectiveness of using anthocyanin profile, color image analysis, and NIR-HSI to differentiate between different grape varieties, with the aid of Stepwise Linear Discriminant Analysis (SLDA) that was created for each dataset to differentiate grapes based on their variety.

Spectroscopic methods provide a valuable means to assess the quality of food by determining the presence and quantity of important components such as flavonoids and antioxidants. The combination of anthocyanin profiling, color image analysis, and near-infrared hyperspectral imaging (NIR-HSI), along with the aid of dataset-specific Stepwise Linear Discriminant Analysis (SLDA), proves effective in distinguishing between different grape varieties based on their specific characteristics. In addition, it can be concluded that NIR spectroscopy holds substantial promise for the non-destructive determination of total phenolics and flavonoids [209]. The content of flavonoids in food can be influenced by various factors, including seasonality, food maturity, and preparation methods employed. In this sense, the spectroscopy methods such as Visible/NIR were used to determine the levels of flavonoids in food [210]. These techniques are employed by various industries including the milk, meat, coffee, and wine sectors to analyze the composition of food [211]. Assessing the quality of food based solely on sensory evaluation may not suffice, particularly for intricate evaluations. Advanced spectroscopy equipment with chemometrics is highly recommended in such scenarios as it offers advantages that are not attainable through manual inspection alone.

Traditional methods of food quality detection can be cumbersome, repetitive, destructive, and take up a lot of time. Non-destructive methods, on the other hand, offer a more efficient way of gaining both quantitative and qualitative data without destroying the sample. The recent developments in non-destructive food quality assessment techniques include imaging, spectroscopy, and cutting-edge approaches such as electronic nose, electronic tongue, dielectric, and acoustic methods [212]. For instance, conventional methods of analysis and detection, for instance, thin-layer chromatography and high-performance liquid chromatography (HPLC) are yet largely used in the basic food industry to detect food quality. Nevertheless, these procedures are damaging, laborious, and time-consuming. Consequently, spectroscopic techniques were demonstrated their use and importance for food quality control including visible/infrared (VIS/IR) [212,213], Raman spectroscopy [214], NMR [215], and HSI [216].

In contrast, the field of food authentication poses a major challenge that can be addressed using advanced spectroscopic tools. Adulteration in food is often accomplished by substituting ingredients with cheaper options that are not easily distinguishable by either consumers or conventional analytical techniques [217]. To confirm the geographical origin, storage conditions, and processing methods listed on food labels, it is necessary to analyze specific components in samples. Hence, the development of trustworthy and effective analytical methods is vital for creating new policies, programs, and techniques to verify food authentication.

The scope of the proposed paragraph is to emphasize the importance of advanced spectroscopy techniques in food analysis, including their non-destructive nature, ability to provide detailed chemical information, and the role of chemometric tools in analyzing complex spectral data. The endpoints are to highlight the use of spectroscopy techniques for assessing food quality, detecting adulteration, and verifying food authentication. However, it should be noted that the paragraph does not delve into the specific methodologies or algorithms associated with spectroscopy techniques and chemometric data analysis. However, there are limitations to consider when using spectroscopy techniques in food analysis. These limitations include the influence of various factors on the content of flavonoids in food, the need for dataset-specific analysis methods, and the challenges associated with distinguishing food adulteration using conventional analytical techniques.

In summary, the integration of different spectroscopy techniques either separately or in a combination or fusion can also enhance the overall performance of food analysis. This can be achieved by using chemometric data analysis approaches.

5. Food Application and Aspects

Spectroscopic methods involve the use of electromagnetic radiation through absorption, transmission, and emission, and are based on the wavelength of the radiation. Unlike traditional techniques, spectroscopic methods have become essential tools for determining food properties through real-time monitoring and non-invasive techniques. Recently, advanced spectroscopic techniques combined with chemometric approaches have been developed and applied in various food applications and areas such as sensory analysis, adulteration detection, chemical analysis, mycotoxin detection, parasitic infection detection, internal physiological analysis, and others.

5.1. Sensory Attributes

The sensory characteristics of basic foods, such as structure, color, toughness, texture, and outside defects, are crucial elements of food quality. The color of a food is generated by the reflection of distinct wavelengths in the visible light section. The structure of a product, which is often determined by its size, weight, or volume, can influence consumer preferences and final consumption. Hardness is a primary indicator of food property, indicating the texture and moisture content of food. External defects, which occur during or after harvest, are another common sensory attribute that greatly impacts food quality. Consequently, precise, and timely forecasting of these sensory characteristics is a primary concern for the farming and food industry. The scope of the proposed paragraph is to discuss the importance of sensory characteristics in food quality and the role of spectroscopy techniques combined with chemometrics in evaluating these characteristics. It highlights the significance of texture, color, hardness, and external defects in determining food quality and consumer preferences.

The sensory characteristics of texture and color were established using VIS and NIR spectroscopy. The VIS/NIR spectroscopy was shown to be a reliable method for evaluating the quality of dry beans during the canning process [218]. A discriminatory linear model was applied to classify the canned beans into two categories: "acceptable" and "unacceptable". The model achieved an ordinary classification accuracy of 72.60%. As sensors and instruments improve, it is expected that VIS/NIR spectroscopy will meet the necessary requirements for its use. The evaluation of basic foods using HSI systems mainly focuses on color, hardness, and external defects. In a recent study [219], the color of fresh soybeans was determined using an active contour model to segment the spectral images. This HSI technique was found to be more effective in acquiring the mean reflectance and entropy parameters of the image. A PLSR model was established to detect the color of processed soybeans with a good R^2 of 0.74 and used HSI in the range of 400–1000 nm to discriminate rice samples based on color and shape. Five shape features (minor axis length, major axis length, perimeter, length-width ratio, and eccentricity) and one-color feature (degree of chalkiness) were utilized as inputs for a back propagation neural network (BPNN) model, resulting in a high classification accuracy of 94.45%. The performance of PLSR and PCR in evaluating the hardness of wheat samples using HSI in the wavelength range of 960–1700 nm and found that PLSR outperformed PCR. Additionally, hyperspectral imaging was also significantly used to determine the surface defects of potatoes [220]. The effectiveness of the proposed method in identifying kernels where the germination process has begun was demonstrated through experiments involving three wheat cultivars. The results achieved 100% accuracy for the samples utilized in this study [221]. The surface defects of potatoes were also determined using the HSI technique. Moreover, the results were combined with the SVM classification tool to achieve a high-accuracy performance in determining the surface defects of potatoes [222]. The endpoints of the paragraph include the effectiveness of spectroscopy techniques, such as VIS/NIR and HSI, in determining the color, hardness, and external defects of food samples. It also highlights the successful application of discriminatory models and neural networks in classifying and predicting sensory characteristics based on spectroscopic data.

Spectroscopic techniques are useful in food sensory analysis to measure properties such as color, texture, and flavor. The advantages of these techniques include objective measurement, non-destructiveness, and rapid results. They can provide a detailed analysis of food chemical composition, helping to identify changes due to processing or storage.

However, limitations include the need for calibration and reference materials, the possibility of interference from other compounds, and the complexity of data analysis. The cost of equipment and expertise required may also be a barrier for smaller food businesses.

5.2. Adulteration Attributes

Adulteration refers to the practice of mixing inferior quality substances with superior original substances and or/by adding ingredients of lower quality, which can negatively impact the completeness and nutritional value of food products [217]. Often, the presence of contaminated materials is very comparable to that of the initial products, making it complicated to differentiate them when blended. Additionally, establishing the species and source of staple foods is important in protecting consumers from potential fraud, as food geographical indications cannot be confirmed solely from food labels. Thus, reliable analysis of adulteration is required to confirm the quality of the product. The scope of the proposed paragraph is to discuss the problem of food adulteration and the role of spectroscopic techniques with chemometrics in detecting and differentiating adulterated food samples. It highlights the challenges posed by adulteration, including the difficulty in differentiating contaminated materials from the original products and the importance of confirming the species and source of staple foods.

Spectroscopic techniques, such as Transmission Raman spectroscopy, have been used to achieve this goal. In a recent study, Transmission Raman spectroscopy was used to differentiate rice samples based on their geographic origin using PCA and LDA [223]. Similarly, Feng et al. (2013) used a combination of Raman spectroscopy and multivariate data analysis techniques to differentiate rice samples from various regions of China, with an overall accuracy of over 90% [224]. Different types of rice samples and their geographical origin were effectively discriminated using ^{1}H-NMR spectroscopy and multivariate data analysis. The accuracy of wheat flour sample discrimination using a simple linear model was 80% [225]. The study by Esteve Agelet et al. (2012) tested the feasibility of NIR spectroscopy to discriminate viable-germinating corn and soybeans from dead seeds and found that dead corn kernels could be discriminated with an accuracy of 99% using partial least squares discriminant analysis (PLS-DA) [226], Haughey et al. (2013) used NIR spectroscopy (833–2632 nm) to detect adulteration of soybeans with melamine and achieved R^2 values ranging from 0.89 to 0.99 using PLSR and PCR algorithms [227]. The use of Fourier transform mid-infrared (FT-MIR) spectroscopy and discriminant analysis was successful in detecting adulterated potato and sweet potato starch by Liu et al. (2013) [228]. Similarly, FT-IR spectroscopy and discriminant analysis were combined for the geographical differentiation of dried lentil seeds [229]. For imaging spectroscopy, it was examined to identify different types of wheat kernels, and found that better results were obtained by selecting three specific wavelength intervals. The ability of NIR spectroscopy to detect adulteration of soybeans by melamine was demonstrated, with R^2 values of 0.89 to 0.99 using PLSR and PCR algorithms [230]. NIR spectroscopy was also used to distinguish viable-germinating corn and soybeans from dead seeds, with perfect accuracy based on PCA and PLS-DA [231]. The NIR and HSI in combination with PCA to detect residues of 0.10% peanuts in wheat flour, with an R^2 of 0.95 [232]. All these spectroscopic techniques demonstrated very great accomplishments in detecting different forms of food fraud and authentication. The end points of the paragraph include the effectiveness of spectroscopic techniques in detecting and differentiating adulterated food samples based on their chemical composition. It mentions the advantages of spectroscopic techniques, such as high sensitivity, specificity, and non-destructiveness, which allow for the identification of small changes in the food samples.

Spectroscopic techniques detect food adulteration by analyzing the chemical composition of food samples. Advantages include high sensitivity, specificity, and non-destructiveness, which can identify small changes in the food sample. Limitations include the need for calibration and reference materials, interference from other compounds, limited applicability, and cost of equipment and expertise.

5.3. Chemical Attributes

The chemical composition of food plays a crucial role in determining its nutritional value and consumer acceptance. The scope of the proposed paragraph is to discuss the chemical composition of food, focusing on cereals and legumes, and highlight the importance of accurate measurement and control of their chemical components for determining nutritional value and food quality. The main chemical components in cereals and legumes are carbohydrates, protein, moisture, fiber, fat, and ash. Starch, a type of carbohydrate, is composed of helical amylose and branched amylopectin. The nutritional value and consumer acceptance of food are significantly influenced by its chemical composition. Cereals and legumes primarily consist of carbohydrates, protein, moisture, fibre, fat, and ash. For instance, maize can range from 20% to 36%, sorghum from 21% to 35%, wheat from 17% to 29%, barley from 11% to 26%, rice from 8% to 37%, pea from 34% to 37%, and potatoes from 18% to 23% [233]. Among these components, starch, a type of carbohydrate, is composed of amylose and amylopectin. The relative proportions of amylose and amylopectin in starch have a notable influence on the quality of food products such as bread and noodles [234]. The content of those chemicals can vary greatly which can influence the food quality properties and high-performance techniques needed to control their concentrations. Staple foods are primarily composed of proteins, which contribute to their structural and functional properties, affecting their quality and taste. Moisture content is a major factor that influences the shelf life and germination success of staple foods. The fiber in these foods, made up of cellulose, hemicellulose, lignin, pectin, and gums, can help lower cholesterol levels but does not provide energy. Lipids are the most energy-dense component, providing more energy per gram than carbohydrates and proteins. Ash content, which is the inorganic residue after organic matter is burned, indicates the presence of minerals in the food sample. Moisture content is commonly measured through drying with an oven or Karl Fisher titration, both of which have low efficiency [235]. The content of crude fat in staple foods can be measured by extracting the dried material using ether or petroleum ether. The crude protein content is typically determined using Kjeldahl or Dumas methods, which quantify the organic nitrogen content [236]. However, these traditional methods are destructive, time-consuming, and require a long preparation time. To overcome these limitations, the use of non-destructive and rapid detection methods is desired. The VIS/IR spectroscopy techniques hold the potential for determining the chemical components in staple foods. Nie et al., demonstrated that VIS/NIR spectroscopy in the range of 400 to 1000 nm can be used non-destructively to determine the presence of the poisonous phytohemagglutinin in beans by monitoring the boiling time of yard-long beans [237]. Both NIR and MIR techniques were used to verify the chemical characteristics, as well as protein, lipid, moisture, and ash amounts in soybean, while PLSR models showed good performance [238]. The preparation and determination of staple food samples using VIS/IR spectroscopy techniques take less than five minutes, compared to the 10–16 h required by traditional methods. FT-NIR spectroscopy has been shown to estimate the concentration of moisture, protein, lipid, ash, and carbohydrate in Brazilian soybeans with high accuracy [239]. The protein and moisture content had the best results with correlation coefficients of 0.81 and 0.80, respectively. NIR spectroscopy has also been used to accurately predict the crude protein content in potatoes, with high correlation coefficients ranging from 0.86 to 0.95 using PLSR for calibration [240]. The VIS/NIR (446–1125 nm) was applied for chemical components in two varieties of potato tubers [241]. The chemical and enzymatic compositions of staple foods, including phenolics, flavonoids, anthocyanins, carotenoids, dioscin, and catalase, have been screened using NIR or MIR spectroscopy and analyzed using chemometric methods namely HCA, PCA, SVM and PLS-DA, producing good predictive ability [241–243]. The endpoints of this section include the advantages of spectroscopic techniques in offering high sensitivity, specificity, and non-destructiveness for food chemical analysis. It mentions their ability to quickly identify small changes in food composition and measure parameters such as acidity, pH, and moisture content.

Spectroscopic techniques offer high sensitivity, specificity, and non-destructiveness for food chemical analysis. They can quickly identify small changes in food composition and measure parameters such as acidity, pH, and moisture content. However, limitations include the need for calibration and reference materials, interference from other compounds, and the complexity of data analysis. The cost of equipment and expertise required may also be a barrier for smaller food businesses.

5.4. Mycotoxin Attributes

The presence of mold and its associated toxins during post-harvest storage can result in a decrease in food quality, leading to losses in nutrients and market value, as well as pose serious food safety risks. The scope of the proposed paragraph is to discuss the presence of mold and mycotoxins in post-harvest storage and their negative impact on food quality, safety, and market value. It emphasizes the harmful effects of mycotoxins, such as aflatoxins and Fusarium toxins, which are known to be carcinogenic and linked to liver and lung cancer in humans.

Mycotoxins, like aflatoxins and Fusarium toxins, are harmful byproducts produced by mold and are recognized as carcinogenic, linked to liver and lung cancer in humans. According to estimates, as much as 25% of crops grown for both animal feed and human consumption globally may be contaminated with mycotoxins [244]. Mycotoxin amounts for essential food products have been specifically restricted by the EU [245]. In the USA, 20 ppb (parts per billion) of aflatoxin amounts in food and 100 ppb in feed are permitted for the administrative market [246]. Giving to the FAO, one billion metric tons of food are rotten worldwide yearly caused by mycotoxins [247]. Hence, the ability to accurately detect various levels of fungal contamination can greatly aid in controlling plant diseases and reducing food safety hazards. Currently, the methods for identifying and measuring toxins consist primarily of thin-layer chromatography and high-performance liquid chromatography, but these methods are both expensive and time-consuming and involve the destruction of samples [248]. The detection of mycotoxins in staple foods can be performed quickly and easily using IR spectroscopy. The NIR spectroscopy in the range of 950–1650 nm was used in conjunction with PLSR to detect total fungal and yellow-green Aspergillus flavus infections in rice [249]. However, the accuracy levels for both total fungi and yellow-green A. flavus infections were not high. The PLS was based on full cross-validation to detect fumonisin contamination in maize through NIR spectroscopy, achieving a high R^2 of 0.91. This demonstrates that NIR spectroscopy is a viable alternative tool for detecting fumonisin infections [250]. The FT-MIR spectroscopy in the range of 2500–16,000 nm with an attenuated total reflectance unit to differentiate peanut kernels contaminated with aflatoxin and non-aflatoxin strains. The "Acceptable" stream (aflatoxin $\leq$ 20 ppb) was separated from "Mildly" (20 $<$ aflatoxin $<$ 300 ppb), "Highly Toxic" (300 $<$ aflatoxin $<$ 1200 ppb), and "Highly Moldy" (aflatoxin $>$ 1200 ppb) through classification. The fingerprint region (5556–12,500 nm) was utilized to predict the A. flavus and A. parasiticus species with varying levels of contamination based on PLS regression, achieving an R^2 of 99.98% [251]. The utilization of the Raman technique in combination with PCA allows for the non-invasive detection of deoxynivalenol (commonly known as vomitoxin) in contaminated wheat and barley [252]. The benefits of this approach consist of the utilization of a NIR laser excitation (1064 nm), which minimized disruption from the fluorescence of biological substances. The endpoints of the paragraph highlight the advantages of spectroscopic techniques in food mycotoxin analysis, including high sensitivity, specificity, and non-destructiveness. It emphasizes the ability of these techniques with chemometric models to detect even small amounts of mycotoxins in food samples and provide rapid measurements of various chemical parameters.

Spectroscopic techniques offer advantages in food mycotoxin analysis such as high sensitivity, specificity, and non-destructiveness. These techniques can detect even small amounts of mycotoxins in food samples and provide rapid measurements of various chemical parameters. However, limitations include the need for calibration and reference materials, the interference of other compounds in the sample, and the limited applicability of certain techniques to specific mycotoxins.

5.5. Parasitic Contamination

Staple foods can be degraded and lose market value due to contamination from parasitic insects. The scope here is to discuss the contamination of staple foods by parasitic insects and the detrimental effects they have on food quality and market value. The activities of parasitic insects, both external and internal, and their impact are highlighted.

The insects not only feed directly on the food, but also create heat and moisture through their metabolic activity, leading to localized hotspots and spoilage. This results in weight loss, nutrient depletion, reduced germination ability, and increased risk of contamination during storage. External insects like the Oryzaephilus surinamensis and internal insects such as Sitophilus granarius can both harm stored products. Some insects like the sweet potato weevil (Cylas formicarius elegantulus) do most of the damage inside the food without significant external changes. Others, like the rice weevil (Sitophilus oryzae) and lesser grain borer (Rhyzopertha dominica), lay eggs, larvae, or pupae in seeds, continuing their destructive activity for up to 7 weeks, until the adult insects emerge and leave an exit hole, making the damage visible [253].

The detection of parasitic insect infestations in staple foods is crucial for the food industry. Such infestations can decrease the quality and value of the food. Internal infestations are more challenging to detect and require effective methods for inspection. HSI, a spectral information technique, has the potential to provide information about internal infestations through reflectance or absorbance measurements. A study by Singh et al. (2009) used LW-NIR HSI (900–1700 nm) to differentiate between insect-infested and healthy wheat kernels, achieving an accuracy of over 85% with LDA and QDA classifiers [254]. The endpoints of the paragraph emphasize the importance of detecting parasitic insect infestations in staple foods to mitigate the negative effects on food quality and value. Internal infestations are highlighted as being more challenging to detect and requiring effective inspection methods.

Spectroscopic techniques detect parasitic contamination in food with high sensitivity, specificity, and non-destructiveness. These techniques rapidly and precisely measure various chemical parameters, but limitations include the need for calibration and reference materials, the possibility of inaccurate results due to other compounds in the sample, and limitations in detectability for certain parasites. The cost and expertise required may also be a barrier for smaller food applications.

5.6. Internal Functional Characteristics

Functional disturbances within staple crop plants are linked to irregular growth patterns brought on by less-than-ideal environmental factors. The scope here is to discuss briefly some examples of the internal functional characteristics disturbances within staple crop plants such as those caused by environmental factors.

The functional disturbances include variations in temperature, moisture, oxygen, nutrients, the presence of toxic gases, and a deficiency in growth regulators. The symptoms of these internal conditions appear to be root tuber disorders such as hollow heart, black heart, and internal brown spot. Overuse of nitrogen fertilizer during the growth phase can cause hollow heart in tubers, a condition where the tuber's core dies or splits, creating a cavity [255]. This disorder is commonly triggered by the rapid growth of the tuber and results in reduced storage life, poorer quality chips, and an unattractive appearance. Blackheart is a condition that arises when there's insufficient oxygen during the tuber's growth or storage period. Tubers cultivated in excessively damp areas or those exposed to severe temperatures are more prone to this condition. Internal brown spot refers to the internal death of the tuber's central tissue, which significantly diminishes its culinary worth. Other physiological damage, such as freezing or chilling injuries, occurs due to extended exposure to freezing temperatures post-harvest. The symptoms of these injuries manifest as grey or black patches, or a brown discoloration around the tuber's vascular ring. Even though these internal physiological disorders may not be immediately visible, they greatly affect the quality and value of root and tuber crops. X-ray examination has been reported to successfully detect hollow hearts in potatoes but attempts to use acoustic methods for detection have been unsuccessful [256], although unsuccessful using an acoustics method [257]. Traditional methods are also heavily dependent on the orientation of the potato. However, spectroscopic techniques have been shown to be effective for the non-destructive prediction of internal physiological disorders, including black heart, hollow heart, and internal brown spot in root tubers. The VIS/NIR transmission spectroscopy (513–850 nm) used to compare three different morphological correction methods combined with PLS-DA and PCA to detect black hearts in potatoes. The best performance was found with height-corrected transmittance. Using six wavelengths (839, 817, 741, 711, 698 and 678 nm), the overall validation classification rate for the black heart was 96.53% [258]. The HSI in the 1000–1700 nm range was applied to distinguish hollow heart and internal brown spots in tubers. Specifically, the HSI technique was applied to detect the presence of hollow hearts in potato tubers [259]. An accurate recognition rate of 89.10% was achieved by combining support vector machine (SVM) with various image processing techniques. Similarly, time-resolved reflectance spectroscopy (540–900 nm) was used for non-destructive external measurement of internal brown spots in potato tubers [260]. By internal detection of healthy tissue and black spots, the most sensitive wavelength for detection was found to be 690 nm. The endpoints of the paragraph emphasize the negative effects of these internal physiological disorders on the quality and value of root and tuber crops.

Spectroscopic techniques offer insights into food's internal functional characteristics, such as composition, structure, and functional properties, with high sensitivity, specificity, and non-destructiveness. They allow for rapid and precise measurements of various chemical and physical parameters. However, challenges arise from the need for calibration and reference materials, potential interference from other compounds, and limitations in the applicability of certain techniques. Moreover, the cost of equipment and expertise required may be an obstacle.

6. Conclusions

In conclusion, advancing spectroscopy techniques offer a game-changing solution for food composition analysis, presenting a non-destructive, rapid, cost-effective, and eco-friendly alternative to traditional methods. These techniques provide valuable insights into food quality, chemical components, composition, structure-function relationships, and sensory attributes, eliminating the need for extensive sample preparation in many cases, thus saving time and resources. Coupled with appropriate chemometric multivariate methods, spectroscopy enables comprehensive and accurate analyses of various food materials.

This review has highlighted recent studies showcasing the effectiveness of spectroscopy techniques, such as hyperspectral and multispectral imaging, NMR, IR, Raman, X-ray-based methods, fluorescence, and UV-visible, in conjunction with chemometric approaches for diverse food analysis applications. From determining the chemical composition and identifying geographical origin to ensuring food safety and traceability, monitoring storage and preservation, assessing sensory characteristics, evaluating microbial quality, and detecting food spoilage, these techniques have shown immense potential and versatility in the food industry.

Looking ahead, the future of spectroscopic techniques in food analysis appears highly promising. As technology continues to advance, these techniques are expected to become even more powerful, accurate, and efficient, opening new avenues of research and application. It is strongly recommended that food businesses and researchers embrace spectroscopy in their analyses and consider integrating these techniques into their standard practices to enhance the overall efficiency and quality of food analysis.

However, it is crucial to acknowledge the challenges and limitations that may arise with spectroscopic methods, such as issues related to sample heterogeneity, sensitivity, accuracy, and the need for proper calibration and reference materials. Chemometric methods, while invaluable, require careful handling to avoid issues like overfitting and to ensure reliable results, potential drawbacks of chemometric methods, such as overfitting or the need for extensive data processing and model optimization should be also considered.

As the field progresses, further research and development are essential to optimize and expand the capabilities of spectroscopic techniques in food analysis. By addressing these challenges and pushing the boundaries of innovation, spectroscopy can revolutionize the food industry, bolstering food safety, quality control, and product innovation in a sustainable and impactful manner.

Author Contributions: Conceptualization, methodology, and validation, M.K.; formal analysis and investigation, M.K., M.A.M., M.T. and H.Y.; resources, M.K.; writing—original draft preparation, M.K., M.A.M., M.T. and H.Y.; writing—review and editing, M.K.; supervision, project administration and funding acquisition, M.K. All authors have read and agreed to the published version of the manuscript.

Funding: Open access funding provided by University of Helsinki.

Data Availability Statement: The data used to support the findings of this study can be made available by the corresponding author upon request.

Conflicts of Interest: The authors declare no conflict of interest.

References

1. Lakshmi, V.; Pradesh, A. Food Adulteration. *Int. J. Sci. Invent. Today* **2012**, *1*, 106–113.
2. The Lancet. Melamine and Food Safety in China. *Lancet* **2009**, *373*, 353. [CrossRef] [PubMed]
3. Ramesh, R.; Jha, S.; Lawrence, F.; Dodd, V. From Mumbai to Your Supermarket: On the Murky Trail of Britain's Biggest Food Scandal. *Guardian* **2005**, *23*.
4. Wu, D.; Shi, H.; He, Y.; Yu, X.; Bao, Y. Potential of Hyperspectral Imaging and Multivariate Analysis for Rapid and Non-Invasive Detection of Gelatin Adulteration in Prawn. *J. Food Eng.* **2013**, *119*, 680–686. [CrossRef]
5. Lohumi, S.; Lee, S.; Lee, H.; Cho, B.-K. A Review of Vibrational Spectroscopic Techniques for the Detection of Food Authenticity and Adulteration. *Trends Food Sci. Technol.* **2015**, *46*, 85–98. [CrossRef]
6. Yen, T.-H.; Lin-Tan, D.-T.; Lin, J.-L. Food Safety Involving Ingestion of Foods and Beverages Prepared with Phthalate-Plasticizer-Containing Clouding Agents. *J. Formos. Med. Assoc.* **2011**, *110*, 671–684. [CrossRef]
7. Yang, J.; Hauser, R.; Goldman, R.H. Taiwan Food Scandal: The Illegal Use of Phthalates as a Clouding Agent and Their Contribution to Maternal Exposure. *Food Chem. Toxicol.* **2013**, *58*, 362–368. [CrossRef]
8. Schneider, A. Asian Honey, Banned in Europe, Is Flooding US Grocery Shelves. *Am. Bee J.* **2011**, *151*, 933.
9. Baeten, V.; Dardenne, P. Spectroscopy: Developments in Instrumentation and Analysis. *Grasas Y Aceites* **2002**, *53*, 45–63. [CrossRef]
10. Joshi, V.C.; Srinivas, P.V.; Khan, I.A. Rapid and Easy Identification of Illicium Verum Hook. f. and Its Adulterant Illicium Anisatum Linn. by Fluorescent Microscopy and Gas Chromatography. *J. AOAC Int.* **2005**, *88*, 703–706. [CrossRef]
11. Gemperline, P. *Practical Guide to Chemometrics*; CRC Press: Boca Raton, FL, USA, 2006; ISBN 0429119569.
12. Otto, M. *Chemometrics: Statistics and Computer Application in Analytical Chemistry*; John Wiley & Sons: Hoboken, NJ, USA, 2016; ISBN 3527340971.
13. Folch-Fortuny, A.; Arteaga, F.; Ferrer, A. Missing Data Imputation Toolbox for MATLAB. *Chemom. Intell. Lab. Syst.* **2016**, *154*, 93–100. [CrossRef]
14. Folch-Fortuny, A.; Arteaga, F.; Ferrer, A. Assessment of Maximum Likelihood PCA Missing Data Imputation. *J. Chemom.* **2016**, *30*, 386–393. [CrossRef]
15. Elcoroaristizabal, S.; Bro, R.; García, J.A.; Alonso, L. PARAFAC Models of Fluorescence Data with Scattering: A Comparative Study. *Chemom. Intell. Lab. Syst.* **2015**, *142*, 124–130. [CrossRef]
16. Tomasi, G.; Bro, R. PARAFAC and Missing Values. *Chemom. Intell. Lab. Syst.* **2005**, *75*, 163–180. [CrossRef]
17. Wu, Y.; Peng, S.; Xie, Q.; Han, Q.; Zhang, G.; Sun, H. An Improved Weighted Multiplicative Scatter Correction Algorithm with the Use of Variable Selection: Application to near-Infrared Spectra. *Chemom. Intell. Lab. Syst.* **2019**, *185*, 114–121. [CrossRef]
18. Mishra, P.; Nordon, A.; Roger, J.-M. Improved Prediction of Tablet Properties with Near-Infrared Spectroscopy by a Fusion of Scatter Correction Techniques. *J. Pharm. Biomed. Anal.* **2021**, *192*, 113684. [CrossRef]
19. Zhang, Z.-M.; Chen, S.; Liang, Y.-Z. Baseline Correction Using Adaptive Iteratively Reweighted Penalized Least Squares. *Analyst* **2010**, *135*, 1138–1146. [CrossRef]
20. Skogholt, J.; Liland, K.H.; Indahl, U.G. Baseline and Interferent Correction by the Tikhonov Regularization Framework for Linear Least Squares Modeling. *J. Chemom.* **2018**, *32*, e2962. [CrossRef]
21. Li, Z.; Wang, Y.; Yu, J.; Guo, Y.; Cao, W. Deep Learning Based Radiomics (DLR) and Its Usage in Noninvasive IDH1 Prediction for Low Grade Glioma. *Sci. Rep.* **2017**, *7*, 5467. [CrossRef]

22. Li, Z.; Kim, S.; Zhong, S.; Zhong, Z.; Kato, I.; Zhang, X. Coherent Point Drift Peak Alignment Algorithms Using Distance and Similarity Measures for Two-dimensional Gas Chromatography Mass Spectrometry Data. *J. Chemom.* **2020**, *34*, e3236. [CrossRef]
23. Deng, B.; Kim, S.; Li, H.; Heath, E.; Zhang, X. Global Peak Alignment for Comprehensive Two-Dimensional Gas Chromatography Mass Spectrometry Using Point Matching Algorithms. *J. Bioinform. Comput. Biol.* **2016**, *14*, 1650032. [CrossRef] [PubMed]
24. Yu, H.; Bro, R.; Gallagher, N.B. PARASIAS: A New Method for Analyzing Higher-Order Tensors with Shifting Profiles. *Anal. Chim. Acta* **2022**, *1238*, 339848. [CrossRef] [PubMed]
25. Mishra, P.; Biancolillo, A.; Roger, J.M.; Marini, F.; Rutledge, D.N. New Data Preprocessing Trends Based on Ensemble of Multiple Preprocessing Techniques. *TrAC Trends Anal. Chem.* **2020**, *132*, 116045. [CrossRef]
26. Andersen, C.M.; Bro, R. Variable Selection in Regression—A Tutorial. *J. Chemom.* **2010**, *24*, 728–737. [CrossRef]
27. Li, Y.; Fang, T.; Zhu, S.; Huang, F.; Chen, Z.; Wang, Y. Detection of Olive Oil Adulteration with Waste Cooking Oil via Raman Spectroscopy Combined with IPLS and SiPLS. *Spectrochim. Acta A Mol. Biomol. Spectrosc.* **2018**, *189*, 37–43. [CrossRef] [PubMed]
28. Zhang, Z.; Trevino, V.; Hoseini, S.S.; Belciug, S.; Boopathi, A.M.; Zhang, P.; Gorunescu, F.; Subha, V.; Dai, S. Variable Selection in Logistic Regression Model with Genetic Algorithm. *Ann. Transl. Med.* **2018**, *6*, 45. [CrossRef] [PubMed]
29. Mao, S.; Lu, C.; Li, M.; Ye, Y.; Wei, X.; Tong, H. Identification of Key Aromatic Compounds in Congou Black Tea by Partial Least-square Regression with Variable Importance of Projection Scores and Gas Chromatography–Mass Spectrometry/Gas Chromatography–Olfactometry. *J. Sci. Food Agric.* **2018**, *98*, 5278–5286. [CrossRef]
30. Kvalheim, O.M. Variable Importance: Comparison of Selectivity Ratio and Significance Multivariate Correlation for Interpretation of Latent-variable Regression Models. *J. Chemom.* **2020**, *34*, e3211. [CrossRef]
31. Azizan, A.; Lee, A.X.; Abdul Hamid, N.A.; Maulidiani, M.; Mediani, A.; Abdul Ghafar, S.Z.; Zolkeflee, N.K.Z.; Abas, F. Potentially Bioactive Metabolites from Pineapple Waste Extracts and Their Antioxidant and α-Glucosidase Inhibitory Activities by 1H NMR. *Foods* **2020**, *9*, 173. [CrossRef]
32. Pearson, K. LIII. On Lines and Planes of Closest Fit to Systems of Points in Space. *Lond. Edinb. Dublin Philos. Mag. J. Sci.* **1901**, *2*, 559–572. [CrossRef]
33. Hotelling, H. Analysis of a Complex of Statistical Variables into Principal Components. *J. Educ. Psychol.* **1933**, *24*, 417. [CrossRef]
34. Bro, R.; Smilde, A.K. Principal Component Analysis. *Anal. Methods* **2014**, *6*, 2812–2831. [CrossRef]
35. Granato, D.; Santos, J.S.; Escher, G.B.; Ferreira, B.L.; Maggio, R.M. Use of Principal Component Analysis (PCA) and Hierarchical Cluster Analysis (HCA) for Multivariate Association between Bioactive Compounds and Functional Properties in Foods: A Critical Perspective. *Trends Food Sci. Technol.* **2018**, *72*, 83–90. [CrossRef]
36. Tleis, M.; Callieris, R.; Roma, R. Segmenting the Organic Food Market in Lebanon: An Application of k-Means Cluster Analysis. *Br. Food J.* **2017**, *119*, 1423–1441. [CrossRef]
37. Moro, M.K.; dos Santos, F.D.; Folli, G.S.; Romao, W.; Filgueiras, P.R. A Review of Chemometrics Models to Predict Crude Oil Properties from Nuclear Magnetic Resonance and Infrared Spectroscopy. *Fuel* **2021**, *303*, 121283. [CrossRef]
38. Lee, G.; Lee, K. Feature Selection Using Distributions of Orthogonal PLS Regression Vectors in Spectral Data. *BioData Min.* **2021**, *14*, 1–16. [CrossRef]
39. Vong, R.; Geladi, P.; Wold, S.; Esbensen, K. Source Contributions to Ambient Aerosol Calculated by Discriminat Partial Least Squares Regression (PLS). *J. Chemom.* **1988**, *2*, 281–296. [CrossRef]
40. Bro, R. Multiway Calibration. Multilinear Pls. *J. Chemom.* **1996**, *10*, 47–61. [CrossRef]
41. Yu, H.; Guo, L.; Kharbach, M.; Han, W. Multi-Way Analysis Coupled with near-Infrared Spectroscopy in Food Industry: Models and Applications. *Foods* **2021**, *10*, 802. [CrossRef]
42. Rocha, W.F.d.C.; do Prado, C.B.; Blonder, N. Comparison of Chemometric Problems in Food Analysis Using Non-Linear Methods. *Molecules* **2020**, *25*, 3025. [CrossRef]
43. Ausati, S.; Amanollahi, J. Assessing the Accuracy of ANFIS, EEMD-GRNN, PCR, and MLR Models in Predicting PM 2.5. *Atmos. Environ.* **2016**, *142*, 465–474. [CrossRef]
44. Li, L.-N.; Liu, X.-F.; Yang, F.; Xu, W.-M.; Wang, J.-Y.; Shu, R. A Review of Artificial Neural Network Based Chemometrics Applied in Laser-Induced Breakdown Spectroscopy Analysis. *Spectrochim. Acta Part B Spectrosc.* **2021**, *180*, 106183. [CrossRef]
45. Lee, L.C.; Liong, C.-Y.; Jemain, A.A. Partial Least Squares-Discriminant Analysis (PLS-DA) for Classification of High-Dimensional (HD) Data: A Review of Contemporary Practice Strategies and Knowledge Gaps. *Analyst* **2018**, *143*, 3526–3539. [CrossRef] [PubMed]
46. Branden, K.V.; Hubert, M. Robust Classification in High Dimensions Based on the SIMCA Method. *Chemom. Intell. Lab. Syst.* **2005**, *79*, 10–21. [CrossRef]
47. Bylesjö, M.; Rantalainen, M.; Cloarec, O.; Nicholson, J.K.; Holmes, E.; Trygg, J. OPLS Discriminant Analysis: Combining the Strengths of PLS-DA and SIMCA Classification. *J. Chemom. A J. Chemom. Soc.* **2006**, *20*, 341–351. [CrossRef]
48. Xia, J.; Zhang, J.; Zhao, Y.; Huang, Y.; Xiong, Y.; Min, S. Fourier Transform Infrared Spectroscopy and Chemometrics for the Discrimination of Paper Relic Types. *Spectrochim. Acta A Mol. Biomol. Spectrosc.* **2019**, *219*, 8–14. [CrossRef]
49. Luts, J.; Ojeda, F.; Van de Plas, R.; De Moor, B.; Van Huffel, S.; Suykens, J.A.K. A Tutorial on Support Vector Machine-Based Methods for Classification Problems in Chemometrics. *Anal. Chim. Acta* **2010**, *665*, 129–145. [CrossRef]
50. De Juan, A.; Jaumot, J.; Tauler, R. Multivariate Curve Resolution (MCR). Solving the Mixture Analysis Problem. *Anal. Methods* **2014**, *6*, 4964–4976. [CrossRef]

51. Jaumot, J.; Tauler, R. MCR-BANDS: A User Friendly MATLAB Program for the Evaluation of Rotation Ambiguities in Multivariate Curve Resolution. *Chemom. Intell. Lab. Syst.* **2010**, *103*, 96–107. [CrossRef]
52. Bro, R. PARAFAC. Tutorial and Applications. *Chemom. Intell. Lab. Syst.* **1997**, *38*, 149–171. [CrossRef]
53. Yu, H.; Bro, R. PARAFAC2 and Local Minima. *Chemom. Intell. Lab. Syst.* **2021**, *219*, 104446. [CrossRef]
54. ten Berge, J.M.F.; Kiers, H.A.L. Some Uniqueness Results for PARAFAC2. *Psychometrika* **1996**, *61*, 123–132. [CrossRef]
55. Yu, H.; Augustijn, D.; Bro, R. Accelerating PARAFAC2 Algorithms for Non-Negative Complex Tensor Decomposition. *Chemom. Intell. Lab. Syst.* **2021**, *214*, 104312. [CrossRef]
56. Gamela, R.R.; Pereira-Filho, E.R.; Pereira, F.M.V. Minimal-Invasive Analytical Method and Data Fusion: An Alternative for Determination of Cu, K, Sr, and Zn in Cocoa Beans. *Food Anal. Methods* **2021**, *14*, 545–551. [CrossRef]
57. Speranҫa, M.A.; Mayorquín-Guevara, J.E.; da Cruz, M.C.P.; de Almeida Teixeira, G.H.; Pereira, F.M.V. Biofortification Quality in Bananas Monitored by Energy-Dispersive X-ray Fluorescence and Chemometrics. *Food Chem.* **2021**, *362*, 130172. [CrossRef] [PubMed]
58. Priyashantha, H.; Höjer, A.; Saedén, K.H.; Lundh, Å.; Johansson, M.; Bernes, G.; Geladi, P.; Hetta, M. Use of Near-Infrared Hyperspectral (NIR-HS) Imaging to Visualize and Model the Maturity of Long-Ripening Hard Cheeses. *J. Food Eng.* **2020**, *264*, 109687. [CrossRef]
59. Darnay, L.; Králik, F.; Oros, G.; Koncz, Á.; Firtha, F. Monitoring the Effect of Transglutaminase in Semi-Hard Cheese during Ripening by Hyperspectral Imaging. *J. Food Eng.* **2017**, *196*, 123–129. [CrossRef]
60. Lu, B.; Sun, J.; Yang, N.; Hang, Y. Fluorescence Hyperspectral Image Technique Coupled with HSI Method to Predict Solanine Content of Potatoes. *J. Food Process. Preserv.* **2019**, *43*, e14198. [CrossRef]
61. Xiao, Q.; Bai, X.; He, Y. Rapid Screen of the Color and Water Content of Fresh-Cut Potato Tuber Slices Using Hyperspectral Imaging Coupled with Multivariate Analysis. *Foods* **2020**, *9*, 94. [CrossRef]
62. Tian, X.-Y.; Aheto, J.H.; Bai, J.-W.; Dai, C.; Ren, Y.; Chang, X. Quantitative Analysis and Visualization of Moisture and Anthocyanins Content in Purple Sweet Potato by Vis–NIR Hyperspectral Imaging. *J. Food Process. Preserv.* **2021**, *45*, e15128. [CrossRef]
63. Li, B.; Cobo-Medina, M.; Lecourt, J.; Harrison, N.; Harrison, R.J.; Cross, J. V Application of Hyperspectral Imaging for Nondestructive Measurement of Plum Quality Attributes. *Postharvest Biol. Technol.* **2018**, *141*, 8–15. [CrossRef]
64. He, H.-J.; Chen, Y.; Li, G.; Wang, Y.; Ou, X.; Guo, J. Hyperspectral Imaging Combined with Chemometrics for Rapid Detection of Talcum Powder Adulterated in Wheat Flour. *Food Control* **2023**, *144*, 109378. [CrossRef]
65. Kim, G.; Lee, H.; Baek, I.; Cho, B.-K.; Kim, M.S. Quantitative Detection of Benzoyl Peroxide in Wheat Flour Using Line-Scan Short-Wave Infrared Hyperspectral Imaging. *Sens. Actuators B Chem.* **2022**, *352*, 130997. [CrossRef]
66. Sun, M.; Zhang, D.; Liu, L.; Wang, Z. How to Predict the Sugariness and Hardness of Melons: A near-Infrared Hyperspectral Imaging Method. *Food Chem.* **2017**, *218*, 413–421. [CrossRef] [PubMed]
67. Wang, H.; Lv, D.; Dong, N.; Wang, S.; Liu, J. Application of Near-Infrared Spectroscopy for Screening the Potato Flour Content in Chinese Steamed Bread. *Food Sci. Biotechnol.* **2019**, *28*, 955–963. [CrossRef]
68. Du, C.; Sun, L.; Bai, H.; Zhao, Z.; Li, X.; Gai, Z. Quantitative Detection of Talcum Powder in Wheat Flour Based on Near-Infrared Spectroscopy and Hybrid Feature Selection. *Infrared Phys. Technol.* **2022**, *123*, 104185. [CrossRef]
69. Kandpal, L.M.; Mouazen, A.M.; Masithoh, R.E.; Mishra, P.; Lohumi, S.; Cho, B.-K.; Lee, H. Sequential Data-Fusion of near-Infrared and Mid-Infrared Spectroscopy Data for Improved Prediction of Quality Traits in Tuber Flours. *Infrared Phys. Technol.* **2022**, *127*, 104371. [CrossRef]
70. Kamboj, U.; Guha, P.; Mishra, S. Comparison of PLSR, MLR, SVM Regression Methods for Determination of Crude Protein and Carbohydrate Content in Stored Wheat Using near Infrared Spectroscopy. *Mater. Today Proc.* **2022**, *48*, 576–582. [CrossRef]
71. Liang, P.-S.; Haff, R.P.; Hua, S.-S.T.; Munyaneza, J.E.; Mustafa, T.; Sarreal, S.B.L. Nondestructive Detection of Zebra Chip Disease in Potatoes Using Near-Infrared Spectroscopy. *Biosyst. Eng.* **2018**, *166*, 161–169. [CrossRef]
72. Jiang, H.; Liu, T.; Chen, Q. Quantitative Detection of Fatty Acid Value during Storage of Wheat Flour Based on a Portable Near-Infrared (NIR) Spectroscopy System. *Infrared Phys. Technol.* **2020**, *109*, 103423. [CrossRef]
73. Ning, H.; Wang, J.; Jiang, H.; Chen, Q. Quantitative Detection of Zearalenone in Wheat Grains Based on Near-Infrared Spectroscopy. *Spectrochim. Acta A Mol. Biomol. Spectrosc.* **2022**, *280*, 121545. [CrossRef] [PubMed]
74. Câmara, A.B.F.; de Oliveira, K.G.; Santos, M.C.D.; de Lima, R.R.S.; de Lima, K.M.G.; de Carvalho, L.S. Multivariate Assessment for Predicting Antioxidant Activity from Clove and Pomegranate Extracts by MCR-ALS and PLS Models Combined to IR Spectroscopy. *Food Chem.* **2022**, *384*, 132321. [CrossRef] [PubMed]
75. Castro, R.C.; Ribeiro, D.S.M.; Santos, J.L.M.; Páscoa, R.N.M.J. Comparison of near Infrared Spectroscopy and Raman Spectroscopy for the Identification and Quantification through MCR-ALS and PLS of Peanut Oil Adulterants. *Talanta* **2021**, *230*, 122373. [CrossRef]
76. Castro, R.C.; Ribeiro, D.S.M.; Santos, J.L.M.; Páscoa, R.N.M.J. Near Infrared Spectroscopy Coupled to MCR-ALS for the Identification and Quantification of Saffron Adulterants: Application to Complex Mixtures. *Food Control* **2021**, *123*, 107776. [CrossRef]
77. Li, S.; Xing, B.; Lin, D.; Yi, H.; Shao, Q. Rapid Detection of Saffron (*Crocus sativus* L.) Adulterated with Lotus Stamens and Corn Stigmas by near-Infrared Spectroscopy and Chemometrics. *Ind. Crops. Prod.* **2020**, *152*, 112539. [CrossRef]
78. Li, S.; Shao, Q.; Lu, Z.; Duan, C.; Yi, H.; Su, L. Rapid Determination of Crocins in Saffron by Near-Infrared Spectroscopy Combined with Chemometric Techniques. *Spectrochim. Acta A Mol. Biomol. Spectrosc.* **2018**, *190*, 283–289. [CrossRef]

79. Liu, P.; Wang, J.; Li, Q.; Gao, J.; Tan, X.; Bian, X. Rapid Identification and Quantification of Panax Notoginseng with Its Adulterants by near Infrared Spectroscopy Combined with Chemometrics. *Spectrochim. Acta A Mol. Biomol. Spectrosc.* **2019**, *206*, 23–30. [CrossRef]
80. Liu, Y.; Wang, Y.; Xia, Z.; Wang, Y.; Wu, Y.; Gong, Z. Rapid Determination of Phytosterols by NIRS and Chemometric Methods. *Spectrochim. Acta A Mol. Biomol. Spectrosc.* **2019**, *211*, 336–341. [CrossRef]
81. Joshi, R.; Baek, I.; Joshi, R.; Kim, M.S.; Cho, B.-K. Detection of Fabricated Eggs Using Fourier Transform Infrared (FT-IR) Spectroscopy Coupled with Multivariate Classification Techniques. *Infrared Phys. Technol.* **2022**, *123*, 104163. [CrossRef]
82. Mazivila, S.J.; Páscoa, R.N.M.J.; Castro, R.C.; Ribeiro, D.S.M.; Santos, J.L.M. Detection of Melamine and Sucrose as Adulterants in Milk Powder Using Near-Infrared Spectroscopy with DD-SIMCA as One-Class Classifier and MCR-ALS as a Means to Provide Pure Profiles of Milk and of Both Adulterants with Forensic Evidence: A Short Communication. *Talanta* **2020**, *216*, 120937.
83. Novianty, I.; Gilang Baskoro, R.; Iqbal Nurulhaq, M.; Achirul Nanda, M. Empirical Mode Decomposition of Near-Infrared Spectroscopy Signals for Predicting Oil Content in Palm Fruits. *Inf. Process. Agric.* 2022; *in press*.
84. Başar, B.; Özdemir, D. Determination of Honey Adulteration with Beet Sugar and Corn Syrup Using Infrared Spectroscopy and Genetic-algorithm-based Multivariate Calibration. *J. Sci. Food Agric.* **2018**, *98*, 5616–5624. [CrossRef] [PubMed]
85. Qin, J.; Kim, M.S.; Chao, K.; Gonzalez, M.; Cho, B.-K. Quantitative Detection of Benzoyl Peroxide in Wheat Flour Using Line-Scan Macroscale Raman Chemical Imaging. *Appl. Spectrosc.* **2017**, *71*, 2469–2476. [CrossRef]
86. Yuan, H.; Liu, M.; Huang, S.; Zhao, J.; Tao, J. Classification and Detection of Testosterone Propionate and Nandrolone Residues in Duck Meat Using Surface-Enhanced Raman Spectroscopy Coupled with Multivariate Analysis. *Poult. Sci.* **2021**, *100*, 296–301. [CrossRef]
87. Ning, X.; Mu-Hua, L.; Hai-Chao, Y.; Shuang-Gen, H.; Xiao, W.; Jin-Hui, Z.; Jian, C.; Ting, W.; Wei, H.; Yi-Xin, S. Classification of Sulfadimidine and Sulfapyridine in Duck Meat by Surface Enhanced Raman Spectroscopy Combined with Principal Component Analysis and Support Vector Machine. *Anal. Lett.* **2020**, *53*, 1513–1524. [CrossRef]
88. Hara, R.; Ishigaki, M.; Kitahama, Y.; Ozaki, Y.; Genkawa, T. Excitation Wavelength Selection for Quantitative Analysis of Carotenoids in Tomatoes Using Raman Spectroscopy. *Food Chem.* **2018**, *258*, 308–313. [CrossRef] [PubMed]
89. de Oliveira Mendes, T.; Manzolli Rodrigues, B.V.; Simas Porto, B.L.; Alves da Rocha, R.; de Oliveira, M.A.L.; de Castro, F.K.; dos Anjos, V.d.C.; Bell, M.J.V. Raman Spectroscopy as a Fast Tool for Whey Quantification in Raw Milk. *Vib. Spectrosc.* **2020**, *111*, 103150. [CrossRef]
90. Czaja, T.; Baranowska, M.; Mazurek, S.; Szostak, R. Determination of Nutritional Parameters of Yoghurts by FT Raman Spectroscopy. *Spectrochim. Acta A Mol. Biomol. Spectrosc.* **2018**, *196*, 413–417. [CrossRef]
91. Tian, H.; Chen, S.; Li, D.; Lou, X.; Chen, C.; Yu, H. Simultaneous Detection for Adulterations of Maltodextrin, Sodium Carbonate, and Whey in Raw Milk Using Raman Spectroscopy and Chemometrics. *J. Dairy. Sci.* **2022**, *105*, 7242–7252. [CrossRef]
92. Bērziņš, K.; Harrison, S.D.L.; Leong, C.; Fraser-Miller, S.J.; Harper, M.J.; Diana, A.; Gibson, R.S.; Houghton, L.A.; Gordon, K.C. Qualitative and Quantitative Vibrational Spectroscopic Analysis of Macronutrients in Breast Milk. *Spectrochim. Acta A Mol. Biomol. Spectrosc.* **2021**, *246*, 118982. [CrossRef]
93. de Sá Oliveira, K.; de Souza Callegaro, L.; Stephani, R.; Almeida, M.R.; de Oliveira, L.F.C. Analysis of Spreadable Cheese by Raman Spectroscopy and Chemometric Tools. *Food Chem.* **2016**, *194*, 441–446. [CrossRef]
94. Puertas, G.; Vázquez, M. Cholesterol Determination in Egg Yolk by UV-VIS-NIR Spectroscopy. *Food Control* **2019**, *100*, 262–268. [CrossRef]
95. Wang, X.; Esquerre, C.; Downey, G.; Henihan, L.; O'Callaghan, D.; O'Donnell, C. Development of Chemometric Models Using Vis-NIR and Raman Spectral Data Fusion for Assessment of Infant Formula Storage Temperature and Time. *Innov. Food Sci. Emerg. Technol.* **2021**, *67*, 102551. [CrossRef]
96. Valinger, D.; Longin, L.; Grbeš, F.; Benković, M.; Jurina, T.; Gajdoš Kljusurić, J.; Jurinjak Tušek, A. Detection of Honey Adulteratio—The Potential of UV-VIS and NIR Spectroscopy Coupled with Multivariate Analysis. *LWT* **2021**, *145*, 111316. [CrossRef]
97. Galvan, D.; Tanamati, A.A.C.; Casanova, F.; Danieli, E.; Bona, E.; Killner, M.H.M. Compact Low-Field NMR Spectroscopy and Chemometrics Applied to the Analysis of Edible Oils. *Food Chem.* **2021**, *365*, 130476. [CrossRef]
98. Barreto, M.C.; Braga, R.G.; Lemos, S.G.; Fragoso, W.D. Determination of Melamine in Milk by Fluorescence Spectroscopy and Second-Order Calibration. *Food Chem.* **2021**, *364*, 130407. [CrossRef]
99. Gu, H.; Lv, R.; Huang, X.; Chen, Q.; Dong, Y. Rapid Quantitative Assessment of Lipid Oxidation in a Rapeseed Oil-in-Water (o/w) Emulsion by Three-Dimensional Fluorescence Spectroscopy. *J. Food Compos. Anal.* **2022**, *114*, 104762. [CrossRef]
100. Tarhan, İ. A Comparative Study of ATR-FTIR, UV–Visible and Fluorescence Spectroscopy Combined with Chemometrics for Quantification of Squalene in Extra Virgin Olive Oils. *Spectrochim. Acta A Mol. Biomol. Spectrosc.* **2020**, *241*, 118714. [CrossRef]
101. Wu, X.; Bian, X.; Lin, E.; Wang, H.; Guo, Y.; Tan, X. Weighted Multiscale Support Vector Regression for Fast Quantification of Vegetable Oils in Edible Blend Oil by Ultraviolet-Visible Spectroscopy. *Food Chem.* **2021**, *342*, 128245. [CrossRef]
102. Zhang, W.; Li, N.; Feng, Y.; Su, S.; Li, T.; Liang, B. A Unique Quantitative Method of Acid Value of Edible Oils and Studying the Impact of Heating on Edible Oils by UV–Vis Spectrometry. *Food Chem.* **2015**, *185*, 326–332. [CrossRef]
103. Ríos-Reina, R.; Azcarate, S.M.; Camiña, J.; Callejón, R.M. Assessment of UV–Visible Spectroscopy as a Useful Tool for Determining Grape-Must Caramel in High-Quality Wine and Balsamic Vinegars. *Food Chem.* **2020**, *323*, 126792. [CrossRef]
104. Cavdaroglu, C.; Ozen, B. Prediction of Vinegar Processing Parameters with Chemometric Modelling of Spectroscopic Data. *Microchem. J.* **2021**, *171*, 106886. [CrossRef]

105. Santos, P.M.; Pereira-Filho, E.R.; Colnago, L.A. Detection and Quantification of Milk Adulteration Using Time Domain Nuclear Magnetic Resonance (TD-NMR). *Microchem. J.* **2016**, *124*, 15–19. [CrossRef]
106. Liu, J.; Zhang, J.; Tan, Z.; Hou, Q.; Liu, R. Detecting the Content of the Bright Blue Pigment in Cream Based on Deep Learning and Near-Infrared Spectroscopy. *Spectrochim. Acta A Mol. Biomol. Spectrosc.* **2022**, *270*, 120757. [CrossRef]
107. Sun, Q.; Zhang, M.; Yang, P. Combination of LF-NMR and BP-ANN to Monitor Water States of Typical Fruits and Vegetables during Microwave Vacuum Drying. *LWT* **2019**, *116*, 108548. [CrossRef]
108. Hajjar, G.; Haddad, L.; Rizk, T.; Akoka, S.; Bejjani, J. High-Resolution 1H NMR Profiling of Triacylglycerols as a Tool for Authentication of Food from Animal Origin: Application to Hen Egg Matrix. *Food Chem.* **2021**, *360*, 130056. [CrossRef]
109. Jiang, L.; Mehedi Hassan, M.; Jiao, T.; Li, H.; Chen, Q. Rapid Detection of Chlorpyrifos Residue in Rice Using Surface-Enhanced Raman Scattering Coupled with Chemometric Algorithm. *Spectrochim. Acta A Mol. Biomol. Spectrosc.* **2021**, *261*, 119996. [CrossRef]
110. Haddad, L.; Francis, J.; Rizk, T.; Akoka, S.; Remaud, G.S.; Bejjani, J. Cheese Characterization and Authentication through Lipid Biomarkers Obtained by High-Resolution 1H NMR Profiling. *Food Chem.* **2022**, *383*, 132434. [CrossRef]
111. Wang, T.; Wu, H.-L.; Long, W.-J.; Hu, Y.; Cheng, L.; Chen, A.-Q.; Yu, R.-Q. Rapid Identification and Quantification of Cheaper Vegetable Oil Adulteration in Camellia Oil by Using Excitation-Emission Matrix Fluorescence Spectroscopy Combined with Chemometrics. *Food Chem.* **2019**, *293*, 348–357. [CrossRef]
112. Richardson, P.I.C.; Muhamadali, H.; Ellis, D.I.; Goodacre, R. Rapid Quantification of the Adulteration of Fresh Coconut Water by Dilution and Sugars Using Raman Spectroscopy and Chemometrics. *Food Chem.* **2019**, *272*, 157–164. [CrossRef]
113. Galvan, D.; de Andrade, J.C.; Effting, L.; Lelis, C.A.; Melquiades, F.L.; Bona, E.; Conte-Junior, C.A. Energy-Dispersive X-ray Fluorescence Combined with Chemometric Tools Applied to Tomato and Sweet Pepper Classification. *Food Control* **2023**, *143*, 109326. [CrossRef]
114. Scatigno, C.; Festa, G. A First Elemental Pattern and Geo-Discrimination of Italian EVOO by Energy Dispersive X-ray Fluorescence and Chemometrics. *Microchem. J.* **2021**, *171*, 106863. [CrossRef]
115. Panebianco, S.; Mazzoleni, P.; Barone, G.; Musumarra, A.; Pellegriti, M.G.; Pulvirenti, A.; Scordino, A.; Cirvilleri, G. Feasibility Study of Tomato Fruit Characterization by Fast XRF Analysis for Quality Assessment and Food Traceability. *Food Chem.* **2022**, *383*, 132364. [CrossRef] [PubMed]
116. Allegretta, I.; Squeo, G.; Gattullo, C.E.; Porfido, C.; Cicchetti, A.; Caponio, F.; Cesco, S.; Nicoletto, C.; Terzano, R. TXRF Spectral Information Enhanced by Multivariate Analysis: A New Strategy for Food Fingerprint. *Food Chem.* **2023**, *401*, 134124. [CrossRef] [PubMed]
117. Vitali Čepo, D.; Karoglan, M.; Borgese, L.; Depero, L.E.; Marguí, E.; Jablan, J. Application of Benchtop Total-Reflection X-ray Fluorescence Spectrometry and Chemometrics in Classification of Origin and Type of Croatian Wines. *Food Chem. X* **2022**, *13*, 100209. [CrossRef]
118. Li, J.; Chen, L.; Huang, W. Detection of Early Bruises on Peaches (*Amygdalus persica* L.) Using Hyperspectral Imaging Coupled with Improved Watershed Segmentation Algorithm. *Postharvest Biol. Technol.* **2018**, *135*, 104–113. [CrossRef]
119. He, P.; Wu, Y.; Wang, J.; Ren, Y.; Ahmad, W.; Liu, R.; Ouyang, Q.; Jiang, H.; Chen, Q. Detection of Mites *Tyrophagus putrescentiae* and *Cheyletus eruditus* in Flour Using Hyperspectral Imaging System Coupled with Chemometrics. *J. Food Process. Eng.* **2020**, *43*, e13386. [CrossRef]
120. Al-Sarayreh, M.; Reis, M.M.; Yan, W.Q.; Klette, R. Potential of Deep Learning and Snapshot Hyperspectral Imaging for Classification of Species in Meat. *Food Control* **2020**, *117*, 107332. [CrossRef]
121. Pan, L.; Zhang, Q.; Zhang, W.; Sun, Y.; Hu, P.; Tu, K. Detection of Cold Injury in Peaches by Hyperspectral Reflectance Imaging and Artificial Neural Network. *Food Chem.* **2016**, *192*, 134–141. [CrossRef]
122. Sun, Y.; Gu, X.; Sun, K.; Hu, H.; Xu, M.; Wang, Z.; Tu, K.; Pan, L. Hyperspectral Reflectance Imaging Combined with Chemometrics and Successive Projections Algorithm for Chilling Injury Classification in Peaches. *LWT* **2017**, *75*, 557–564. [CrossRef]
123. Babellahi, F.; Paliwal, J.; Erkinbaev, C.; Amodio, M.L.; Chaudhry, M.M.A.; Colelli, G. Early Detection of Chilling Injury in Green Bell Peppers by Hyperspectral Imaging and Chemometrics. *Postharvest Biol. Technol.* **2020**, *162*, 111100. [CrossRef]
124. Cen, H.; Lu, R.; Zhu, Q.; Mendoza, F. Nondestructive Detection of Chilling Injury in Cucumber Fruit Using Hyperspectral Imaging with Feature Selection and Supervised Classification. *Postharvest Biol. Technol.* **2016**, *111*, 352–361. [CrossRef]
125. Carreiro Soares, S.F.; Medeiros, E.P.; Pasquini, C.; de Lelis Morello, C.; Harrop Galvão, R.K.; Ugulino Araújo, M.C. Classification of Individual Cotton Seeds with Respect to Variety Using Near-Infrared Hyperspectral Imaging. *Anal. Methods* **2016**, *8*, 8498–8505. [CrossRef]
126. Fan, S.; Li, C.; Huang, W.; Chen, L. Detection of Blueberry Internal Bruising over Time Using NIR Hyperspectral Reflectance Imaging with Optimum Wavelengths. *Postharvest Biol. Technol.* **2017**, *134*, 55–66. [CrossRef]
127. Sun, Y.; Pessane, I.; Pan, L.; Wang, X. Hyperspectral Characteristics of Bruised Tomatoes as Affected by Drop Height and Fruit Size. *LWT* **2021**, *141*, 110863. [CrossRef]
128. Susič, N.; Žibrat, U.; Širca, S.; Strajnar, P.; Razinger, J.; Knapič, M.; Vončina, A.; Urek, G.; Gerič Stare, B. Discrimination between Abiotic and Biotic Drought Stress in Tomatoes Using Hyperspectral Imaging. *Sens. Actuators B Chem.* **2018**, *273*, 842–852. [CrossRef]
129. Zhao, X.; Que, H.; Sun, X.; Zhu, Q.; Huang, M. Hybrid Convolutional Network Based on Hyperspectral Imaging for Wheat Seed Varieties Classification. *Infrared Phys. Technol.* **2022**, *125*, 104270. [CrossRef]

130. Zhao, Y.; Zhu, S.; Zhang, C.; Feng, X.; Feng, L.; He, Y. Application of Hyperspectral Imaging and Chemometrics for Variety Classification of Maize Seeds. *RSC Adv.* **2018**, *8*, 1337–1345. [CrossRef]
131. Tsouvaltzis, P.; Babellahi, F.; Amodio, M.L.; Colelli, G. Early Detection of Eggplant Fruit Stored at Chilling Temperature Using Different Non-Destructive Optical Techniques and Supervised Classification Algorithms. *Postharvest Biol. Technol.* **2020**, *159*, 111001. [CrossRef]
132. Huang, F.; Song, H.; Guo, L.; Guang, P.; Yang, X.; Li, L.; Zhao, H.; Yang, M. Detection of Adulteration in Chinese Honey Using NIR and ATR-FTIR Spectral Data Fusion. *Spectrochim. Acta A Mol. Biomol. Spectrosc.* **2020**, *235*, 118297. [CrossRef]
133. De Girolamo, A.; von Holst, C.; Cortese, M.; Cervellieri, S.; Pascale, M.; Longobardi, F.; Catucci, L.; Porricelli, A.C.R.; Lippolis, V. Rapid Screening of Ochratoxin A in Wheat by Infrared Spectroscopy. *Food Chem.* **2019**, *282*, 95–100. [CrossRef]
134. Chen, H.; Tan, C.; Lin, Z. Non-Destructive Identification of Native Egg by near-Infrared Spectroscopy and Data Driven-Based Class-Modeling. *Spectrochim. Acta A Mol. Biomol. Spectrosc.* **2019**, *206*, 484–490. [CrossRef] [PubMed]
135. Marquetti, I.; Link, J.V.; Lemes, A.L.G.; Scholz, M.B.d.S.; Valderrama, P.; Bona, E. Partial Least Square with Discriminant Analysis and near Infrared Spectroscopy for Evaluation of Geographic and Genotypic Origin of Arabica Coffee. *Comput. Electron. Agric.* **2016**, *121*, 313–319. [CrossRef]
136. Miao, X.; Miao, Y.; Tao, S.; Liu, D.; Chen, Z.; Wang, J.; Huang, W.; Yu, Y. Classification of Rice Based on Storage Time by Using near Infrared Spectroscopy and Chemometric Methods. *Microchem. J.* **2021**, *171*, 106841. [CrossRef]
137. Rovira, G.; Miaw, C.S.W.; Martins, M.L.C.; Sena, M.M.; de Souza, S.V.C.; Callao, M.P.; Ruisánchez, I. One-Class Model with Two Decision Thresholds for the Rapid Detection of Cashew Nuts Adulteration by Other Nuts. *Talanta* **2023**, *253*, 123916. [CrossRef]
138. Visconti, L.G.; Rodríguez, M.S.; Di Anibal, C. V Determination of Grated Hard Cheeses Adulteration by near Infrared Spectroscopy (NIR) and Multivariate Analysis. *Int. Dairy J.* **2020**, *104*, 104647. [CrossRef]
139. Xie, L.-H.; Tang, S.-Q.; Wei, X.-J.; Sheng, Z.-H.; Shao, G.-N.; Jiao, G.-A.; Hu, S.-K.; Wang-Lin; Hu, P.-S. Simultaneous Determination of Apparent Amylose, Amylose and Amylopectin Content and Classification of Waxy Rice Using near-Infrared Spectroscopy (NIRS). *Food Chem.* **2022**, *388*, 132944. [CrossRef]
140. Ziegler, J.U.; Leitenberger, M.; Longin, C.F.H.; Würschum, T.; Carle, R.; Schweiggert, R.M. Near-Infrared Reflectance Spectroscopy for the Rapid Discrimination of Kernels and Flours of Different Wheat Species. *J. Food Compos. Anal.* **2016**, *51*, 30–36. [CrossRef]
141. Rozali, N.L.; Azizan, K.A.; Singh, R.; Syed Jaafar, S.N.; Othman, A.; Weckwerth, W.; Ramli, U.S. Fourier Transform Infrared (FTIR) Spectroscopy Approach Combined with Discriminant Analysis and Prediction Model for Crude Palm Oil Authentication of Different Geographical and Temporal Origins. *Food Control* **2022**, *146*, 109509. [CrossRef]
142. Márquez, C.; López, M.I.; Ruisánchez, I.; Callao, M.P. FT-Raman and NIR Spectroscopy Data Fusion Strategy for Multivariate Qualitative Analysis of Food Fraud. *Talanta* **2016**, *161*, 80–86. [CrossRef]
143. Unuvar, A.; Boyaci, I.H.; Koksel, H. A Novel Approach for Rapid Discrimination of Common and Durum Wheat Flours Using Spectroscopic Analyses Combined with Chemometrics. *J. Cereal Sci.* **2021**, *100*, 103269. [CrossRef]
144. Amjad, A.; Ullah, R.; Khan, S.; Bilal, M.; Khan, A. Raman Spectroscopy Based Analysis of Milk Using Random Forest Classification. *Vib. Spectrosc.* **2018**, *99*, 124–129. [CrossRef]
145. de Oliveira Mendes, T.; Porto, B.L.S.; Almeida, M.R.; Fantini, C.; Sena, M.M. Discrimination between Conventional and Omega-3 Fatty Acids Enriched Eggs by FT-Raman Spectroscopy and Chemometric Tools. *Food Chem.* **2019**, *273*, 144–150. [CrossRef] [PubMed]
146. Nieuwoudt, M.K.; Holroyd, S.E.; McGoverin, C.M.; Simpson, M.C.; Williams, D.E. Raman Spectroscopy as an Effective Screening Method for Detecting Adulteration of Milk with Small Nitrogen-Rich Molecules and Sucrose. *J. Dairy Sci.* **2016**, *99*, 2520–2536. [CrossRef] [PubMed]
147. Robert, C.; Fraser-Miller, S.J.; Jessep, W.T.; Bain, W.E.; Hicks, T.M.; Ward, J.F.; Craigie, C.R.; Loeffen, M.; Gordon, K.C. Rapid Discrimination of Intact Beef, Venison and Lamb Meat Using Raman Spectroscopy. *Food Chem.* **2021**, *343*, 128441. [CrossRef]
148. Tian, F.; Tan, F.; Li, H. An Rapid Nondestructive Testing Method for Distinguishing Rice Producing Areas Based on Raman Spectroscopy and Support Vector Machine. *Vib. Spectrosc.* **2020**, *107*, 103017. [CrossRef]
149. Wu, X.; Xu, B.; Ma, R.; Gao, S.; Niu, Y.; Zhang, X.; Du, Z.; Liu, H.; Zhang, Y. Botanical Origin Identification and Adulteration Quantification of Honey Based on Raman Spectroscopy Combined with Convolutional Neural Network. *Vib. Spectrosc.* **2022**, *123*, 103439. [CrossRef]
150. Yao, S.; Li, T.; Li, J.; Liu, H.; Wang, Y. Geographic Identification of Boletus Mushrooms by Data Fusion of FT-IR and UV Spectroscopies Combined with Multivariate Statistical Analysis. *Spectrochim. Acta A Mol. Biomol. Spectrosc.* **2018**, *198*, 257–263. [CrossRef]
151. Antônio, D.C.; de Assis, D.C.S.; Botelho, B.G.; Sena, M.M. Detection of Adulterations in a Valuable Brazilian Honey by Using Spectrofluorimetry and Multiway Classification. *Food Chem.* **2022**, *370*, 131064. [CrossRef]
152. Fang, H.; Wu, H.-L.; Wang, T.; Long, W.-J.; Chen, A.-Q.; Ding, Y.-J.; Yu, R.-Q. Excitation-Emission Matrix Fluorescence Spectroscopy Coupled with Multi-Way Chemometric Techniques for Characterization and Classification of Chinese Lager Beers. *Food Chem.* **2021**, *342*, 128235. [CrossRef]
153. Jiménez-Carvelo, A.M.; Lozano, V.A.; Olivieri, A.C. Comparative Chemometric Analysis of Fluorescence and near Infrared Spectroscopies for Authenticity Confirmation and Geographical Origin of Argentinean Extra Virgin Olive Oils. *Food Control* **2019**, *96*, 22–28. [CrossRef]

154. Meng, X.; Yin, C.; Yuan, L.; Zhang, Y.; Ju, Y.; Xin, K.; Chen, W.; Lv, K.; Hu, L. Rapid Detection of Adulteration of Olive Oil with Soybean Oil Combined with Chemometrics by Fourier Transform Infrared, Visible-near-Infrared and Excitation-Emission Matrix Fluorescence Spectroscopy: A Comparative Study. *Food Chem.* **2023**, *405*, 134828. [CrossRef]
155. Yuan, L.; Meng, X.; Xin, K.; Ju, Y.; Zhang, Y.; Yin, C.; Hu, L. A Comparative Study on Classification of Edible Vegetable Oils by Infrared, near Infrared and Fluorescence Spectroscopy Combined with Chemometrics. *Spectrochim. Acta A Mol. Biomol. Spectrosc.* **2022**, *288*, 122120. [CrossRef] [PubMed]
156. Uncu, O.; Ozen, B. A Comparative Study of Mid-Infrared, UV–Visible and Fluorescence Spectroscopy in Combination with Chemometrics for the Detection of Adulteration of Fresh Olive Oils with Old Olive Oils. *Food Control* **2019**, *105*, 209–218. [CrossRef]
157. Gonçalves, T.R.; Rosa, L.N.; Gonçalves, R.P.; Torquato, A.S.; Março, P.H.; Marques Gomes, S.T.; Matsushita, M.; Valderrama, P. Monitoring the Oxidative Stability of Monovarietal Extra Virgin Olive Oils by UV–Vis Spectroscopy and MCR–ALS. *Food Anal. Methods* **2018**, *11*, 1936–1943. [CrossRef]
158. Suhandy, D.; Yulia, M. Peaberry Coffee Discrimination Using UV-Visible Spectroscopy Combined with SIMCA and PLS-DA. *Int. J. Food Prop.* **2017**, *20*, S331–S339. [CrossRef]
159. Torrecilla, J.S.; Aroca-Santos, R.; Cancilla, J.C.; Matute, G. Linear and Non-Linear Modeling to Identify Vinegars in Blends through Spectroscopic Data. *LWT-Food Sci. Technol.* **2016**, *65*, 565–571. [CrossRef]
160. Cavdaroglu, C.; Ozen, B. Detection of Vinegar Adulteration with Spirit Vinegar and Acetic Acid Using UV–Visible and Fourier Transform Infrared Spectroscopy. *Food Chem.* **2022**, *379*, 132150. [CrossRef]
161. Kucharska-Ambrożej, K.; Martyna, A.; Karpińska, J.; Kiełtyka-Dadasiewicz, A.; Kubat-Sikorska, A. Quality Control of Mint Species Based on UV-VIS and FTIR Spectral Data Supported by Chemometric Tools. *Food Control* **2021**, *129*, 108228. [CrossRef]
162. Botoran, O.R.; Ionete, R.E.; Miricioiu, M.G.; Costinel, D.; Radu, G.L.; Popescu, R. Amino Acid Profile of Fruits as Potential Fingerprints of Varietal Origin. *Molecules* **2019**, *24*, 4500. [CrossRef]
163. Consonni, R.; Polla, D.; Cagliani, L.R. Organic and Conventional Coffee Differentiation by NMR Spectroscopy. *Food Control* **2018**, *94*, 284–288. [CrossRef]
164. de Moura Ribeiro, M.V.; Boralle, N.; Redigolo Pezza, H.; Pezza, L.; Toci, A.T. Authenticity of Roasted Coffee Using1H NMR Spectroscopy. *J. Food Compos. Anal.* **2017**, *57*, 24–30. [CrossRef]
165. da Silva, L.A.; Flumignan, D.L.; Tininis, A.G.; Pezza, H.R.; Pezza, L. Discrimination of Brazilian Lager Beer by 1H NMR Spectroscopy Combined with Chemometrics. *Food Chem.* **2019**, *272*, 488–493. [CrossRef] [PubMed]
166. Gougeon, L.; da Costa, G.; Guyon, F.; Richard, T. 1H NMR Metabolomics Applied to Bordeaux Red Wines. *Food Chem.* **2019**, *301*, 125257. [CrossRef]
167. Marseglia, A.; Acquotti, D.; Consonni, R.; Cagliani, L.R.; Palla, G.; Caligiani, A. HR MAS 1H NMR and Chemometrics as Useful Tool to Assess the Geographical Origin of Cocoa Beans–Comparison with HR 1H NMR. *Food Res. Int.* **2016**, *85*, 273–281. [CrossRef]
168. Milani, M.I.; Rossini, E.L.; Catelani, T.A.; Pezza, L.; Toci, A.T.; Pezza, H.R. Authentication of Roasted and Ground Coffee Samples Containing Multiple Adulterants Using NMR and a Chemometric Approach. *Food Control* **2020**, *112*, 107104. [CrossRef]
169. Rachineni, K.; Rao Kakita, V.M.; Awasthi, N.P.; Shirke, V.S.; Hosur, R.V.; Chandra Shukla, S. Identifying Type of Sugar Adulterants in Honey: Combined Application of NMR Spectroscopy and Supervised Machine Learning Classification. *Curr. Res. Food Sci.* **2022**, *5*, 272–277. [CrossRef]
170. Shi, T.; Zhu, M.; Chen, Y.; Yan, X.; Chen, Q.; Wu, X.; Lin, J.; Xie, M. 1H NMR Combined with Chemometrics for the Rapid Detection of Adulteration in Camellia Oils. *Food Chem.* **2018**, *242*, 308–315. [CrossRef]
171. Zhang, Y.; Zhao, Y.; Shen, G.; Zhong, S.; Feng, J. NMR Spectroscopy in Conjugation with Multivariate Statistical Analysis for Distinguishing Plant Origin of Edible Oils. *J. Food Compos. Anal.* **2018**, *69*, 140–148. [CrossRef]
172. Borgese, L.; Bilo, F.; Tsuji, K.; Fernández-Ruiz, R.; Margui, E.; Streli, C.; Pepponi, G.; Stosnach, H.; Yamada, T.; Vandenabeele, P.; et al. First Total Reflection X-ray Fluorescence Round-Robin Test of Water Samples: Preliminary Results. *Spectrochim. Acta Part B Spectrosc.* **2014**, *101*, 6–14. [CrossRef]
173. Xue, X.; Chen, Z.; Wu, H.; Gao, H. Identification of Guiboutia Species by NIR-HSI Spectroscopy. *Sci. Rep.* **2022**, *12*, 11507. [CrossRef]
174. Tao, F.; Ngadi, M. Recent Advances in Rapid and Nondestructive Determination of Fat Content and Fatty Acids Composition of Muscle Foods. *Crit. Rev. Food Sci. Nutr.* **2018**, *58*, 1565–1593. [CrossRef] [PubMed]
175. Wu, D.; Sun, D.-W. Advanced Applications of Hyperspectral Imaging Technology for Food Quality and Safety Analysis and Assessment: A Review—Part I: Fundamentals. *Innov. Food Sci. Emerg. Technol.* **2013**, *19*, 1–14. [CrossRef]
176. Sha, M.; Zhang, D.; Zhang, Z.; Wei, J.; Chen, Y.; Wang, M.; Liu, J. Improving Raman Spectroscopic Identification of Rice Varieties by Feature Extraction. *J. Raman Spectrosc.* **2020**, *51*, 702–710. [CrossRef]
177. Alamprese, C.; Casale, M.; Sinelli, N.; Lanteri, S.; Casiraghi, E. Detection of Minced Beef Adulteration with Turkey Meat by UV–Vis, NIR and MIR Spectroscopy. *LWT-Food Sci. Technol.* **2013**, *53*, 225–232. [CrossRef]
178. Tammer, M.G. *Sokrates: Infrared and Raman Characteristic Group Frequencies: Tables and Charts*; Wiley: Chichester, UK, 2004.
179. Mehrnia, M.-A.; Jafari, S.-M.; Makhmal-Zadeh, B.S.; Maghsoudlou, Y. Rheological and Release Properties of Double Nano-Emulsions Containing Crocin Prepared with Angum Gum, Arabic Gum and Whey Protein. *Food Hydrocoll.* **2017**, *66*, 259–267. [CrossRef]

180. Cai, C.; Huang, J.; Zhao, L.; Liu, Q.; Zhang, C.; Wei, C. Heterogeneous Structure and Spatial Distribution in Endosperm of High-Amylose Rice Starch Granules with Different Morphologies. *J. Agric. Food Chem.* **2014**, *62*, 10143–10152. [CrossRef]
181. Jääskeläinen, A.-S.; Holopainen-Mantila, U.; Tamminen, T.; Vuorinen, T. Endosperm and Aleurone Cell Structure in Barley and Wheat as Studied by Optical and Raman Microscopy. *J. Cereal Sci.* **2013**, *57*, 543–550. [CrossRef]
182. Piot, O.; Autran, J.-C.; Manfait, M. Spatial Distribution of Protein and Phenolic Constituents in Wheat Grain as Probed by Confocal Raman Microspectroscopy. *J. Cereal Sci.* **2000**, *32*, 57–71. [CrossRef]
183. Nakajima, S.; Kuroki, S.; Ikehata, A. Selective Detection of Starch in Banana Fruit with Raman Spectroscopy. *Food Chem.* **2023**, *401*, 134166. [CrossRef]
184. Czaja, T.; Mazurek, S.; Szostak, R. Quantification of Gluten in Wheat Flour by FT-Raman Spectroscopy. *Food Chem.* **2016**, *211*, 560–563. [CrossRef]
185. Tay, M.; Fang, G.; Chia, P.L.; Li, S.F.Y. Rapid Screening for Detection and Differentiation of Detergent Powder Adulteration in Infant Milk Formula by LC–MS. *Forensic Sci. Int.* **2013**, *232*, 32–39. [CrossRef] [PubMed]
186. Taylan, O.; Cebi, N.; Tahsin Yilmaz, M.; Sagdic, O.; Bakhsh, A.A. Detection of Lard in Butter Using Raman Spectroscopy Combined with Chemometrics. *Food Chem.* **2020**, *332*, 127344. [CrossRef] [PubMed]
187. Wang, Q.; Lonergan, S.M.; Yu, C. Rapid Determination of Pork Sensory Quality Using Raman Spectroscopy. *Meat Sci.* **2012**, *91*, 232–239. [CrossRef]
188. Fowler, S.M.; Hopkins, D.L.; Torley, P.J.; Gill, H.; Blanch, E.W. Investigation of Chemical Composition of Meat Using Spatially Off-Set Raman Spectroscopy. *Analyst* **2019**, *144*, 2618–2627.
189. Karunathilaka, S.R.; Yakes, B.J.; He, K.; Brückner, L.; Mossoba, M.M. First Use of Handheld Raman Spectroscopic Devices and On-Board Chemometric Analysis for the Detection of Milk Powder Adulteration. *Food Control* **2018**, *92*, 137–146. [CrossRef]
190. Aykas, D.P.; Shotts, M.-L.; Rodriguez-Saona, L.E. Authentication of Commercial Honeys Based on Raman Fingerprinting and Pattern Recognition Analysis. *Food Control* **2020**, *117*, 107346. [CrossRef]
191. Kelis Cardoso, V.G.; Poppi, R.J. Cleaner and Faster Method to Detect Adulteration in Cassava Starch Using Raman Spectroscopy and One-Class Support Vector Machine. *Food Control* **2021**, *125*, 107917. [CrossRef]
192. Figueiredo, L.P.; Borém, F.M.; Almeida, M.R.; de Oliveira, L.F.C.; de Carvalho Alves, A.P.; dos Santos, C.M.; Rios, P.A. Raman Spectroscopy for the Differentiation of Arabic Coffee Genotypes. *Food Chem.* **2019**, *288*, 262–267. [CrossRef]
193. Jiménez-Carvelo, A.M.; Osorio, M.T.; Koidis, A.; González-Casado, A.; Cuadros-Rodríguez, L. Chemometric Classification and Quantification of Olive Oil in Blends with Any Edible Vegetable Oils Using FTIR-ATR and Raman Spectroscopy. *LWT* **2017**, *86*, 174–184. [CrossRef]
194. Jin, H.; Li, H.; Yin, Z.; Zhu, Y.; Lu, A.; Zhao, D.; Li, C. Application of Raman Spectroscopy in the Rapid Detection of Waste Cooking Oil. *Food Chem.* **2021**, *362*, 130191. [CrossRef]
195. Trimigno, A.; Marincola, F.C.; Dellarosa, N.; Picone, G.; Laghi, L. Definition of Food Quality by NMR-Based Foodomics. *Curr. Opin. Food Sci.* **2015**, *4*, 99–104. [CrossRef]
196. Ullah, R.; Khan, S.; Ali, H.; Bilal, M. Potentiality of Using Front Face Fluorescence Spectroscopy for Quantitative Analysis of Cow Milk Adulteration in Buffalo Milk. *Spectrochim. Acta A Mol. Biomol. Spectrosc.* **2020**, *225*, 117518. [CrossRef]
197. Liu, H.; Chen, Y.; Shi, C.; Yang, X.; Han, D. FT-IR and Raman Spectroscopy Data Fusion with Chemometrics for Simultaneous Determination of Chemical Quality Indices of Edible Oils during Thermal Oxidation. *LWT* **2020**, *119*, 108906. [CrossRef]
198. Li, Y.; Huang, Y.; Xia, J.; Xiong, Y.; Min, S. Quantitative Analysis of Honey Adulteration by Spectrum Analysis Combined with Several High-Level Data Fusion Strategies. *Vib. Spectrosc.* **2020**, *108*, 103060. [CrossRef]
199. Wang, W.; Paliwal, J. Near-Infrared Spectroscopy and Imaging in Food Quality and Safety. *Sens. Instrum. Food Qual. Saf.* **2007**, *1*, 193–207. [CrossRef]
200. Kademi, H.I.; Ulusoy, B.H.; Hecer, C. Applications of Miniaturized and Portable near Infrared Spectroscopy (NIRS) for Inspection and Control of Meat and Meat Products. *Food Rev. Int.* **2019**, *35*, 201–220. [CrossRef]
201. McVey, C.; Elliott, C.T.; Cannavan, A.; Kelly, S.D.; Petchkongkaew, A.; Haughey, S.A. Portable Spectroscopy for High Throughput Food Authenticity Screening: Advancements in Technology and Integration into Digital Traceability Systems. *Trends Food Sci. Technol.* **2021**, *118*, 777–790. [CrossRef]
202. Herrero, A.M. Raman Spectroscopy a Promising Technique for Quality Assessment of Meat and Fish: A Review. *Food Chem.* **2008**, *107*, 1642–1651. [CrossRef]
203. Özbalci, B.; Boyaci, İ.H.; Topcu, A.; Kadılar, C.; Tamer, U. Rapid Analysis of Sugars in Honey by Processing Raman Spectrum Using Chemometric Methods and Artificial Neural Networks. *Food Chem.* **2013**, *136*, 1444–1452. [CrossRef]
204. Prieto, N.; Dugan, M.E.R.; Juárez, M.; López-Campos, Ó.; Zijlstra, R.T.; Aalhus, J.L. Using Portable Near-Infrared Spectroscopy to Predict Pig Subcutaneous Fat Composition and Iodine Value. *Can. J. Anim. Sci.* **2017**, *98*, 221–229. [CrossRef]
205. Mishra, P.; Sytsma, M.; Chauhan, A.; Polder, G.; Pekkeriet, E. All-in-One: A Spectral Imaging Laboratory System for Standardised Automated Image Acquisition and Real-Time Spectral Model Deployment. *Anal. Chim. Acta* **2022**, *1190*, 339235. [CrossRef] [PubMed]
206. Mishra, P.; Nordon, A.; Asaari, M.S.M.; Lian, G.; Redfern, S. Fusing Spectral and Textural Information in Near-Infrared Hyperspectral Imaging to Improve Green Tea Classification Modelling. *J. Food Eng.* **2019**, *249*, 40–47. [CrossRef]

207. Bwambok, D.K.; Siraj, N.; Macchi, S.; Larm, N.E.; Baker, G.A.; Pérez, R.L.; Ayala, C.E.; Walgama, C.; Pollard, D.; Rodriguez, J.D. QCM Sensor Arrays, Electroanalytical Techniques and NIR Spectroscopy Coupled to Multivariate Analysis for Quality Assessment of Food Products, Raw Materials, Ingredients and Foodborne Pathogen Detection: Challenges and Breakthroughs. *Sensors* **2020**, *20*, 6982. [CrossRef] [PubMed]
208. Beć, K.B.; Grabska, J.; Huck, C.W. Miniaturized NIR Spectroscopy in Food Analysis and Quality Control: Promises, Challenges, and Perspectives. *Foods* **2022**, *11*, 1465. [CrossRef]
209. Hassan, H.; Fan, M.; Zhang, T.; Yang, K. Prediction of Total Phenolics and Flavonoids Contents in Chinese Wild Rice (*Zizania latifolia*) Using FT-NIR Spectroscopy. *Am. J. Food Technol.* **2015**, *10*, 109–117. [CrossRef]
210. Fu, Y.; Liu, W.; Soladoye, O. Towards innovative food processing of flavonoid compounds: Insights into stability and bioactivity. *Lwt Food Sci. Technol.* **2021**, *150*, 111968. [CrossRef]
211. Power, A.C.; Chapman, J.; Chandra, S.; Cozzolino, D. Ultraviolet-Visible Spectroscopy for Food Quality Analysis. In *Evaluation Technologies for Food Quality*; Woodhead Publishing: Sawston, UK, 2019; pp. 91–104.
212. Ali, M.M.; Hashim, N. Non-Destructive Methods for Detection of Food Quality. In *Future Foods*; Elsevier: Amsterdam, The Netherlands, 2022; pp. 645–667.
213. Xu, W.; Liu, X.; Xie, L.; Ying, Y. Comparison of Fourier Transform Near-Infrared, Visible near-Infrared, Mid-Infrared, and Raman Spectroscopy as Non-Invasive Tools for Transgenic Rice Discrimination. *Trans. ASABE* **2014**, *57*, 141–150.
214. Corvucci, F.; Nobili, L.; Melucci, D.; Grillenzoni, F.-V. The Discrimination of Honey Origin Using Melissopalynology and Raman Spectroscopy Techniques Coupled with Multivariate Analysis. *Food Chem.* **2015**, *169*, 297–304. [CrossRef]
215. Ohtsuki, T.; Sato, K.; Abe, Y.; Sugimoto, N.; Akiyama, H. Quantification of Acesulfame Potassium in Processed Foods by Quantitative 1H NMR. *Talanta* **2015**, *131*, 712–718. [CrossRef]
216. Nogales-Bueno, J.; Rodríguez-Pulido, F.J.; Heredia, F.J.; Hernández-Hierro, J.M. Comparative Study on the Use of Anthocyanin Profile, Color Image Analysis and near-Infrared Hyperspectral Imaging as Tools to Discriminate between Four Autochthonous Red Grape Cultivars from La Rioja (Spain). *Talanta* **2015**, *131*, 412–416. [CrossRef]
217. Marvin, H.J.P.; Bouzembrak, Y.; Janssen, E.M.; van der Fels-Klerx, H.J.; van Asselt, E.D.; Kleter, G.A. A Holistic Approach to Food Safety Risks: Food Fraud as an Example. *Food Res. Int.* **2016**, *89*, 463–470. [CrossRef] [PubMed]
218. Mendoza, F.A.; Cichy, K.; Lu, R.; Kelly, J.D. Evaluation of Canning Quality Traits in Black Beans (*Phaseolus vulgaris* L.) by Visible/near-Infrared Spectroscopy. *Food Bioproc. Tech.* **2014**, *7*, 2666–2678. [CrossRef]
219. Wang, L.; Liu, D.; Pu, H.; Sun, D.-W.; Gao, W.; Xiong, Z. Use of Hyperspectral Imaging to Discriminate the Variety and Quality of Rice. *Food Anal. Methods* **2015**, *8*, 515–523. [CrossRef]
220. Mahesh, S.; Jayas, D.S.; Paliwal, J.; White, N.D.G. Comparison of Partial Least Squares Regression (PLSR) and Principal Components Regression (PCR) Methods for Protein and Hardness Predictions Using the near-Infrared (NIR) Hyperspectral Images of Bulk Samples of Canadian Wheat. *Food Bioproc. Tech.* **2015**, *8*, 31–40. [CrossRef]
221. Barbedo, J.; Guarienti, E.; Tibola, C. Detection of sprout damage in wheat kernels using NIR hyperspectral imaging. *Biosyst. Eng.* **2018**, *175*, 124–132. [CrossRef]
222. Ji, Y.; Sun, L.; Li, Y.; Li, J.; Liu, S.; Xu, X.; Yuantong, X. Non-destructive classification of defective potatoes based on hyperspectral imaging and support vector machine. *Infrared Phys. Technol.* **2019**, *99*, 71–79. [CrossRef]
223. Li, F.; Wang, J.; Xu, L.; Wang, S.; Zhou, M.; Yin, J.; Lu, A. Rapid Screening of Cadmium in Rice and Identification of Geographical Origins by Spectral Method. *Int. J. Environ. Res. Public Health* **2018**, *15*, 312. [CrossRef]
224. Lamanna, R.; Cattivelli, L.; Miglietta, M.L.; Troccoli, A. Geographical Origin of Durum Wheat Studied by 1H-NMR Profiling. *Magn. Reson. Chem.* **2011**, *49*, 1–5. [CrossRef]
225. Monakhova, Y.B.; Rutledge, D.N.; Roßmann, A.; Waiblinger, H.; Mahler, M.; Ilse, M.; Kuballa, T.; Lachenmeier, D.W. Determination of Rice Type by 1H NMR Spectroscopy in Combination with Different Chemometric Tools. *J. Chemom.* **2014**, *28*, 83–92. [CrossRef]
226. Agelet, L.E.; Ellis, D.D.; Duvick, S.; Goggi, A.S.; Hurburgh, C.R.; Gardner, C.A. Feasibility of near Infrared Spectroscopy for Analyzing Corn Kernel Damage and Viability of Soybean and Corn Kernels. *J. Cereal. Sci.* **2012**, *55*, 160–165. [CrossRef]
227. Haughey, S.A.; Graham, S.F.; Cancouët, E.; Elliott, C.T. The Application of Near-Infrared Reflectance Spectroscopy (NIRS) to Detect Melamine Adulteration of Soya Bean Meal. *Food Chem.* **2013**, *136*, 1557–1561. [CrossRef] [PubMed]
228. Liu, J.; Wen, Y.; Dong, N.; Lai, C.; Zhao, G. Authentication of Lotus Root Powder Adulterated with Potato Starch and/or Sweet Potato Starch Using Fourier Transform Mid-Infrared Spectroscopy. *Food Chem.* **2013**, *141*, 3103–3109. [CrossRef] [PubMed]
229. Kouvoutsakis, G.; Mitsi, C.; Tarantilis, P.A.; Polissiou, M.G.; Pappas, C.S. Geographical Differentiation of Dried Lentil Seed (*Lens culinaris*) Samples Using Diffuse Reflectance Fourier Transform Infrared Spectroscopy (DRIFTS) and Discriminant Analysis. *Food Chem.* **2014**, *145*, 1011–1014. [CrossRef]
230. Serranti, S.; Cesare, D.; Bonifazi, G. The Development of a Hyperspectral Imaging Method for the Detection of Fusarium-Damaged, Yellow Berry and Vitreous Italian Durum Wheat Kernels. *Biosyst. Eng.* **2013**, *115*, 20–30. [CrossRef]
231. Panero, J.; Silva, H.; Panero, P.; Smiderle, O.; Panero, F.; Faria, F.; Rodríguez, A. Separation of Cultivars of Soybeans by Chemometric Methods Using Near Infrared Spectroscopy. *J. Agric. Sci.* **2018**, *10*, 351–360. [CrossRef]

232. Mishra, P.; Herrero-Langreo, A.; Barreiro, P.; Roger, J.M.; Diezma, B.; Gorretta, N.; Lleó, L. Detection and Quantification of Peanut Traces in Wheat Flour by near Infrared Hyperspectral Imaging Spectroscopy Using Principal-Component Analysis. *J. Near Infrared Spectrosc.* **2015**, *23*, 15–22. [CrossRef]
233. Zhang, Z.; Li, E.; Fan, X.; Yang, C.; Ma, H.; Gilbert, R. The effects of the chain-length distributions of starch molecules on rheological and thermal properties of wheat flour paste. *Food Hydrocoll.* **2020**, *101*, 105563. [CrossRef]
234. Li, Q.; Li, C.; Li, E.; Gilbert, R.; Xu, B. A molecular explanation of wheat starch physicochemical properties related to noodle eating quality. *Food Hydrocoll.* **2020**, *108*, 106035. [CrossRef]
235. Corpaş, L.; Hădărugă, N.G.; David, I.; Pîrşan, P.; Hădărugă, D.I.; Isengard, H.-D. Karl Fischer Water Titration– Principal Component Analysis Approach on Wheat Flour. *Food Anal. Methods* **2014**, *7*, 1353–1358. [CrossRef]
236. Dreher, J.; Blach, C.; Terjung, N.; Gibis, M.; Weiss, J. Influence of protein content on plant-based emulsified and crosslinked fat crystal networks to mimic animal fat tissue. *Food Hydrocoll.* **2020**, *106*, 105864. [CrossRef]
237. Nie, P.; Wu, D.; Yang, Y.; He, Y. Fast Determination of Boiling Time of Yardlong Bean Using Visible and near Infrared Spectroscopy and Chemometrics. *J. Food Eng.* **2012**, *109*, 155–161. [CrossRef]
238. Ferreira, D.S.; Galão, O.F.; Pallone, J.A.L.; Poppi, R.J. Comparison and Application of Near-Infrared (NIR) and Mid-Infrared (MIR) Spectroscopy for Determination of Quality Parameters in Soybean Samples. *Food Control* **2014**, *35*, 227–232. [CrossRef]
239. Ferreira, D.S.; Pallone, J.A.L.; Poppi, R.J. Fourier Transform Near-Infrared Spectroscopy (FT-NIRS) Application to Estimate Brazilian Soybean [Glycine Max (L.) Merril] Composition. *Food Res. Int.* **2013**, *51*, 53–58. [CrossRef]
240. López, A.; Arazuri, S.; Jarén, C.; Mangado, J.; Arnal, P.; de Galarreta, J.I.R.; Riga, P.; López, R. Crude Protein Content Determination of Potatoes by NIRS Technology. *Procedia Technol.* **2013**, *8*, 488–492. [CrossRef]
241. Rady, A.M.; Guyer, D.E.; Kirk, W.; Donis-González, I.R. The Potential Use of Visible/near Infrared Spectroscopy and Hyperspectral Imaging to Predict Processing-Related Constituents of Potatoes. *J. Food Eng.* **2014**, *135*, 11–25. [CrossRef]
242. Uarrota, V.G.; Moresco, R.; Coelho, B.; da Costa Nunes, E.; Peruch, L.A.M.; de Oliveira Neubert, E.; Rocha, M.; Maraschin, M. Metabolomics Combined with Chemometric Tools (PCA, HCA, PLS-DA and SVM) for Screening Cassava (*Manihot esculenta* Crantz) Roots during Postharvest Physiological Deterioration. *Food Chem.* **2014**, *161*, 67–78. [CrossRef]
243. Sanchez, T.; Ceballos, H.; Dufour, D.; Ortiz, D.; Morante, N.; Calle, F.; Zum Felde, T.; Dominguez, M.; Davrieux, F. Prediction of Carotenoids, Cyanide and Dry Matter Contents in Fresh Cassava Root Using NIRS and Hunter Color Techniques. *Food Chem.* **2014**, *151*, 444–451. [CrossRef]
244. Kovalsky, P.; Kos, G.; Nährer, K.; Schwab, C.; Jenkins, T.; Schatzmayr, G.; Sulyok, M.; Krska, R. Co-Occurrence of Regulated, Masked and Emerging Mycotoxins and Secondary Metabolites in Finished Feed and Maize—An Extensive Survey. *Toxins* **2016**, *8*, 363. [CrossRef]
245. Cheli, F.; Battaglia, D.; Gallo, R.; Dell'Orto, V. EU Legislation on Cereal Safety: An Update with a Focus on Mycotoxins. *Food Control* **2014**, *37*, 315–325. [CrossRef]
246. Yao, H.; Hruska, Z.; Kincaid, R.; Brown, R.L.; Bhatnagar, D.; Cleveland, T.E. Hyperspectral Image Classification and Development of Fluorescence Index for Single Corn Kernels Infected with Aspergillus Flavus. *Trans. ASABE* **2013**, *56*, 1977–1988.
247. Twarużek, M.; Skrzydlewski, P.; Kosicki, R.; Grajewski, J. Mycotoxins survey in feed materials and feedingstuffs in years 2015–2020. *Toxicon Off. J. Int. Soc. Toxinology* **2021**, *202*, 27–39. [CrossRef]
248. Ahmed, N.; Khan, S.H.; Anjum, M.A.; Rehman, A. A Cost Effective Preparative Thin Layer Chromatography Cleanup Method for High Performance Liquid Chromatography Analysis of Aflatoxins B1, B2 and G2. *Adv. Life Sci.* **2014**, *2*, 1–4.
249. Sirisomboon, C.D.; Putthang, R.; Sirisomboon, P. Application of near Infrared Spectroscopy to Detect Aflatoxigenic Fungal Contamination in Rice. *Food Control* **2013**, *33*, 207–214. [CrossRef]
250. Della Riccia Giacomo, D.Z.S. A Multivariate Regression Model for Detection of Fumonisins Content in Maize from near Infrared Spectra. *Food Chem.* **2013**, *141*, 4289–4294. [CrossRef] [PubMed]
251. Kaya-Celiker, H.; Mallikarjunan, P.K.; Schmale III, D.; Christie, M.E. Discrimination of Moldy Peanuts with Reference to Aflatoxin Using FTIR-ATR System. *Food Control* **2014**, *44*, 64–71. [CrossRef]
252. Kim, K.; Shim, W.; Kim, J.; Chung, D. Development of a simultaneous lateral flow strip test for the rapid and simple detection of deoxynivalenol and zearalenone. *J. Food Sci.* **2014**, *79*, M2048–M2055. [CrossRef] [PubMed]
253. Singh, C.B.; Jayas, D.S.; Paliwal, J.; White, N.D.G. Identification of Insect-Damaged Wheat Kernels Using Short-Wave near-Infrared Hyperspectral and Digital Colour Imaging. *Comput. Electron. Agric.* **2010**, *73*, 118–125. [CrossRef]
254. Singh, C.B.; Jayas, D.S.; Paliwal, J.; White, N.D.G. Detection of Insect-Damaged Wheat Kernels Using near-Infrared Hyperspectral Imaging. *J. Stored Prod. Res.* **2009**, *45*, 151–158. [CrossRef]
255. Jenni, S.; Truco, M.; Michelmore, R. Quantitative trait loci associated with tipburn, heat stress-induced physiological disorders, and maturity traits in crisphead lettuce. *Theor. Appl. Genet.* **2013**, *126*, 3065–3079. [CrossRef]
256. Bamberg, J.; Lombard, K.; Palta, J.; Workmaster, B.; Atucha, A. Survival of Solanum jamesii Tubers at Freezing Temperatures. *Am. J. Potato Res.* **2020**, *97*, 497–504. [CrossRef]
257. Elbatawi, I.E. An Acoustic Impact Method to Detect Hollow Heart of Potato Tubers. *Biosyst. Eng.* **2008**, *100*, 206–213. [CrossRef]
258. Zhou, Z.; Zeng, S.; Li, X.; Zheng, J. Nondestructive Detection of Blackheart in Potato by Visible/near Infrared Transmittance Spectroscopy. *J. Spectrosc.* **2015**, *2015*, 786709. [CrossRef]

259. Dacal-Nieto, A.; Formella, A.; Carrión, P.; Vazquez-Fernandez, E.; Fernández-Delgado, M. Non–Destructive Detection of Hollow Heart in Potatoes Using Hyperspectral Imaging. In Proceedings of the International Conference on Computer Analysis of Images and Patterns, Seville, Spain, 29–31 August 2011; Springer: Berlin/Heidelberg, Germany, 2011; pp. 180–187.
260. Vanoli, M.; Rizzolo, A.; Spinelli, L.; Parisi, B.; Torricelli, A. Non Destructive Detection of Internal Brown Spot in Potato Tubers by Time-Resolved Reflectance Spectroscopy: Preliminary Results on a Susceptible Cultivar. In Proceedings of the International Conference of Agricultural Engineering, CIGRAgeng 2012, Valencia, Spain, 8–12 July 2012.

MDPI
St. Alban-Anlage 66
4052 Basel
Switzerland
www.mdpi.com

Foods Editorial Office
E-mail: foods@mdpi.com
www.mdpi.com/journal/foods

www.ingramcontent.com/pod-product-compliance
Lightning Source LLC
Chambersburg PA
CBHW061506210926
51850CB00008B/1836
9783036585802